Professor Tetsuro Yamamoto

G. Alefeld
X. Chen (eds.)

Topics in Numerical Analysis

With Special Emphasis
on Nonlinear Problems

Computing
Supplement 15

Springer-Verlag Wien GmbH

Professor Dr. Goetz Alefeld
Institut für Angewandte Mathematik,
Universität Karlsruhe,
Karlsruhe, Germany

Professor Dr. Xiaojun Chen
Department of Mathematical Sciences,
Shimane University,
Matsue, Japan

Originally published by Springer-Verlag Wien New York in 2001

Typesetting: Scientific Publishing Services (P) Ltd., Madras

Printed on acid-free and chlorine-free bleached paper
SPIN: 10779596

With 29 Figures

ISSN 0344-8029
ISBN 978-3-211-83673-6 ISBN 978-3-7091-6217-0 (eBook)
DOI 10.1007/978-3-7091-6217-0

Preface

This volume contains eighteen papers submitted in celebration of the sixty-fifth birthday of Professor Tetsuro Yamamoto of Ehime University. Professor Yamamoto was born in Tottori, Japan on January 4, 1937. He obtained his B.S. and M.S. in mathematics from Hiroshima University in 1959 and 1961, respectively. In 1966, he took a lecturer position in the Department of Mathematics, Faculty of General Education, Hiroshima University and obtained his Ph.D. degree from Hiroshima University two years later. In 1969, he moved to the Department of Applied Mathematics, Faculty of Engineering, Ehime University as an associate professor and he has been a full professor of the Department of Mathematics (now Department of Mathematical Sciences), Faculty of Science, since 1975. At the early stage of his study, he was interested in algebraic eigenvalue problems and linear iterative methods. He published some papers on these topics in high level international journals. After moving to Ehime University, he started his research on Newton's method and Newton-like methods for nonlinear operator equations. He published many papers on error estimates of the methods. He established the remarkable result that all the known error bounds for Newton's method under the Kantorovich assumptions follow from the Newton-Kantorovich theorem, which put a period to the race of finding sharper error bounds for Newton's method. This paper was written while he was visiting the Mathematics Research Center, University of Wisconsin-Madison, March–October 1985, where Professor Rall had asserted the necessity of interval computation through his seminar and papers. Professor Yamamoto had also recognized the importance of verified computation in the late 1970's. Recently he has been interested in superconvergence and related properties of finite difference methods for Dirichlet problems.

Professor Yamamoto has been a member of the editorial boards of four international journals. In particular, he was a principal editor of the Journal of Computational and Applied Mathematics during 1992–2000. He organized a number of domestic and international conferences on computational and applied mathematics. Among others, the meeting "International Symposium on Computational Mathematics" held in Matsuyama in 1990 was the first international meeting having more than 40 overseas participators on computational mathematics held in Japan. He was also elected as dean of the Faculty of Science, Ehime University, during 1991–1995 and as vice president of Japan SIAM in 1997.

We thank the authors for their creative contributions and the referees for their prompt and careful reviews. Thanks also go to Professor Hackbusch, the chief editor of *Computing*, and to Springer-Verlag for their supports.

Finally we wish Professor Yamamoto continued health, happiness, and rich scientific work.

Goetz Alefeld, Karlsruhe University
Xiaojun Chen, Shimane University

Selected Publications of Professor Tetsuro Yamamoto

1. A computational method for the dominant root of a non-negative irreducible matrix. Numer. Math. *8*, 324–333 (1966).
2. On the extreme values of the roots of matrices. J. Math. Soc. Japan *19*, 173–178 (1967).
3. On the eigenvalues of compact operators in a Hilbert space. Numer. Math. *11*, 211–219 (1968).
4. On Lanczos' algorithm for tri-diagonalization. J. Sci. Hiroshima Univ. Ser. A-I Math. *32*, 259–284 (1968).
5. A note on the spectral mapping theorem. SIAM J. Math. Anal. *2*, 49–51 (1971).
6. The Rayleigh-Ritz and Weinstein-Bazley methods applied to a class of ordinary differential equations of the second order. Math. Japon. *19*, 95–111 (1974).
7. The Rayleigh-Ritz and Weinstein-Bazley methods applied to a class of ordinary differential equations of the second order. II, SIAM J. Numer. Anal. *12*, 428–438 (1975).
8. On the distribution of errors in the iterative solution of a system of linear equations. Numer. Math. *24*, 71–79 (1975).
9. On the distribution of errors in the iterative solution of a system of linear equations. II. Numer. Math. *25*, 461–463 (1976).
10. (with Y. Ikebe) Inversion of band matrices. Linear Alg. Appl. *24*, 105–111 (1979).
11. Componentwise error estimates for approximate solutions of nonlinear equations. J. Inf. Proc. *2*, 121–126 (1979).
12. Error bounds for computed eigenvalues and eigenvectors. Numer. Math. *34*, 189–199 (1980).
13. Componentwise error estimates for approximate solutions of systems of equations. Lecture Notes Num. Anal. *3*, 1–22 (1981).
14. Error bounds for computed eigenvalues and eigenvectors. II. Numer. Math. *40*, 201–206 (1982).
15. An existence theorem of solution to boundary value problems and its application to error estimates. Math. Japon. *27*, 301–308 (1982).
16. A posteriori componentwise error estimate for a computed solution of a system of linear equations. Pub. RIMS *18*, 522–537 (1982).
17. The Symm-Wilkinson method for improving an approximate eigenvalue and its associated eigenvector. Computing *33*, 179–184 (1984).
18. Error bounds for approximate solutions of systems of equations. Japan J. Appl. Math. *1*, 157–171 (1984).
19. Error bounds for Newton's process derived from the Kantorovich theorem. Japan J. Appl. Math. *2*, 285–292 (1985).

20. A unified derivation of several error bound for Newton's process. J. Comp. Appl. Math. *12* & *13*, 179–191 (1985).
21. A convergence theorem for Newton's method in Banach spaces. Japan J. Appl. Math. *3*, 37–52 (1986).
22. Error bounds for Newton-like methods under Kantorovich type assumptions. Japan J. Appl. Math. *3*, 295–313 (1986).
23. Error bounds for Newton's iterates derived from the Kantorovich theorem. Numer. Math. *48*, 91–98 (1986).
24. A method for finding sharp error bounds for Newton's method under the Kantorovich assumptions. Numer. Math. *49*, 203–220 (1986).
25. A convergence theorem for Newton-like methods in Banach spaces. Numer. Math. *51*, 545–557 (1987).
26. A note on a posteriori error bound of Zabrejko and Nguen for Zincenko's iteration. Numer. Funct. Anal. Optimiz. *9*, 987–994 (1987).
27. On the method of tangent hyperbolas in Banach spaces. J. Comp. Appl. Math. *21*, 75–86 (1988).
28. Newton's method and it neighborhood. Sugaku Expositions *1*, 219–238 (1988).
29. Uniqueness of the solution in a Kantorovich-type theorem of Haussler for the Gauss-Newton method. Japan J. Appl. Math. *6*, 77–81 (1989).
30. (with X. Chen) Convergence domain of certain iterative methods for solving nonlinear equations. Numer. Funct. Anal. Optimiz. *10*, 37–48 (1989).
31. (with X. Chen) Ball-convergence theorems and error estimates for certain iterative methods for nonlinear equations. Jap. J. Appl. Math. *7*, 131–143 (1990).
32. (with X. Chen) An existence and nonexistence theorem for solutions of nonlinear systems and its application to algebraic equations. J. Comp. Appl. Math. *30*, 87–97 (1990).
33. (with X. Chen) PCG methods applied to a system of nonlinear equations. J. Comp. Appl. Math. *38*, 61–75 (1991).
34. (with X. Chen) On the convergence of some quasi-Newton methods for nonlinear equations with nondifferentiable operators. Computing *49*, 87–94 (1992).
35. (with X. Chen) A convergence ball for multistep simplified Newton-like methods. Numer. Funct. Anal. Optimiz. *14*, 15–24 (1993).
36. (with L. Atanassova and S. Kanno) Validated computation of polynomial zeros by the Durand-Kerner method, in Topics in Validated Computation (J. Herzberger, Ed.), North-Holland, 1994, 27–53.
37. (with X. Chen) Newton-like methods for solving underdetermined non-linear equations with nondifferentiable terms. J. Comp. Appl. Math. *55*, 311–324 (1994).
38. Thirty years of numerical analysis in Japan. Math. Japon. *44*, 201–208 (1996).
39. (with S. Kanno and N.V. Kjurkchiev) On some methods for the simultaneous determination of polynomial zeros. Jap. J. Indust. Appl. Math. *13*, 267–288 (1996).

40. On nonlinear SOR-like methods. I-Application to simultaneous methods for polynomial zeros. Jap. J. Indust. Appl. Math. *14*, 87–97 (1997).
41. (with K. Ishihara and Y. Muroya) On nonlinear SOR-like methods. II-Convergence of the SOR-Newton method for mildly nonlinear equations. Jap. J. Indust. Appl. Math. *14*, 99–110 (1997).
42. (with K. Ishihara) On nonlinear SOR-like methods. III-Global convergence of SOR, SSOR and USSOR methods for convex problems. Japan J. Indust. Appl. Math. *15*, 135–145 (1998).
43. (with N. Matsunaga) Convergence of Swartztrauber-Sweet's approximation for the Poisson-type equation on a disk. Numer. Funct. Anal. Optimiz. *20*, 917–928 (1999).
44. (with X. Chen and N. Matsunaga) Smoothing Newton method for nonsmooth Dirichlet problems, in Reformulation: Nonsmooth, Piecewise Smooth, Semismooth and Smoothing Methods (M. Fukushima and L. Qi, Eds.), Kluwer Academic Pub., 1999, 65–79.
45. (with N. Matsunaga) Superconvergence of the Shortley-Weller approximation for Dirichlet problems. J. Comp. Appl. Math. *116*, 263–273 (2000).
46. Historical development in convergence analysis for Newton's and Newton-like methods. J. Comp. Appl. Math. *124*, 1–23 (2000).
47. (with Q. Fang) Superconvergence of finite difference approximations for convection-diffusion problems. Numer. Linear Alg. Appl. *8*, 99–110 (2001).
48. (with Q. Fang and X. Chen) Superconvergence and nonsuperconvergence of the Shortley-Weller approximations for Dirichlet problems. Numer. Funct. Anal. Optimiz. (to appear).
49. Inversion formulas for tridiagonal matrices with applications to boundary value problems. Numer. Funct. Anal. Appl. (to appear).
50. (with Q. Fang and T. Tsuchiya) Finite difference, finite element and finite volume methods applied to two-point boundary value problems. J. Comp. Appl. Math. (to appear).

Contents

Computing [Suppl] 15, 1–7 (2001)

Computing

A Unified Approach for Bounding the Positive Root of Certain Classes of Polynomials with Applications

A. Abels, Leinfelden-Echterdingen, and **J. Herzberger,** Oldenburg

Dedicated to Professor Tetsuro Yamamoto on the occasion of his 65th birthday

Abstract

The paper shows that by means of a simple variable transformation many formulas given in the literature for bounds including the positive root of certain polynomials can already be derived out of those for special simple cases of these polynomials. In the same way some more complicated cases can now be treated with the help of known results. These polynomials are occurring in mathematics of finance and in the theory of the determination of the order of convergence of iterative numerical processes.

AMS Subject Classifications: 26C05, 65H05, 91B28.

Key Words: Polynomial roots, mathematics of finance, order of convergence.

1. Introduction

In classical Mathematics of Finance often polynomial equations of certain types crop up, the positive root q^* of which determine the internal rate of return i^* of an investment by $q^* = 1 + i^*$.

For an ordinary annuity for which A denotes the periodic payments (of equal amounts), S the present value of the annuity and n the term of the annuity (which is according to the US-rule also the total number of payments), we get the equation

$$S \cdot q^n - A \cdot \sum_{j=1}^{n} q^{n-j} = 0$$

or written in another form

$$q^n - a \cdot \sum_{j=1}^{n} q^{n-j} = 0 \quad \text{with } a = A/S. \tag{1}$$

If the periodic payments are growing with a growth factor h, which means that the payments are showing geometric growth defined by

$$A_j = h^{j-1} \cdot A, \quad 1 \leq j \leq n,$$

then we get the similar equation

$$q^n - a \cdot \sum_{j=0}^{n-1} h^j \cdot q^{n-1-j} = 0 \quad \text{with } a = A/S. \tag{2}$$

In both cases the effective rate of the annuity i^* is given by $i^* = q^* - 1$ where q^* is the unique positive root of Eq. (1) or (2), resp., by Descartes' rule of signs.

A closer look at the Eqs. (1) and (2) reveals that since some decades nearly the same type of polynomials have been investigated in numerical analysis with respect to the derivation of bounds for the positive root q^* determining the rate of convergence of some iterative processes for the solution of nonlinear equations (see Traub [9]). Equation (1) with the condition $n \cdot a > 1$ was treated by Traub with quite satisfactory bounds whereas for Eq. (2) corresponding results can be found in Herzberger [2], M.S. Petković and L.D. Petković [8] and more recently in Kjurkchiev [6].

For example, in Traub [9] we can find the formulas for Eq. (1)

$$\frac{n}{n+1} \cdot (a+1) < q^* < a+1 \tag{3}$$

or more accurate

$$a+1-\frac{a}{(a+1)^n} \cdot \left(1+\frac{1}{n}\right)^n < q^* < a+1-\frac{a}{(a+1)^n} \tag{4}$$

both analytically derived for the case $n \cdot a > 1$.

The corresponding results in Herzberger [2] for Eq. (2) read as

$$\frac{n}{n+1} \cdot (a+h) < q^* < a+h \tag{5}$$

and more precisely in Kjurkchiev [6] as

$$a+h-\frac{a \cdot h^n}{(a+h)^n} \cdot \left(1+\frac{1}{n}\right)^n < q^* < a+h-\frac{a \cdot q^n}{(a+h)^n} \tag{6}$$

under the assumption that $n > h/a$ holds.

The proof of the last two formulas (5) and (6) follows the way first given in Traub [9] and the results are similar to those for Eq. (1). The formulas of M. Petković and Lj. Petković are of quite different structure, require no restriction on n unlike (3)–(6) – which is important for a variety of applications in finance – and are derived from the estimation of the Perron-root using linear algebra.

It should be mentioned that there exist some more results for more generalized polynomial equations, like in Herzberger [3].

2. Application of Variable Transformation

As pointed out by Ivanov [5] Eq. (1) can be transformed into the Eq. (2) by means of a simple variable transformation

$$u = q/h. \tag{T}$$

This should be demonstrated by starting with Eq. (2) where we apply transformation (T) and get

$$S \cdot u^n \cdot h^n = A \cdot \sum_{j=0}^{n-1} h^j \cdot (u \cdot h)^{n-1-j} = 0$$

and after dividing the whole equation by h^n

$$S \cdot u^n - (A/h) \cdot \sum_{j=0}^{n-1} u^{n-1-j} = 0$$

which finally can be written in the form

$$S \cdot u^n - \widetilde{a} \cdot \sum_{j=1}^{n} u^{n-j} = 0, \quad \widetilde{a} = A/h \tag{1*}$$

equivalently to Eq. (1).

The conclusion now is that the bounds derived for Eq. (1) given by (3) and (4) can be transformed into those given by (5) and (6) by a mere application of the transformation (T) to the formulas of the original bounds. It can be shown that the assumptions in the analytic proofs of (3) and (4) are not affected by this transformation.

The following scheme shall symbolize this very simple but nevertheless effective process:

$$\text{(3) of Traub} \overset{(T)}{\Longrightarrow} \text{(5) Herzberger [2]},$$

$$\text{(4) of Traub} \overset{(T)}{\Longrightarrow} \text{(6) Kjurkchiev [6]}.$$

Another application can be that bounds, derived for a generalized polynomial of type (2) in Herzberger [3], now are proved for the simpler case $h = 1$. Then using the same process and after a transformation (T) applied to these results we get the more complicated bounds in Abels [1].

We now demonstrate that by means of our approach even some more complicated cases not yet considered in the literature, can easily be treated. For example, we want to mention the following problem (communicated by Lichtenberg [7]).

The conditions of an ordinary simple annuity are changed in such a way that the amount A of the periodic payments is reduced to A/k $(k \geq 2)$ *and in return the total number of payments is increased by the factor k (by reducing the length of the payment intervals by a factor of* $1/k$*). How about the value of the new effective rate* j^* *of the changed annuity?*

The changed annuity leads to the polynomial equation

$$q^{n \cdot k} - a/k \cdot \sum_{j=1}^{n \cdot k} q^{n \cdot k - j} = 0 \tag{7}$$

and its unique positive root q^* defines by $j^* = q^* - 1$ the effective rate of the changed annuity.

The mathematical question is now, what is the relation between the unique positive root q^* of the polynomial Eq. (1) and q_k^* of Eq. (7)?

In Herzberger [4] the following bounds expressed by the known effective rate i of the original annuity leading to Eq. (1) are derived:

$$j^* > i/k + \frac{(a-i)/k}{1 - \frac{1}{(1+i/k)^{k \cdot n}(1-a/i)+a/i}} \tag{8}$$

$$j^* < i/k - \frac{(1+i/k)^{k \cdot n} \cdot (1-a/i) + a/i}{(1+i/k)^{k \cdot n-1}\left[k \cdot n - a/k \cdot \frac{(k \cdot n-1)i/k-1}{(i/k)^2}\right] - \frac{a/k}{(i/k)^2}}. \tag{9}$$

We are now considering the Eq. (2) and in connection with this the changed equation

$$p_k(q) = q^{n \cdot k} - a/k \sum_{j=1}^{n \cdot k} h^{n \cdot k - j} \cdot q^{j-1} = 0 \tag{10}$$

which is derived from (2) using the same procedure (which leads from (1) to (7)).

For this equation we are now able to prove the following lemma.

Lemma. *For the polynomial Eq.* (10) *one gets in case of* $h < q^*$ *the following bounds for its uniquely existing positive root* q^*

$$q^* > h + \alpha + \frac{\frac{a-p+h}{k}}{1 - \frac{1}{\left(1+\frac{p-h}{k \cdot h}\right)^{k \cdot n} \cdot \left(1-\frac{a}{p-h}\right) + \frac{a}{p-h}}} \tag{11}$$

$$q^* < h + \alpha - \frac{(h+\alpha)^{k\cdot n} \cdot \left(1 - \frac{a}{p-h}\right) + h^{k\cdot n} \cdot \frac{a}{p-h}}{(h+\alpha)^{k\cdot n-1} \cdot \left[k \cdot n - \frac{a}{k} \cdot \frac{(k\cdot n-1)\cdot\alpha - h}{\alpha^2}\right] - h^{k\cdot n} \cdot \frac{\frac{a}{k}}{\alpha^2}} \tag{12}$$

with $\alpha = \frac{p-h}{k}$, $p = 1 + i$.

In case of $h > q^*$ *only the upper bound* (12) *can be proved.*

Proof: We first apply again the transformation (T) to the Eq. (10) and thus finally get the transformed equation

$$u^{n\cdot k} - \tilde{a}/k \cdot \sum_{j=1}^{k\cdot n} u^{j-1} = 0 \quad \text{with } \tilde{a} = a/h.$$

For the quantity a we get by a simple calculation the expression

$$a = \tilde{a} \cdot h = \frac{q^* - h}{1 - (h/q^*)^n}.$$

If we now follow the proof of the bounds (8) and (9) like in Herzberger [4], we have to apply one step of the secant-method and Newton-method respectively to the simple to verify initial bounds, which are here transformed by (T) into the bounds

$$1 + \frac{\tilde{q}^* - 1}{k} < u^* < \tilde{a}/k \quad \text{with } \tilde{q}^* = q^*/h.$$

However, the technique used in Herzberger [4] works only if we can make sure that these are really lower and upper bounds for u^*. For this it is sufficient to show that for the polynomial p_k the inequalities

$$p_k(h + (q^* - h)/k) < 0 \quad \text{and} \quad p_k(h + a/k) > 0$$

hold true.

In order to prove these inequalities we can first verify the following expressions:

$$p_k(h + (q - h)/k) = \frac{(h + (q-h)/k)^{n\cdot k} \cdot (h/q)^n - h^{n\cdot k}}{(h/q)^n - 1}$$

and

$$p_k(h + a/k) = h^{n\cdot k}.$$

Since we always have $h > 0$ the second value of p_k at $h + a/k$ is always positive ensuring that $h + a/k$ is always an upper bound for u^*.

But the second value of p_k at $h + (q - h)/k$ is only negative if $h < q^*$ as a simple analysis shows. Only in this case it can be ensured that $h + (q - h)/k$ is a lower bound for u^*. Otherwise it will be transformed into another upper bound for u^*.

Since the polynomial p_k can be proved to be convex in this region, applying one step of the secant-method in the first case, an improved lower bound for u^* is calculated. In both cases one step of the Newton-method will result in a new upper bound.

The rest of the proof of our statement is a mere technical matter and only has to verify that the transformation (T) applied to the bounds (8) and (9) results in the bounds (11) and (12). The elementary details will be omitted here. □

Let us demonstrate the accuracy of the derived bounds in (11) and (12) with two numerical examples:

Example.

(a) Case $h < q^*$.
$n = 5$, $k = 2$, $i = 0.06$ ($p = 6\%$), $h = 1.05$ (5% growth rate) exact solution:
$j^* = 0.055446743$ (≈ 0.0554) bounds: $0.055249362 < j^* < 0.055447882$

(b) Case $h > q^*$.
$n = 8$, $k = 2$, $i = 0.04$ ($p = 4\%$), $h = 1.09$ (9% growth rate) exact solution:
$j^* = 0.063328562$ (≈ 0.0633) bounds: $j^* < 0.063353991$

3. Conclusions

The proposed and applied principle of a proper variable transformation proved to be very successful in deriving bounds for the positive root of polynomial equations of type (2) out of those already proved for the solution of Eq. (1).

Even new equations can be treated applying this transformation to bounds for already considered equations but it has to be verified that the transformation does not destroy the assumption made for deriving the known bounds.

Last but not least it has to be mentioned that the bounds derived by M. Petković and Lj. Petković for a more general polynomial equation than (2) proved to be resistant against our approach up to now. A closer inspection reveals that the variable transformation make the main steps of the proof break down. This shows that our proposed approach also has some limits in its application.

References

[1] Abels, A.: Schranken für positive Wurzeln bestimmter Polynomklassen (mit Anwendungen). Diploma-Thesis, Universität Oldenburg, 1999.
[2] Herzberger, J.: On the R-order of some recurrences with application to inclusion-methods. Computing *36*, 175–180 (1986a).
[3] Herzberger, J.: Bounds for the R-order of certain iterative numerical processes. BIT *26*, 259–262 (1986b).
[4] Herzberger, J.: Bounds for the positive root of a certain class of polynomials with applications. BIT *39*, 366–372 (1999).

[5] Ivanov, K.: Private communication 1998.
[6] Kjurkchiev, N.: Note on the estimation of the order of convergence of some iterative processes. BIT *32*, 525–528 (1992).
[7] Lichtenberg, H.: Private communication 1998.
[8] Petković, M. S., Petković, L. D.: On bounds of the R-order of some iterative methods. Z. Angew. Math. Mech. *69*, T 197–T 198 (1989).
[9] Traub, J. F.: Iterative methods for the solution of equations. Englewood-Cliffs: Prentice Hall, 1964.

Andreas Abels
COR Gessner GmbH
Karlsruher Str. 3
70771 Leinfelden-Echterdingen
Germany
e-mail: a.abels@corag.de

Jürgen Herzberger
Fachbereich Mathematik
Universität Oldenburg
26111 Oldenburg
Germany
e-mail: meyer@mathematik.uni-oldenburg.de

Computing [Suppl] 15, 9–19 (2001)

Computing

Numerical Verifications of Solutions for Obstacle Problems

R. P. Agarwal, Singapore, and **C. S. Ryoo**, Taegu

Dedicated to Professor Tetsuro Yamamoto on the occasion of his 65th birthday

Abstract

In this paper, we consider a numerical technique which enables us to verify the existence of solutions for some simple obstacle problems. Using the finite element approximations for obstacle problems, the explicit a priori error estimates and the Krawczyk type operator, we construct, in computer, a set of solutions which satisfies the hypothesis of Schauder's fixed point theorem for compact map on a certain Sobolev space. Further, the conditions of verifiability by this method are considered and a numerical example is presented.

AMS Subject Classifications: 65N15, 65G20.

Key Words: Obstacle problems, numerical verification, error estimates, finite element method.

1. Introduction

We propose a numerical method to verify the existence of solutions to obstacle problems for two dimensional case. The basic approach of this method consists of the fixed point formulation of obstacle problem and construction of the function set, in computer, satisfying the validation condition of a certain infinite dimensional fixed point theorem. For a fixed point formulation, a candidate set which possibly contains a solution is considered. In order to get such a candidate set, we divide the verification procedure into two phases; one is the computation of a projection into closed convex subset of some finite dimensional subspace (rounding), the other is the estimation of the error for the projection (rounding error).

Combining them with some iterative technique, the exact solution can be enclosed by sum of rounding parts, which is a subset of finite dimensional space, and the rounding error, which is indicated by a nonnegative real number. These two concepts enable us to treat the infinite dimensional problems as finite procedures, i.e., by computer. They are based on the finite dimensional fixed point theorem, i.e. Schauder's fixed point theorem and the explicit error estimates for finite element approximations. In Section 2, we describe the obstacle problem considered and a computer algorithm to construct the set satisfying the verification conditions is presented. In our numerical verification methods, the magnitude of the constants in the a priori error estimates for finite element solutions plays an essential role. In Section 3, explicit a priori error bound is presented for approximation by piecewise linear elements over triangles. A numerical example is illustrated in the last section.

2. Basic Concepts and Verification Method

In this paper, we will use the classical notation concerning the Sobolev space: $W^{k,p}(\Omega)$, k integer, $1 \leq k < \infty$, is the Sobolev spaces of real valued L^p functions on Ω, such that all their derivatives up to order k belong to $L^p(\Omega)$. If $p = 2$, we write $H^k(\Omega) = W^{k,2}(\Omega)$. We write (u, v), $\|u\|_{L^2(\Omega)}$, the scalar product and the norm in $L^2(\Omega)$. Let Ω be a convex bounded domain in R^2 with a smooth boundary $\partial\Omega$, and

$$a(u, v) = \int_\Omega \nabla u \cdot \nabla v \, dx,$$

where

$$\nabla u \cdot \nabla v = \frac{\partial u}{\partial x_1}\frac{\partial v}{\partial x_1} + \frac{\partial u}{\partial x_2}\frac{\partial v}{\partial x_2}.$$

Also set $H_0^1(\Omega) \equiv \{v \in H^1(\Omega) : v = 0 \text{ on } \partial\Omega\}$ with scalar product $(\nabla u, \nabla v)$ for $u, v \in H_0^1(\Omega)$. The H_0^1 norm and H^2 seminorm are defined by

$$\|u\|_{H_0^1(\Omega)} \equiv \|\nabla u\|_{L^2(\Omega)} \quad \text{and} \quad |u|^2_{H^2(\Omega)} \equiv \sum_{i,j=1}^{2} \left\| \frac{\partial^2 u}{\partial x_i \partial x_j} \right\|^2_{L^2(\Omega)},$$

respectively.

Next, we define $K = \{v : v \in H_0^1(\Omega), v \geq 0 \text{ a.e. on } \Omega\}$.
We now suppose the following assumptions on nonlinear function $f(\cdot)$.
A1. f is a continuous map from $H_0^1(\Omega)$ to $L^2(\Omega)$.
A2. For each bounded subset $W \subset H_0^1(\Omega)$, $f(W)$ is also bounded in $L^2(\Omega)$.

We consider the following obstacle problem.

$$\begin{cases} \text{Find } u \in K \text{ such that} \\ a(u, v - u) \geq (f(u), v - u), \quad \forall v \in K. \end{cases} \tag{2.1}$$

Now we describe a numerical method to verify the existence of solutions of (2.1). For each $u \in H_0^1(\Omega)$, by the Riesz representation theorem for Hilbert spaces, there exists a unique $F(u) \in H_0^1(\Omega)$ such that

$$(\nabla F(u), \nabla v) = (f(u), v), \quad \forall v \in H_0^1(\Omega). \tag{2.2}$$

Then the map $F : H_0^1(\Omega) \to H_0^1(\Omega)$ is a compact operator. By (2.2), problem (2.1) is equivalent to that of finding $u \in H_0^1(\Omega)$ such that

$$a(u, v - u) \geq a(F(u), v - u), \quad \forall v \in K \tag{2.3}$$

By (2.3) and the definition of P_K, we have the following fixed point problem for compact operator

$$u = P_K F(u),$$

where P_K denotes the projection operator from $H_0^1(\Omega)$ to K.

In the following, we shall follow the arguments by [11]. We determine a set V for a bounded, convex and closed subset $U \subset H_0^1(\Omega)$ as

$$V = \{v \in H_0^1(\Omega) : v = P_K F(u), u \in U\}. \tag{2.4}$$

From Schauder's fixed point theorem, if $V \subset U$ holds, then there exists a solution of (2.1) in the set U. A procedure to verify $V \subset U$ using a computer is as follows.

First, we consider the auxiliary problem: given $g \in L^2(\Omega)$,

$$\begin{cases} \text{find } u \in K \text{ such that} \\ a(u, v-u) \geq (g, v-u), \quad \forall v \in K. \end{cases} \tag{2.5}$$

We note that, by well known result [5], for any $g \in L^2(\Omega)$, the problem (2.5) has a unique solution $u \in H_0^1(\Omega) \cap H^2(\Omega)$, and the estimate

$$\|\Delta u\|_{L^2(\Omega)} \leq \|g\|_{L^2(\Omega)}$$

holds.

We now approximate the solution of (2.5) by means of a finite element approximation. Henceforth we shall assume that Ω is a bounded convex polygonal domain in R^2. We consider a classical triangulation of $\mathscr{T}_h$ of Ω and

$$\bar{\Omega} = \bigcup_{T \in \mathscr{T}_h} T.$$

We introduce $\sum_h = \{p : p \in \bar{\Omega}, p \text{ is a vertex of } T \in \mathscr{T}_h\}$ and define the approximation S_h of $H_0^1(\Omega)$ by

$$S_h = \{v_h : v_h \in H_0^1(\Omega) \cap C^0(\bar{\Omega}), v_h|_T \in P_1, \quad \forall T \in \mathscr{T}_h\},$$

where $v_h|_T$ denotes the restriction of v_h to T and P_1 representing the space of polynomials in two variables of degree ≤ 1. We then define K_h, an approximate subset of K, by

$$K_h = S_h \cap K = \{v_h : v_h \in S_h, v_h \geq 0 \text{ on } \bar{\Omega}\}$$

and the dual cone of K_h by $K_h^* = \{w \in H_0^1(\Omega) : a(w, v) \leq 0, \quad \forall v \in K_h\}$.

For given V, we calculate the rounding $R(V) \subset K_h$ and the rounding error $RE(V) \subset K_h^*$ such that

$$V \subset R(V) \oplus RE(V) \tag{2.6}$$

holds. Then it is sufficient to find U which satisfies

$$R(V) \oplus RE(V) \subset U.$$

In order to calculate the rounding $R(V)$, we define the approximate problem corresponding to (2.5) as

$$\begin{cases} \text{find } u_h \in K_h \text{ such that} \\ a(u_h, v_h - u_h) \geq (g, v_h - u_h), \quad \forall v_h \in K_h, \end{cases} \tag{2.7}$$

and (2.7) admits one and only one solution (see [5]).

Now, as one of the approximation properties of K_h, assume that, for each $u \in H_0^1(\Omega) \cap H^2(\Omega)$, there exists a positive constant C, independent of u and h, such that

$$\|u - u_h\|_{H_0^1(\Omega)} \leq Ch\|g\|_{L^2(\Omega)}. \tag{2.8}$$

When $\Omega = (0, 1)$, we may take $C = \frac{\sqrt{5}}{\pi}$ (see [11]). The two-dimensional case, is described in Section 3.

We propose computer algorithm to obtain the set $R(V)$. Following [5], (2.7) is equivalent to the quadratic programming problem

$$\min_{v_h \in K_h} \left[\frac{1}{2} a(v_h, v_h) - (g, v_h) \right]. \tag{2.9}$$

As parameters to describe a function $v_h \in S_h$ we choose the values $v_h(A_i)$ of v_h at nodes $A_i, i = 1, \ldots, M$, of $\mathscr{T}_h$. The corresponding basis functions $\phi_j \in S_h$, $j = 1, \ldots, M$, are defined by

$$\phi_j(A_i) = \delta_{ij} \equiv \begin{cases} 1, & i = j, \\ 0, & i \neq j. \end{cases}$$

A function $v_h \in S_h$ now has the representation

$$v_h(t) = \sum_{j=1}^{M} z_j \phi_j(t), \quad z_j = v_h(A_j), \quad \text{for } t \in \bar{\Omega}.$$

Then we can represent the above quadratic programming problem (2.9) in the form

$$\min_{z \geq 0} \left[\frac{1}{2} z^T D z - P^T z \right]. \tag{2.10}$$

Here, $D = (\nabla \phi_i, \nabla \phi_j)$ and $1 \leq i, j \leq M$, and z is the coefficient vector for $\{\phi_j\}$ corresponding to the function v_h in (2.9). Further, $P \equiv \big((g, \phi_j)\big)$ is an M dimensional vector.

By the Kuhn-Tucker theorem, a vector $z = (z_j) \in R^M$ with $z \geq 0$ is an optimal solution to (2.10) if and only if there exists $w = (w_j) \in R^M$ such that

$$\begin{cases} w - Dz = -P, \\ w_j z_j = 0, \quad w \geq 0, \quad z \geq 0. \end{cases} \tag{2.11}$$

Let R^+ denote the set of nonnegative real numbers. For $\alpha \in R^+$ we set

$$[\alpha] \equiv \{\phi \in K_h^* : \|\phi\|_{H_0^1(\Omega)} \leq \alpha\}. \tag{2.12}$$

Let $a_j (1 \leq j \leq M)$ be intervals on R^1 and let $\sum_{j=1}^M a_j \phi_j$ be a linear combination of $\{\phi_j\}$, i.e., an element of the power set 2^{S_h} in the following sense:

$$\sum_{j=1}^M a_j \phi_j = \left\{ \sum_{j=1}^M \hat{a}_j \phi_j : \ \hat{a}_j \in a_j, 1 \leq j \leq M \right\}.$$

Now, for a given set $U = \sum_{j=1}^M a_j \phi_j + [\alpha]$ and $g = f(U)$ in (2.7), we consider the nonlinear system

$$\begin{cases} w - Dz = -(f(U), \phi_j), & 1 \leq j \leq M, \\ w_j z_j = 0, & w \geq 0, \quad z \geq 0. \end{cases} \tag{2.13}$$

Many algorithms for solving (2.13) are designed via an equivalent system of nonlinear equations

$$\Psi(x) = 0, \tag{2.14}$$

where $\Psi : R^n \longrightarrow R^n$ is continuous. The equivalence means that x^* solves (2.13) if and only if x^* solves (2.14). In order to solve (2.14) with guaranteed accuracy, we effectively apply Theorem 2.1.

Theorem 2.1. ([9]) *Let $\Psi : R^n \to R^n$ be a function with continuous first derivative and let $H \in R^{n \times n}$ (real $n \times n$ matrix), $\tilde{x} \in R^n$. Denote the Jacobian matrix of Ψ* by *$\Psi' \in R^{n \times n}$ and for $X \in IR^n$ (real interval vectors with n components) define $\Psi'(X) := \cap\{Y \in IR^n : \Psi'(x) \in Y$ for all $x \in X\}$. If then for some $X \in IR^n$*

$$\tilde{x} - H\Psi(\tilde{x}) + \{I - H\Psi'(\tilde{x} \underline{\cup} X)\}(X - \tilde{x}) \subseteq \overset{\circ}{X},$$

then there exists an $\hat{x} \in \overset{\circ}{X}$ with $\Psi(\hat{x}) = 0$.

We can compute the solution of (2.14) and evaluate the rounding $R(V)$. First, we generate a sequence of sets $\{U^{(i)}\}$, $i = 1, 2, \ldots$, which consists of subsets of $H_0^1(\Omega)$, in the following manner. For $i = 0$, we choose appropriate initial value $u_h^{(0)} \in K_h$ and $\alpha_0 \in R^+$, and define $U^{(0)} \subset H_0^1(\Omega)$ by $U^{(0)} = u_h^{(0)} + [\alpha_0]$. Usually, $u_h^{(0)}$ will be determined as a solution of the following problem:

$$a(u_h^{(0)}, \varphi - u_h^{(0)}) \geq (f(u_h^{(0)}), \varphi - u_h^{(0)}), \quad \forall \varphi \in K_h, \ u_h^{(0)} \in K_h \tag{2.15}$$

which corresponds to the Galerkin approximation for (2.1). The standard selection for α_0 will be $\alpha_0 = 0$. Then we iterate the following procedure:

For $i \geq 0$, set

$$V^{(i)} = \{v^{(i)} \in H_0^1(\Omega) : v^{(i)} = P_K F(u^{(i)}), u^{(i)} \in U^{(i)}\}.$$

In order to enclose $V^{(i)}$, let define $R(V^{(i)})$ as follows. For $u^{(i)} \in U^{(i)}$, we define $v_h^{(i)} \in K_h$ as the solution of the following problem:

$$a(v_h^{(i)}, \varphi - v_h^{(i)}) \geq (f(u^{(i)}), \varphi - v_h^{(i)}), \quad \forall \varphi \in K_h.$$

Note that $R(V^{(i)})$ is determined as the set of $v_h^{(i)}$ for all $u^{(i)} \in U^{(i)}$. This is a subset of K_h and can be enclosed as follows making use of interval coefficients:

$$R(V^{(i)}) \subset \sum_{j=1}^{M} [\underline{a_j}, \overline{a_j}] \phi_j,$$

where ϕ_j $(j = 1, \ldots, M)$ denote the fixed basis of S_h and $a_j = [\underline{a_j}, \overline{a_j}]$.

Secondly, we define $RE(V^{(i)})$ by

$$RE(V^{(i)}) = \{v \in K_h^* : \|v\|_{H_0^1(\Omega)} \leq Ch \sup_{u^{(i)} \in U^{(i)}} \|f(u^{(i)})\|_{L^2(\Omega)}\},$$

where C is the same constant as in (2.8).

Then the verification condition is written as

$$R(V^{(i)}) \oplus RE(V^{(i)}) \subset U^{(i)}.$$

If the verification condition is not satisfied, we take

$$u_h^{(i+1)} = \sum_{j=1}^{M} [\underline{a_j} - \delta, \overline{a_j} + \delta] \phi_j,$$

$$\alpha_{i+1} = Ch \sup_{u^{(i)} \in U^{(i)}} \|f(u^{(i)})\|_{L^2(\Omega)} + \delta,$$

and define $U^{(i+1)}$ by $U^{(i+1)} = u_h^{(i+1)} \oplus [\alpha_{i+1}]$, where

$$[\alpha_{i+1}] = \{v \in K_h^* : \|v\|_{H_0^1(\Omega)} \leq \alpha_{i+1}\}.$$

Then we continue the above iteration process. The method using δ is called δ-inflation. The reader may refer to [7, 11, 12] for the details.

3. Computation of the Constant C

In this section, we give a bound of the constant C of (2.8). In the numerical verification methods, it is very important, because the actual value for C much influences the possibility and the accuracy of verification. Let Ω be a square with side length 1 and let $\mathcal{T}_h$ be the uniform triangulation of Ω according to Fig. 1.

We have the following well known estimation for the Courant's triangles (see [4]).

Lemma 3.1. *For any $\Phi \in H^2(\Omega)$, let $\tilde{\Phi}$ be the unique interpolating polynomial of degree ≤ 1 of Φ. Then we have*

$$\|\tilde{\Phi} - \Phi\|_{H_0^1(\Omega)} \leq 3\frac{\ell^2}{\rho}|\Phi|_{H^2(\Omega)}$$

and

$$\|\tilde{\Phi} - \Phi\|_{L^2(\Omega)} \leq 3\ell^2|\Phi|_{H^2(\Omega)},$$

where $\ell =$ greatest length of the sides of T, $\rho =$ diameter of the inscribed sphere of T.

Theorem 3.2. *Let $\mathcal{T}_h$ be the uniform triangulation of Ω and let u and u_h be the respective solution of* (2.5) *and* (2.7). *If $g \in L^2(\Omega)$ we then have*

$$\|u_h - u\|_{H_0^1(\Omega)} \leq Ch\|g\|_{L^2(\Omega)},$$

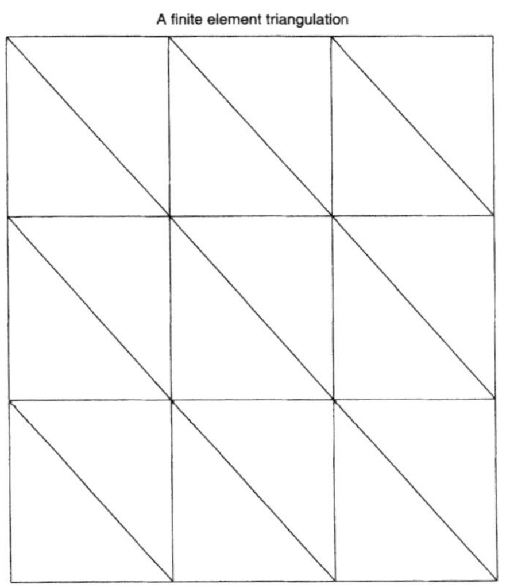

Figure 1. A finite element triangulation

where the constant C can be taken $\sqrt{9(6+4\sqrt{2})+24}$ *in* (2.8) *as in this case and h is uniform mesh size of triangle.*

Proof: Since $u_h \in K_h \subset K$ it is results from (2.5) that

$$a(u, u_h - u) \geq (g, u_h - u). \tag{3.1}$$

We deduce, by adding (2.7) and (3.1), that $\forall v_h \in K_h$,

$$\begin{cases} a(u_h - u, u_h - u) \leq a(v_h - u, u_h - u) \\ +a(u, v_h - u) - (g, v_h - u), \quad \forall v_h \in K_h \end{cases} \tag{3.2}$$

From (3.2) and the inequality

$$2a(u, v) \leq \|u\|^2_{H^1_0(\Omega)} + \|v\|^2_{H^1_0(\Omega)}, \quad \forall u, v \in H^1_0(\Omega),$$

we deduce

$$\tfrac{1}{2}\|u_h - u\|^2_{H^1_0(\Omega)} \leq \tfrac{1}{2}\|v_h - u\|^2_{H^1_0(\Omega)} + a(u, v_h - u) - (g, v_h - u), \quad \forall v_h \in K_h. \tag{3.3}$$

Since $g \in L^2(\Omega)$ implies that $u \in H^1_0(\Omega) \cap H^2(\Omega)$ and hence $-\Delta u - g \in L^2(\Omega)$ and if we put

$$\lambda = -\Delta u - g, \tag{3.4}$$

we get

$$\|\lambda\|_{L^2(\Omega)} \leq \|\Delta u\|_{L^2(\Omega)} + \|g\|_{L^2(\Omega)} \leq 2\|g\|_{L^2(\Omega)}. \tag{3.5}$$

From (3.4) we deduce that

$$a(u, v) = (g + \lambda, v), \quad \forall v \in H^1_0(\Omega), \tag{3.6}$$

and from (3.3), (3.6)

$$\tfrac{1}{2}\|u_h - u\|^2_{H^1_0(\Omega)} \leq \tfrac{1}{2}\|v_h - u\|^2_{H^1_0(\Omega)} + (\lambda, v_h - u), \quad \forall v_h \in K_h,$$

and hence from (3.5)

$$\tfrac{1}{2}\|u_h - u\|^2_{H^1_0(\Omega)} \leq \tfrac{1}{2}\|v_h - u\|^2_{H^1_0(\Omega)} + 2\|g\|_{(\Omega)}\|v_h - u\|_{L^2(\Omega)}, \quad \forall v_h \in K_h. \tag{3.7}$$

To estimate $u_h - u$ we shall use (3.7), choosing a suitable v_h; first we define the interpolation operator $\Pi_h : H^1_0(\Omega) \cap C^0(\Omega) \to S_h$ by

$$\begin{cases} \Pi_h v \in S_h, & \forall v \in H^1_0(\Omega) \cap C^0(\bar{\Omega}) \\ \Pi_h v(p) = v(p), & \forall p \in \sum_h. \end{cases} \tag{3.8}$$

We also have

$$\Pi_h v \in K_h, \quad \forall v \in K \cap C^0(\bar{\Omega}).$$

Replacing v_h by $\Pi_h u$ in (3.7) we then have

$$\tfrac{1}{2}\|u_h - u\|^2_{H^1_0(\Omega)} \le \tfrac{1}{2}\|\Pi_h u - u\|^2_{H^1_0(\Omega)} + 2\|g\|_{L^2(\Omega)}\|\Pi_h u - u\|_{L^2(\Omega)}. \tag{3.9}$$

The regularity property $u \in H^2(\Omega)$ and

$$|u|_{H^2(\Omega)} = \|\Delta u\|_{L^2(\Omega)} \le \|g\|_{L^2(\Omega)},$$

by Lemma 3.1, we have the following estimation:

$$\|\Pi_h u - u\|_{H^1_0(\Omega)} \le 3(2+\sqrt{2})h|u|_{H^2(\Omega)} \le 3(2+\sqrt{2})h\|g\|_{L^2(\Omega)} \tag{3.10}$$

and

$$\|\Pi_h u - u\|_{L^2(\Omega)} \le 6h^2|u|_{H^2(\Omega)} \le 6h^2\|g\|_{L^2(\Omega)}. \tag{3.11}$$

From (3.10)–(3.11), we have

$$\|u_h - u\|_{H^1_0(\Omega)} \le \sqrt{9(6+4\sqrt{2}) + 24}h\|g\|_{L^2(\Omega)}.$$

The problem of obtaining $L^2(\Omega)$ estimates of optimal order (that is, $O(h^2)$) of $u_h - u$ via a generalization of the Aubin-Nitsche method has not yet been completely resolved; for incomplete results in this direction we refer to Natterer [8] and Mosco [6]. But in case two dimension is still open problem.

4. Example of Numerical Verification

With the condition in Section 3, we provide a numerical example of verification in the two dimensional case according to the procedure described in the previous section. We consider the case $f(u) = 0.03u + \sin 2\pi x \sin 2\pi y$. We divide the domain into small triangle with uniform mesh size h, and choose the basis of S_h as the pyramid functions.

Conditions:

$$\begin{cases} \dim S_h = 100. \\ \text{Initial value: } u_h^{(0)} = \text{Galerkin approximation (2.15)}, \alpha_0 = 0, \\ \quad \text{the outline of } u_h^{(0)} \text{ is shown in Fig. 2.} \\ \text{Extension parameter: } \delta = 10^{-3}. \end{cases}$$

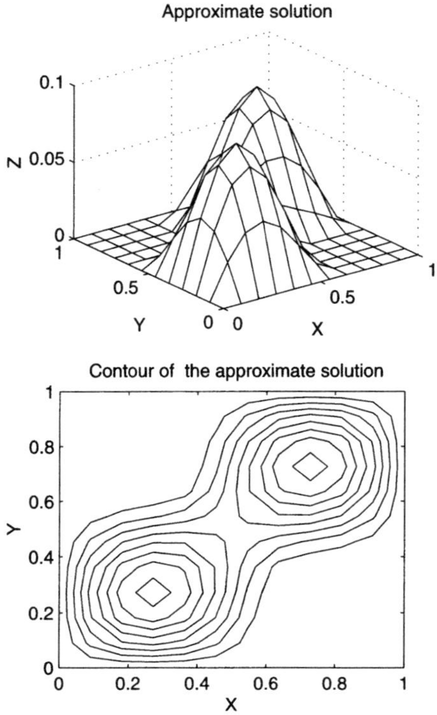

Figure 2. Approximate solution

Results:

$$\begin{cases} \text{Iteration numbers: 4.} \\ H_0^1(\Omega) - \text{error bound: 0.424273.} \\ \text{Maximum width of coefficient intervals in } \{a_j\} = 0.000428. \\ \text{Coefficient intervals: as in Table 1.} \end{cases}$$

Table 1. Coefficient intervals

j	$a_j = [\underline{a_j}, \overline{a_j}]$	j	$a_j = [\underline{a_j}, \overline{a_j}]$
1	[0.020668 0.020772]	12	[0.059283 0.059541]
2	[0.034787 0.034948]	13	[0.067163 0.067472]
3	[0.038938 0.039127]	14	[0.035724 0.036004]
4	[0.032819 0.033011]	15	[0.012006 0.012188]
5	[0.019726 0.019896]	16	[0.000000 0.000000]
6	[0.006010 0.006121]	17	[0.000000 0.000000]
7	[0.000000 0.000000]	18	[0.000000 0.000000]
8	[0.000000 0.000000]	19	[0.000000 0.000000]
9	[0.000000 0.000000]	20	[0.038938 0.039127]
10	[0.000000 0.000000]	21	[0.067163 0.067472]
11	[0.034787 0.034948]	22	[0.077217 0.077591]
	Omitted for 23 $\sim$ 100		

Remark 4.1. In [11], in order to evaluate the rounding $R(V)$, we used the interval linear system which is given in [9]. In this paper we used the Krawczyk type operator instead of using the interval linear system. Comparing the method proposed in this paper with the method presented in [11], there is a little difference between these two results and the results found in this paper are better. That is, the L^2-error bound in this paper is less than in [11]. Moreover, the maximum width of the coefficient interval in this paper is narrower than in [11]. Therefore, using the method presented in this paper, an improvement in the accuracy of the verification can be obtained.

Using computer arithmetic with double precision instead of strict interval computations (e.g., ACRITH, FORTRAN-SC, C-XSC, PROFIL, etc.), the numerical example is computed. So, the round-off errors in this example is neglected. However, it should be sufficient for our present purposes.

Acknowledgement

We are grateful to the refree for his helpful suggestions.

References

[1] Agarwal, R. P., Wong, J. Y.: Explicit error bounds for the derivatives of piecewise-Lidstone interpolation. J. Comp. Appl. Math. *58*, 67–81 (1995).

[2] Agarwal, R. P., Ryoo, C. S.: Numerical inclusion methods of solutions for variational inequalities (to submitted).

[3] Chen, X., Yamamoto, T.: On the convergence of some quasi-Newton methods for nonlinear equations with nondifferentiable operators. Computing *48*, 87–94 (1992).

[4] Ciarlet, P. G.: The finite element method for elliptic problems. Amsterdam: North-Holland 1978.

[5] Glowinski, R.: Numerical methods for nonlinear variational problems. New York: Springer 1984.

[6] Mosco, U.: Error estimate for some variational inequalities. Lecture Notes in Math. *606*, 224–236 (1977).

[7] Nakao, M. T., Yamamoto, N.: Numerical verification of solutions for nonlinear elliptic problems using an L^∞ residual method. J. Math. Anal. Appl. *217*, 246–262 (1998).

[8] Natter, F.: Optimale L^2-Konvergenz finiten Elemente bei Variationsungleichungen. Bonn. Math. Schr. *89*, 1–12 (1976).

[9] Rump, S. M.: Solving algebraic problems with high accuracy. A new approach to scientific computation. New York: Academic Press 1983.

[10] Rump, S. M.: Inclusion of zeros of nowhere differentiable n-dimensional functions. Reliable Computing *3*, 5–16 (1997).

[11] Ryoo, C. S., Nakao, M. T.: Numerical verification of solutions for variational inequalities. Numer. Math. *81*, 305–320 (1998).

[12] Yamamoto, N.: A numerical verification method for solutions of boundary value problems with local uniqueness by Banach's fixed point theorem. SIAM J. Numer. Anal. *35*, 2004–2013 (1998).

R. P. Agarwal
Department of Mathematics
National University of Singapore
Lower Kent Ridge Road
Singapore 119260
e-mail: matravip@nus.edu.sg

C. S. Ryoo
Department of Mathematics
Kyungpook National University
Taegu 702-701
Korea
e-mail: ryoocs@knu.ac.kr

Computing [Suppl] 15, 21–28 (2001)

Computing

On the Existence Theorems of Kantorovich, Moore and Miranda

G. E. Alefeld, Karlsruhe, **F. A. Potra**, Baltimore, and **Z. Shen**, Nanjing

Dedicated to Professor Tetsuro Yamamoto on the occasion of his 65th birthday

Abstract

We show that the assumptions of the well-known Kantorovich theorem imply the assumptions of Miranda's theorem, but not vice versa.

AMS Subject Classifications: 65H10.

Key Words: Nonlinear systems, existence theorems, interval analysis.

1. Introduction

The purpose of this paper is to show that the assumptions of the well-known Kantorovich theorem imply the assumptions of Miranda's theorem. This surprising fact can be concluded from a series of well-known results on the existence of solutions of a nonlinear system of equations in $\mathbf{R}^n$. At the end of the paper we give a short direct proof of this fact.

2. Some Existence Theorems and their Relations

Proving the existence of a solution of a system of nonlinear equations is a fundamental problem in nonlinear analysis. In what follows we will review three well-known theorems related to this problem: The Kantorovich Theorem, Moore's Theorem and Miranda's Theorem. The most famous of them is perhaps the Kantorovich Theorem. We will present it both in its "classical" form (see [8] or [10]), and in its "affine-invariant" form proposed by Deuflhard and Heindl [2]. Although the above mentioned theorems hold in general Banach spaces, for the purpose of this paper we will state them only for the n-dimensional space $\mathbf{R}^n$ endowed with the infinity norm. Also, we will present only the parts of the conclusion of those theorems that are relevant to our paper.

Theorem 1. (*Kantorovich*) *Let $f : D \subset \mathbf{R}^n \to \mathbf{R}^n$ be Frechét differentiable in the open convex set D. Assume that for the point $x^0 \in D$ the Jacobian $f'(x^0)$ is invertible with $||f'(x^0)^{-1}||_\infty \leq \beta$. Let there be a Lipschitz constant κ* for *f' such that*

$$||f'(u) - f'(v)||_\infty \leq \kappa ||u - v||_\infty \quad \textit{for all } u, v \in D.$$

Let

$$\|x^1 - x^0\|_\infty = \|f'(x^0)^{-1} f(x^0)\|_\infty \le \eta.$$

If $h = \beta\kappa\eta \le \frac{1}{2}$ *and* $\bar{B}(x^0, \rho_-) = \{x \in \mathbf{R}^n \mid \|x - x^0\|_\infty \le \rho_-\} \subset D$ *where*

$$\rho_- = \frac{1 - \sqrt{1-2h}}{h}\eta,$$

then f *has a zero* $x^* \in \bar{B}(x^0, \rho_-)$. □

Theorem 2. (*Affine-invariant form of the Kantorovich Theorem*) *Let* $f : D \subset \mathbf{R}^n \to \mathbf{R}^n$ *be Frechét differentiable in the open convex set* D. *Assume that for the point* $x^0 \in D$ *the Jacobian* $f'(x^0)$ *is invertible with*

$$\|f'(x^0)^{-1} f(x^0)\|_\infty \le \eta. \tag{1}$$

Let there be a Lipschitz constant ω for $f'(x^0)^{-1}f'$ *such that*

$$\|f'(x^0)^{-1}(f'(u) - f'(v))\|_\infty \le \omega\|u - v\|_\infty \quad \textit{for all } u, v \in D. \tag{2}$$

If $h = \eta\omega \le \frac{1}{2}$ *and* $\bar{B}(x^0, \rho_-) = \{x \in \mathbf{R}^n \mid \|x - x^0\|_\infty \le \rho_-\} \subset D$, *where*

$$\rho_- = \frac{1 - \sqrt{1-2h}}{\omega}, \tag{3}$$

then f *has a zero* x^* *in* $\bar{B}(x^0, \rho_-)$. □

If the hypothesis of Theorem 1 is satisfied for some constants β, κ and η, then the hypothesis of Theorem 2 is obviously satisfied with $\omega = \beta\kappa$ and η. In many cases the Lipschitz constant ω from Theorem 2 is much smaller than the product of the constants β and κ appearing in Theorem 1. In fact there are examples where Theorem 2 applies but Theorem 1 does not. For a recent application of the affine-invariant form of the Kantorovich Theorem see [9]. The next theorem contains the so-called Moore test [7] which is based on the Krawczyk operator [3].

Theorem 3. (*Moore*) *Let* $f : D \subset \mathbf{R}^n \to \mathbf{R}^n$ *be continuous on the open convex set* D. *Suppose that for* $x^0 \in D$ *and some real vector* $d = (d_i)$ *with* $d_i \ge 0,\ i = 1, \ldots, n$, *the interval vector* $[x] = [x^0 - d, x^0 + d]$ *is contained in* D *and*

$$f(x) - f(x^0) = \delta f(x^0, x)(x - x^0), \quad x \in [x],$$

where the matrix $\delta f(x^0, x)$ *is called slope of* f. *Assume furthermore that for some interval matrix* $\delta f(x^0, [x])$ *we have*

$$\delta f(x^0, x) \in \delta f(x^0, [x]) \quad \textit{for all } x \in [x]. \tag{4}$$

If there is a real nonsingular matrix A such that the Krawczyk operator

$$K([x], x^0, A) := x^0 - Af(x) + (I - A\delta f(x^0, [x]))([x] - x^0)$$

satisfies

$$K([x], x^0, A) \subset [x], \tag{5}$$

then $[x]$ *contains a zero* x^* *of* f. □

There are different possibilities for finding an interval matrix $\delta f(x^0, [x])$ for which (4) holds, such as the interval arithmetic evaluation of the Jacobian, the interval extension of the Jacobian, the interval extension of the slope etc.

In [11] Rall took for $\delta f(x^0, [x])$ the interval extension of the Jacobian of f on $[x]$ and under this assumption he performed a careful comparison of Theorems 1 and 3 coming to the conclusion that Theorem 1 is more general. Furthermore a simple example is given which shows that (5) does not hold for the ball $\bar{B}(x^0, \rho_-)$ constructed using Theorem 1. However, Theorem 3 is easier to apply. For example, no Lipschitz constant is needed for the Moore test.

Neumaier and Shen [7] took for $\delta f(x^0, [x])$ the interval extension of the slope. In contrast to the preceding case they could show that the Moore test (Theorem 3) always works if Theorem 1 can be applied, but not vice versa. Shen and Wolfe [12] performed the same comparison for Theorem 2 and the Moore test (Theorem 3). The result is the same as in [7].

The last existence theorem to be presented in this paper is Miranda's theorem (see [4] and [6]).

Theorem 4. (*Miranda*) *Let* $f : D \subset \mathbf{R}^n \to \mathbf{R}^n$ *be a continuous function. Assume that the interval vector* $[x] = [x^0 - d, x^0 + d], x^0 = (x_i^0), d = (d_i), d_i \geq 0, i = 1, \ldots, n,$ *is contained in D. Let*

$$[x]_i^+ = \{x \in [x], x_i = x_i^0 + d_i\},$$
$$[x]_i^- = \{x \in [x], x_i = x_i^0 - d_i\},$$

be the n pairs of parallel, opposite faces of the interval vector $[x]$. *If*

$$f_i(x) f_i(y) \leq 0 \quad \textit{for all } x \in [x]_i^+, y \in [x]_i^-, \quad i = 1, 2, \ldots, n,$$

then f *has at least one zero* x^* *in* $[x]$. □

Very recently Alefeld and Shen [1] proved that if $f(x)$ satisfies the hypothesis of Theorem 3 then $g(x) = Af(x)$ satisfies the hypothesis of Theorem 4, but not the other way around. Their result holds for any interval matrix $\delta f(x^0, [x])$ satisfying (4) independently of how it is obtained (interval arithmetic evaluation of the Jacobian, interval extension of the Jacobian, interval extension of the slope, etc.) Summarizing the above discussion we obtain the following result.

Theorem 5. *Assume either that the assumptions of Theorem 1 or those of Theorem 2 hold for $f(x)$. Then the assumptions of Theorem 4 hold for $g(x) = f'(x^0)^{-1}f(x)$, but not vice versa.*

This result is quite interesting but the way it was obtained above involves many concepts from interval arithmetic that do not have anything to do with the statement of Theorem 5. It is therefore useful to have a direct proof of this theorem. This will be done in the next section.

3. A Direct Proof

Proof of Theorem 5:

"$\Rightarrow$" As noted in the preceding section, if the hypothesis of Theorem 1 is satisfied then the hypothesis of Theorem 2 is satisfied with $\omega = \beta\kappa$. Therefore in what follows we assume that the hypothesis of Theorem 2 is satisfied. Let e^i denote the i-th unit vector. If one considers for a fixed $i, 1 \le i \le n$, all vectors $u^i = (u^i_j)$ with $(e^i)^T u^i = 1$ and $|u^i_j| \le 1, j = 1, \ldots, n, j \ne i$, then the pairs of opposite faces of the interval vector $[x] = [x^0 - \rho e, x^0 + \rho e], \rho \ge 0, e = (1, \ldots, 1)^T$, can be written as

$$\begin{aligned}[x]_i^+ &= \{x \in [x] \mid x_i = x_i^0 + \rho\} \\ &= \{x = x^0 + \rho u^i \mid (e^i)^T u^i = 1, |u^i_j| \le 1, j = 1, \ldots, n, j \ne i\}\end{aligned}$$

and

$$\begin{aligned}[x]_i^- &= \{x \in [x] \mid x_i = x_i^0 - \rho\} \\ &= \{x = x^0 - \rho u^i \mid (e^i)^T u^i = 1, |u^i_j| \le 1, j = 1, \ldots, n, j \ne i\},\end{aligned}$$

respectively. Let $x = x^0 \pm \rho_- u^i \in [x]_i^\pm$ where ρ_- is defined by (3). We first show that

$$\left| \int_0^1 (e^i)^T \{g'(x^0 \pm t\rho_- u^i) - g'(x^0)\} u^i dt \right| \le \frac{1}{2}\omega\rho_-, \tag{6}$$

where $g(x) = f'(x^0)^{-1}f(x)$. We have

$$\begin{aligned}&\left| \int_0^1 (e^i)^T \{g'(x^0 \pm t\rho_- u^i) - g'(x^0)\} u^i dt \right| \\ &\quad \le \int_0^1 (e^i)^T |g'(x^0 \pm t\rho_- u^i) - g'(x^0)| \cdot |u^i| dt\end{aligned}$$

$$\leq \int_0^1 \sum_{j=1}^n |g'_{ij}(x^0 \pm t\rho_- u^i) - g'_{ij}(x^0)| \cdot |u^i_j| dt$$

$$\leq \int_0^1 \sum_{j=1}^n |g'_{ij}(x^0 \pm t\rho_- u^i) - g'_{ij}(x^0)| dt$$

$$\leq \int_0^1 ||g'(x^0 \pm t\rho_- u^i) - g'(x^0)||_\infty dt$$

$$\leq \omega \int_0^1 || \pm t\rho_- u^i||_\infty dt$$

$$\leq \frac{1}{2}\omega\rho_-.$$

In what follows we use the fact that ρ_- defined by (3) satisfies the quadratic equation

$$\frac{1}{2}\omega\rho_-^2 - \rho_- + \eta = 0. \tag{7}$$

Let $x = x^0 \pm \rho_- u^i \in [x]_i^{\pm}$. By the well-known generalization of the fundamental theorem of calculus we obtain

$$\begin{aligned} g(x^0 \pm \rho_- u^i) &= g(x^0) \pm \rho_- \int_0^1 g'(x^0 \pm t\rho_- u^i) u^i dt \\ &= g(x^0) \pm \rho_- \int_0^1 \{g'(x^0 \pm t\rho_- u^i) - g'(x^0) + g'(x^0)\} u^i dt \\ &= g(x^0) \pm \rho_- g'(x^0) u^i \pm \rho_- \int_0^1 \{g'(x^0 \pm t\rho_- u^i) - g'(x^0)\} u^i dt. \end{aligned}$$

Using $g'(x^0) = I$ and $(e^i)^T u^i = 1$ gives

$$g_i(x^0 \pm \rho_- u^i) = g_i(x^0) \pm \rho_- \pm \rho_- \int_0^1 (e^i)^T \{g'(x^0 \pm t\rho_- u^i) - g'(x^0)\} u^i dt.$$

By taking the + sign and using (1), (2), (6) and (7) we obtain

$$\begin{aligned} g_i(x^0 + \rho_- u^i) & \\ &\geq -|g_i(x^0)| + \rho_- - \rho_- \left| \int_0^1 (e^i)^T \{g'(x^0 + t\rho_- u^i) - g'(x^0)\} u^i dt \right| \\ &\geq -\eta + \rho_- - \frac{1}{2}\omega\rho_-^2 = 0. \end{aligned}$$

Similarly,

$$\begin{aligned} g_i(x^0 - \rho_- u^i) & \\ &\leq |g_i(x^0)| - \rho_- - \rho_- \int_0^1 (e^i)^T \{g'(x^0 - t\rho_- u^i) - g'(x^0)\} u^i dt \\ &\leq \eta - \rho_- + \rho_- \left| \int_0^1 (e^i)^T \{g'(x^0 - t\rho_- u^i) - g'(x^0)\} u^i dt \right| \\ &\leq \eta - \rho_- + \frac{1}{2}\omega\rho_-^2 = 0. \end{aligned}$$

Hence the hypothesis of Miranda's Theorem is satisfied for $g(x)$ and $d = \rho_- e$.

"$\Leftarrow$" This part of the proof is shown by using a simple counter-example. Let

$$f(x) = -x^3 + 6x^2 - 11x + 6 = -(x-1)(x-2)(x-3)$$

for $x \in D = [x^0 - d, x^0 + d]$ with $x^0 = 0$ and some fixed $d > 0$. From

$$f'(x) = -3x^2 + 12x - 11$$

it follows $f'(x^0) = -11$ and therefore we have

$$\begin{aligned} g(x) &= f'(x^0)^{-1} f(x) \\ &= \frac{1}{11}(x^3 - 6x^2 + 11x - 6). \end{aligned}$$

Choosing $d = 1$ it holds that

$$\begin{aligned} g(x^0 - d) &= g(-1) = -\frac{24}{11}, \\ g(x^0 + d) &= g(1) = 0 \end{aligned}$$

and Theorem 4 can be applied to $g(x)$ for $D = [-1, 1]$. On the other hand we obtain immediately

$$|f'(x^0)^{-1}f(x^0)| = \frac{6}{11} =: \eta.$$

Furthermore

$$f'(x^0)^{-1}f'(x) = \frac{3}{11}x^2 - \frac{12}{11}x + 1$$

and therefore

$$\begin{aligned} |f'(x^0)^{-1}(f'(u) - f'(v))| &= \left|\frac{3}{11}(u^2 - v^2) - \frac{12}{11}(u - v)\right| \\ &= \left|\frac{3}{11}(u + v) - \frac{12}{11}\right||u - v|. \end{aligned}$$

If $u, v \in [-d, d]$ then $\frac{3}{11}(u + v) - \frac{12}{11} \in \left[-\frac{6d}{11} - \frac{12}{11}, \frac{6d}{11} - \frac{12}{11}\right]$ and therefore

$$\left|\frac{3}{11}(u + v) - \frac{12}{11}\right| \leq \frac{6}{11}d + \frac{12}{11}.$$

Hence (2) holds with $\omega := \frac{6}{11}d + \frac{12}{11}$.

Since

$$h := \eta\omega = \frac{6}{11}\left(\frac{6}{11}d + \frac{12}{11}\right) = \frac{6}{121}(6d + 12)$$

is bigger than $\frac{1}{2}$ for all positive d we can not apply Theorem 2. □

It is interesting to note that we have proved more than the fact that the hypothesis of the Kantorovich Theorem in the ∞-norm guarantees the applicability of Miranda's Theorem. Actually we have shown that the function g_i has nonpositive values on the "left face" $[x]_i^-$ and nonnegative values on the "right face" $[x]_i^+$ for all $i \in \{1, \ldots, n\}$.

Acknowledgements

The authors are grateful to U. Schäfer for carefully reading the paper. The work of F. A. Poha was supported in part by NSF, grand DMS-9996154. The work of Z. Shen was performed while he was a visiting professor at the University of Karlsruhe. He would like to acknowledge the support by DFG and the University of Karlsruhe.

References

[1] Alefeld, G., Shen, Z.: Miranda's theorem and the verification of linear complementarity problems. Institut für Wissenschaftliches Rechnen und Mathematische Modellbildung. Universität Karlsruhe (2001).

[2] Deuflhard, P., Heindl, G.: Affine invariant convergence theorems for Newton's method and extension to related methods. SIAM J. Numer. Anal. *16*, 1–10 (1980).

[3] Krawczyk, R.: Newton–Algorithmen zur Bestimmung von Nullstellen mit Fehlerschranken. Computing *4*, 187–201 (1969).

[4] Miranda, C.: Un' osservatione su un theorema di Brouwer. Bolletino Unione Mathematica Italiana 2 vol *3*, 5–7 (1940).

[5] Moore, R. E.: A Test for existence of solutions to nonlinear systems. SIAM J. Numer. Anal. *14*, 611–615 (1977).

[6] Moore, R. E., Kioustelidis, J. B.: A simple test for accuracy of approximate solutions to nonlinear (or linear) systems. SIAM J. Numer. Anal. *17*, 521–529 (1980).

[7] Neumaier, A., Shen, Z.: The Krawczyk Operator and Kantorovich's Theorem. J. Math. Anal. Appl. *149*, 437–443 (1990).

[8] Ortega, J. M., Rheinboldt, W. C.: Iterative solution of nonlinear equations in several variables. New York London: Academic Press 1970.

[9] Potra, F. A.: A path-following method for linear complementarity problems based on the affine-invariant Kantorovich Theorem. ZIB-Report 00–30, Konrad-Zuse-Zentrum, Berlin, August 2000.

[10] Rall, L. B.: Computational solution of nonlinear operator equations, New York: John Wiley 1969.

[11] Rall, L. B.: A Comparison of the existence theorems of Kantorovich and Moore. SIAM J. Numer. Anal. *17*, 148–161 (1980).

[12] Shen, Z., Wolfe, M. A.: A note on the comparison of the Kantorovich and Moore theorems. Nonlin. Anal. *15*, 229–232 (1990).

G. E. Alefeld
Institut für Angewandte Mathematik
Universität Karlsruhe
Kaiserstraße 12
D-76128 Karlsruhe
Germany
e-mail: goetz.alefeld@mathematik.uni-karlsruhe.de

F. A. Potra
Department of Mathematics and Statistics
University of Maryland
Baltimore County
1000 Hilltop Circle
Baltimore, MD 21250
USA
e-mail: potra@math.umbc.edu

Z. Shen
Department of Mathematics
Nanjing University
Nanjing, 210093
P. R. China
e-mail: shenzuhe@netra.nju.edu.cn

Computing [Suppl] 15, 29–48 (2001)

Computing

A Survey of Robust Preconditioning Methods

O. Axelsson, Nijmegen

Dedicated to Professor Tetsuro Yamamoto on the occasion of his 65th birthday

Abstract

Iterative solution methods can not yet be considered as having in general a robust behavior w.r.t. problem parameters and shape of elements. To be robust they need a robust preconditioning method. There are three classes of preconditioning methods: (i) Operator based splitting methods, (ii) Element matrix based splitting methods and (iii) Global matrix based splitting methods. Already important results have been achieved but work is still ongoing in the construction of such preconditioners. The paper gives a background to the methods and gives some more detailed examples of some of the methods when used for symmetric and positive definite matrices.

AMS Subject Classifications: 65F10, 65F50, CR: G1.3.

Key Words: Systems of linear equations, conjugate gradient method, preconditioning methods, condition numbers.

1. Classification of Preconditioning Methods

Let A be a given nonsingular operator or matrix. Frequently A is a differential operator for a boundary value problem or a corresponding finite element matrix.

By a preconditioning B we refer to a matrix or operator splitting, $A = B - R$ to be used in an iterative solution method for the equation $Ax = b$, to improve either the spectral radius of $B^{-1}R$, the condition number of the matrix $B^{-1}A$ or, more generally, to improve its distribution of eigenvalues to enable a faster rate of convergence. Here B is an operator which, in some way, is much easier to solve systems with than with A.

There are three major classes of preconditioning or splitting methods:

a) *Operator splitting* or *defect–correction methods.*

Here a simpler but more readily solvable operator is used as preconditioner or corrector for the given operator. Often it suffices to perform a few iteration steps without sacrificing approximation order of the given operator, which is then used in the computation of the defect only.

A well known example of such a method is defect–correction with a monotone correction operator, such as an upwind approximation for convection–diffusion problems. Another example is the preconditioner derived from separate displacement orderings for systems of PDEs as arises for linear elasticity theory problems. See Section 3 for a detailed description.

A similar, but algebraically constructed type of method is based on deletion of positive off-diagonal matrix entries and diagonal compensation of them. The so derived matrix can then be preconditioned more easily by some other method, which requires M-matrices, for instance. See [2, 26, 27] for further details.

b) *Element mesh based preconditioners.*

Here one utilizes a sequence of meshes (sometimes only two levels) in a successive refinement method. The matrix is partitioned in two by two blocks corresponding to the partitioning of nodes on each level in new (added) and old (previous) points. The matrix can be preconditioned by its block diagonal part.

A fundamental quality parameter of the method is the angle between the two corresponding subsets of basisfunctions, see [6, 8]. Another important issue is the preconditioning of the matrix block corresponding to the added node points.

c) *Global matrix based splittings.*

A typical example of such a method is some type of incomplete factorization method, possibly coupled with approximate inverses, or subspace projection based methods. Earlier versions of such methods were restricted to M-matrices or diagonally dominant matrices (see [2, 19]) but a new 'second order' method (see [21]) is applicable for general positive definite matrices. During the years much effort has also been devoted to the construction of algebraic multigrid methods, see [28].

Each splitting can be implemented in a *basic iteration method* or B can be used as a preconditioner in a conjugate gradient method.

We recall first in the next section some convergence properties of such iteration methods and discuss then in the following sections examples of the first two types of preconditioning methods.

The first type of method deals with the splitting on differential operator level. The arising simplified operator B can then, after discretization, be preconditioned itself by some of the algebraic (matrix) preconditioning methods in b) or c).

Methods of the third class (c) will not be discussed in this paper. During the years, much work has already been presented. We refer to [1, 2, 6, 19] for incomplete factorization and approximate inverse methods and references to previous work, to [21] for more recent work on second order robust incomplete factorization methods and to [12, 25] for approximate subspace projection type methods.

2. Some Convergence Properties of Iterative Solution Methods

The basic iterative method to solve the equation $A\hat{x} = b$ takes the form:

begin:
Given an initial approximation x^0, for $k = 0, 1, \ldots$ until convergence
Compute the residual $r^k = b - Ax^k$
Solve $B\delta^k = r^k$
Let $x^{k+1} = x^k + \delta^k$
end

It holds here that

$$\|e^{k+1}\| \leq \|B^{-1}(B - A)e^k\| \leq \|B^{-1}R\| \; \|e^k\| \leq \|B^{-1}R\|^{k+1} \|e^0\|$$

where $e^k = \hat{x} - x^k$, and the method converges monotonically for a general initial approximation if and only if $\|B^{-1}R\| < 1$. The method may converge asymptotically i.e., $\lim_{k\to\infty} \|e^k\| \to 0$, but not necessarily monotonically, if $\rho(B^{-1}R) < 1$. Often the iteration errors may grow unacceptably large for the initial steps if $\|B^{-1}R\| > 1$, even if $\rho(B^{-1}R) < 1$.

As is more common in practice, one uses B as a preconditioner in a (generalized) conjugate gradient method. In this case, the whole spectrum of $B^{-1}A$ plays a decisive role for the rate of convergence. If A and B are symmetric and positive definite and are spectrally equivalent, i.e., there exists constants α, β, with $0 < \alpha \leq \beta < \infty$ such that

$$\alpha \leq \frac{x^* Ax}{x^* Bx} \leq \beta, \quad \text{for all } x \in D(A),$$

where $D(A) = D(B)$ denotes the domain of definition, then it can be shown (see e.g. [2]) that

$$\|e^k\|_{A^{1/2}} \leq 2\left(\frac{1 - \sqrt{\alpha/\beta}}{1 + \sqrt{\alpha/\beta}}\right)^k \|e^0\|_{A^{1/2}} \tag{1}$$

where $\|e^k\|_{A^{1/2}} = (e^{k^*} Ae^k)^{1/2}$.

This estimate depends only on the spectral condition number. If one diagonally scales the matrix so that the largest eigenvalue becomes bounded by β $(\beta < 2)$ then the "negative" norms, $\|u\|_{A^{-\nu}}$, $\nu > 0$ becomes stronger than the "positive" norms, $\|u\|_{A^{\nu}}$, $\nu > 0$. If one uses such different norms, with a weaker norm on the iteration errors than on the initial error then, with $\nu = \frac{1}{2}$, the estimate becomes independent on the smallest eigenvalue (see [2]),

$$\|e^k\|_{A^{1/2}} \le \beta \min_{P_k \in \pi_k^1} \max_{0 \le x \le 1} |xP_k(x)| \, \|e^0\|_{A^{-1/2}}$$
$$= \frac{\beta}{k+1} \tan \frac{\pi}{4(k+1)} \|e^0\|_{A^{-1/2}} \sim \frac{\pi\beta}{4(k+1)^2} \|e^0\|_{A^{-1/2}}$$

This estimate shows an initial decay of the residuals $\|r^k\|_{A^{-1/2}} = \|e^k\|_{A^{1/2}}$ as $(O((k+1)^{-2})$ but the iteration errors may still be large in a stronger norm.

It can be seen that this estimate is of interest mainly when the higher eigenmodes of e^0 are dominating. Such an initial error occurs typically if one computes the initial solution on a coarser mesh and then interpolates it to the finer mesh, see [2, 3].

More generally, one can consider various condition numbers of A w.r.t. the initial error and they show a sublinear rate, say $O(k^{-2})$ of convergence in the initial phase. This phase is normally followed by a more or less pronounced linearly convergent phase as shown by (1). When one considers the whole spectrum it can be seen that this in turn is followed by a superlinearly convergent phase, which is similar to the type of convergence which occurs for operators which are compact perturbations of the identity. Indeed, the conjugate gradient method is applicable also for operators in a general Hilbert (inner product) space.

Finally we remark that the conjugate gradient method when applied in a finite dimensional subspace, in exact arithmetic is always terminating after a number of steps equal to the degree m of the minimal polynomial of A. Hence it can be considered as a direct solution method. However the above shows that in practice we consider it as an iterative method which may converge to a sufficient accuracy after much fewer steps than m.

Operator Splitting Methods as Defect–Correction Methods: If the initial error has a smooth representation, i.e. it is represented mainly by the first harmonics (eigenfunctions) of A, and B is sufficiently close to A in an approximation sense on the subspace spanned by those functions, then the basic iteration method can be implemented by taking very few iteration steps. The method is then normally referred to as a *defect–correction method* which takes the form:

begin:
  Solve $Bx^0 = b$
  Compute the defect (residual) $r^0 = b - Ax^0$
  Solve for the correction, $B\delta^1 = r^0$
  Let $x^1 = x^0 + \delta^1$
  Possibly repeat the defect–correction steps one or a few more times.
end

It holds

$$e^1 = \hat{x} - x^1 = B^{-1}(B - A)e^0,$$

and if $\|B^{-1}\| \leq c$ for some (not too large) constant c, and if

$$\|(B - A)e^0\| \leq \varepsilon \|e^0\|, \quad 0 < \varepsilon \ll 1,$$

where e^0 is smooth, then it may suffice with one or a few steps to reach a required accuracy. An important and well known example of a defect-correction method is the following.

Example 2.1. Singularly perturbed convection–diffusion, see [10] for instance. Let A be a second order convection–diffusion differential operator and let B be the same operator but with stronger $O(h)$ "artificial" diffusion, to make B a monotone operator, $B^{-1} > 0$. Then for the corresponding difference or finite element methods, and for smooth solutions it suffices with two steps to get a second order $O(h^2)$ accuracy, as in this case $\|e^0\| = O(h)$ and $\|(B - A)e^0\| = O(h)\|e^0\|$.

For some more detailed discussions of convergence of iterative methods, see [2].

3. An Operator Splitting Method in Linear Elasticity

During the years, various operator splitting methods have been used in connection with defect–correction methods. An example was given in the previous section. We consider now an important splitting method for the Lamé–Navier equations in linear elasticity, normally used as preconditioner for the conjugate gradient method. For an earlier presentation of the method and references to previous work on the method, see [4].

Problem formulation and Korn's inequalities: We give the problem formulation and discuss the Korn's inequalities which will turn out to be important in the estimates of certain condition numbers.

Consider finding the displacements of an isotropic elastic body $\Omega \subset \mathbb{R}^d$, subject to a body force $\underline{g}$ and surface force $\underline{\gamma}$ on $\Gamma_1 \subset \Gamma$. On $\Gamma_0 = \Gamma \backslash \Gamma_1$ the body is assumed to be fixed and we assume that the surface measure of Γ_0 is positive. Isotropic materials can be described by just two parameters, the Lamé coefficients λ, μ or the *elasticity modules E* and *Poisson ratio* ν. The following relations between the parameters hold,

$$\mu = \frac{E}{2(1+\nu)}, \quad \lambda = \frac{2\mu\nu}{1-2\nu}.$$

Here $E > 0$ and $0 < \nu < \frac{1}{2}$.

The upper bound of ν is taken for incompressible (i.e. fixed density) materials. In general, the material parameters are variable and have jumps at the intersection of two different materials. Let $\underline{u} = (u_1, u_2, u_3)$ be the displacement vector which is assumed to satisfy

$$\underline{u} \in V = \{\underline{u} \in [H^1(\Omega)]^d; \quad \underline{u} = 0 \text{ on } \Gamma_0\},$$

the set of admissible displacements.

The linearized strain tensor $\underline{\varepsilon} = (\varepsilon_{ij}(\underline{u}))$ is related to $\underline{u}$ by

$$\varepsilon_{ij}(\underline{u}) = \frac{1}{2}\left(\frac{\partial u_i}{\partial x_j} + \frac{\partial u_j}{\partial x_i}\right).$$

The Hooke's law (or constitutive equations) for linear elasticity relates the stresses $\underline{\sigma} = (\sigma_{ij}(\underline{u}))$ and strains,

$$\underline{\sigma} = \lambda \operatorname{tr}(\underline{\varepsilon}) I + 2\mu\underline{\varepsilon}.$$

In equilibrium, the displacement vector has the property of minimizing the energy functional

$$\varphi(\underline{u}) = \int_\Omega \left[\sum_{i,j=1}^d \mu(\varepsilon_{ij}(\underline{u}))^2 + \frac{\lambda}{2}(\operatorname{tr}(\underline{\varepsilon}))^2 - \underline{g} \cdot \underline{u}\right] d\Omega - \oint_{\Gamma_1} \underline{\gamma} \cdot \underline{u}\, d\Gamma \tag{2}$$

for all $u \in V$.

The variational formulation of this minimization problem is

$$a(\underline{u}, \underline{v}) = \int_\Omega \sum_{i,j=1}^d \sigma_{ij}(\underline{u})\varepsilon_{ij}(\underline{v}) d\Omega = \int_\Omega \underline{g} \cdot \underline{v}\, d\Omega + \oint_{\Gamma_1} \underline{\gamma} \cdot \underline{v}\, d\Gamma \quad \text{for all } v \in V. \tag{3}$$

Using the divergence theorem one obtains

$$-\sum_{j=1}^d \frac{\partial \sigma_{ij}(\underline{u})}{\partial x_j} = g_i \text{ in } \Omega \; quad, \quad \sum_j \sigma_{ij}(\underline{u})\nu_j = \gamma_i \text{ on } \Gamma_1$$

which are the *equilibrium equations*, balancing the internal and external body and surface forces. Here ν_j are the components of the unit outward pointing normal vector on Γ. If $\Gamma_0 = \emptyset$ (which case we have excluded as there is no unique solution) then we have a *pure traction problem* while if $\Gamma_1 = \emptyset$ we have a *pure displacement problem*.

Let

$$D(\underline{u}) = \frac{1}{2}\int_U \mu \sum_{i=1}^d |\underline{\nabla} u_i|^2 d\Omega, \quad S(\underline{u}) = \int_\Omega \mu \sum_{i,j=1}^d (\varepsilon_{ij}(\underline{u}))^2 d\Omega,$$

$$R(\underline{u}) = \frac{1}{2}\int_\Omega \mu |\underline{\nabla} \times \underline{u}|^2 d\Omega$$

denote the Laplacian operator energy functional, the strain energy and the rotational energy, respectively. Then

$$D(\underline{u}) = \frac{1}{2}\int_\Omega \mu \sum_{i,j=1}^{d} \left(\frac{\partial u_i}{\partial x_j}\right)^2 d\Omega, \quad S(\underline{u}) = \frac{1}{4}\int_\Omega \mu \sum_{i,j=1}^{d} \left(\frac{\partial u_i}{\partial x_j} + \frac{\partial u_j}{\partial x_i}\right)^2 d\Omega,$$

$$R(\underline{u}) = \frac{1}{4}\int_\Omega \mu \sum_{i,j=1}^{d} \left(\frac{\partial u_i}{\partial x_j} - \frac{\partial u_j}{\partial x_i}\right)^2 d\Omega$$

and an elementary computation shows that

$$S(\underline{u}) = 2D(\underline{u}) - R(\underline{u}).$$

Therefore, using also the relations $\mathrm{tr}(\underline{\varepsilon}) = \underline{\nabla} \cdot \underline{u}$, the energy functional $\varphi(\underline{u})$ in (2) can be rewritten as

$$\varphi(\underline{u}) = \int_\Omega \left[\mu \sum_{i=1}^{d} |\nabla u_i|^2 - \frac{\mu}{2}|\underline{\nabla} \times \underline{u}|^2 + \frac{\lambda}{2}(\underline{\nabla} \cdot \underline{u})^2 - \underline{g} \cdot \underline{u}\right] d\Omega - \int_{\Gamma_1} \underline{\gamma} \cdot \underline{u}\, d\Gamma,$$

which involves a kinetic energy, a rotational energy and a potential energy.

Taking variations and using the relation

$$\int_\Omega \underline{\nabla} \times \underline{u} \cdot \underline{\nabla} \times \underline{v}\, d\Omega = \int_\Omega \underline{v} \cdot [\nabla(\underline{\nabla} \cdot \underline{u}) - \Delta \underline{u}]\, d\Omega$$

one obtains the *Lamé–Navier equations*

$$\begin{aligned} &-\mu \,\mathrm{div}\,\mathrm{grad}\, \underline{u} - (\lambda + \mu)\, \mathrm{grad}\,\mathrm{div}\, \underline{u} = \underline{g} \text{ in } \Omega \\ &\lambda(\mathrm{div}\, \underline{u})\underline{v} + 2\mu\underline{\varepsilon} \cdot \underline{v} = \underline{\gamma} \text{ on } \Gamma_1 \\ &\underline{u} = 0 \text{ on } \Gamma_0 \end{aligned}$$

or

$$-(\lambda + 2\mu)\frac{\partial^2 u_i}{\partial x_i^2} - \mu \sum_{j \neq i} \frac{\partial^2 u_i}{\partial x_j^2} - (\lambda + \mu) \sum_{j \neq i} \frac{\partial^2 u_i}{\partial x_i \partial x_j} = g_i, \quad i = 1, 2, 3 \text{ in } \Omega.$$

There are two natural ways of splitting this operator keeping either

$$-(\lambda + 2\mu)\frac{\partial^2 u_i}{\partial x_i^2} - \mu \sum_{j \neq i} \frac{\partial^2 u_i}{\partial x_j^2} \tag{4}$$

or only

$$-(\lambda + 2\mu)\frac{\partial^2 u_i}{\partial x_i^2} \tag{5}$$

as the operator B.

Using Korn's inequality we can establish coercivity of the bilinear form $a(\underline{u}, \underline{v})$ in (3) and, hence, condition number (stability) estimates for the corresponding splittings.

A Korn's inequality (see [4, 17, 18, 22, 24, 27]), asserts that there exists a positive number $k_1 = k(\Omega)$ depending only on the domain Ω, or on the boundary conditions, such that

$$2D(\underline{u}) \leq k_1 S(\underline{u}), \tag{6}$$

for all vectors $\underline{u}$ satisfying certain side conditions (see below). Let V denote the set of such vector functions.

For the estimation of condition numbers we want to relate the quadratic part of the displacement energy functional

$$\varphi(\underline{u}) = S(\underline{u}) + \frac{1}{2}\int_\Omega \lambda(\operatorname{tr}(\underline{\varepsilon}(\underline{u})))^2 d\Omega$$

to $\widetilde{D}(\underline{u})$ and to $D(\underline{u})$, where

$$\begin{aligned}\widetilde{D}(\underline{u}) &= \int_\Omega \sum_{i=1}^{d}\left[\left(\mu+\frac{\lambda}{2}\right)\left(\frac{\partial u_i}{\partial x_i}\right)^2 + \frac{\mu}{2}\sum_{j\neq i}\left(\frac{\partial u_i}{\partial x_j}\right)^2\right] d\Omega \\ &\leq \int_\Omega \left(\mu+\frac{\lambda}{2}\right)\sum_{i,j=1}^{d}\left(\frac{\partial u_i}{\partial x_j}\right)^2 d\Omega\end{aligned}$$

in a spectral sense, i.e. we want to show the existence of positive numbers b_0, b_1, c_0, c_1 such that

$$b_0 D(\underline{u}) \leq \varphi(\underline{u}) \leq b_1 D(\underline{u}), \tag{7}$$

$$c_0 \widetilde{D}(\underline{u}) \leq \varphi(\underline{u}) \leq c_1 \widetilde{D}(\underline{u}), \tag{8}$$

respectively, holds for all $\underline{u} \in V$. Here $\widetilde{D}(\underline{u})$ corresponds to the first splitting (4) and $D(\underline{u})$ to the second (5). To find the upper bounds note first that $2D(\underline{u}) = S(\underline{u}) + R(\underline{u})$ shows that

$$2D(\underline{u}) \geq S(\underline{u}). \tag{9}$$

Further, by an elementary computation,

$$\begin{aligned}\int_\Omega \lambda(\operatorname{tr}(\underline{\varepsilon}(\underline{u})))^2 d\Omega &= \int_\Omega \lambda\left(\sum_{j=1}^{d}\frac{\partial u_j}{\partial x_j}\right)^2 d\Omega \\ &\leq d\int_\Omega \lambda\sum_{j=1}^{d}\left(\frac{\partial u_j}{\partial x_j}\right)^2 d\Omega \leq \left(d\max_\Omega\frac{\lambda}{\mu}\right)D(\underline{u})\end{aligned}$$

so

$$b_1 = 2 + d \max_{\Omega} \frac{\lambda}{\mu} = 2 + d \frac{2\nu_0}{1 - 2\nu_0},$$

where $\nu_0 = \max_{\Omega} \nu$.

To find the lower bound constant b_0, we use Korn's inequality (6), which implies $b_0 = 2/k_1$. Similarly using the relations $\lambda = 2\nu(\lambda + \mu)$ we find

$$\begin{aligned}\varphi(\underline{u}) &\le \int_{\Omega} \left[\sum_{i,j=1}^{d} \mu \left(\frac{\partial u_i}{\partial u_j} \right)^2 + \frac{\lambda}{2} \left(\sum_{i=1}^{d} \frac{\partial u_i}{\partial x_i} \right)^2 \right] d\Omega \\ &\le 2 \int_{\Omega} \left[\sum_{i,j=1}^{d} \frac{\mu}{2} \left(\frac{\partial u_i}{\partial x_j} \right)^2 + d\nu \frac{\lambda + \mu}{2} \sum_{i=1}^{d} \left(\frac{\partial u_i}{\partial x_i} \right)^2 \right] d\Omega,\end{aligned}$$

so

$$\varphi(\underline{u}) \le 2 \max_{\Omega} \{1, d\nu_0\} \widetilde{D}(\underline{u}).$$

Further,

$$\varphi(\underline{u}) \ge \frac{2}{k_1} D(\underline{u}) \ge \frac{2}{k_1} \min_{\Omega} \frac{\mu}{\mu + \lambda/2} \widetilde{D}(\underline{u}) = \frac{2}{k_1} \frac{1 - 2\nu_0}{1 - \nu_0} \widetilde{D}(\underline{u}),$$

i.e.

$$c_0 = \frac{2}{k_1} \frac{1 - 2\nu_0}{1 - \nu_0}, \quad c_1 = 2 \max_{\Omega} \{1, d\nu_0\}.$$

However, being a scaling with the "block diagonal" part of the operator, the upper bound is not greater than 2, so $c_1 = 2$.

If $\underline{u}$ is a pure rotation, i.e. $\partial u_i / \partial x_j = -\partial u_j / \partial x_i$, then $S(\underline{u}) = 0$ but $D(\underline{u}) > 0$ so (6) can not hold in this case. In [17], conditions were given to eliminate the occurrence of such pure rotations. The simplest case is when one assumes that

$$\sum_{i,j=1}^{d} \int_{\Omega} \frac{\partial u_i}{\partial x_j} \frac{\partial u_j}{\partial x_i} \, d\Omega = \sum_{i,j=1}^{d} \int_{\Omega} \frac{\partial u_i}{\partial x_i} \frac{\partial u_j}{\partial x_j} \, d\Omega. \tag{10}$$

Lemma 3.1. *Let $u_i \in H^1(\Omega)$, $i = 1, \ldots, d$. Then*

$$\sum_{i,j=1}^{d} \int_{\Omega} \frac{\partial u_i}{\partial x_j} \frac{\partial u_j}{\partial x_i} \, d\Omega = \sum_{i,j=1}^{d} \int_{\Omega} \frac{\partial u_i}{\partial x_i} \frac{\partial u_j}{\partial x_j} \, d\Omega$$

holds if

$$\oint_\Gamma \left(\sum_{i=1}^{n} (\underline{u} \cdot \nabla u_i) v_i - (\underline{u} \cdot \underline{v})(\nabla \cdot \underline{u}) \right) ds = 0.$$

This boundary condition holds in particular if $\underline{u} = 0$ *on* Γ. *In a 2D problem, it holds if*

$$(u_1 \underline{\tau} \cdot \nabla u_2 - u_2 \underline{\tau} \cdot \nabla u_1)|_\Gamma = 0,$$

where $(\tau_1, \tau_2) = (v_2, -v_1)$ *is the tangent vector.*

Proof: The lemma follows from the divergence theorem, $\int_\Omega \partial f / \partial x_i \, d\Omega = \oint_\Gamma f v_i \, ds$. □

A straightforward computation shows that when (10) holds, the following relation is valid,

$$2S(\underline{u}) = 2D(\underline{u}) + \int_\Omega \mu (\operatorname{div} \underline{u})^2 d\Omega, \tag{11}$$

that is,

$$D(u) \leq S(u), \tag{12}$$

so $k_1 = 2$ in this case. This is the best possible constant because equality holds in (12) when div $\underline{u} = 0$ in Ω, and the above boundary condition holds.

Further, when (10) holds, we have $\int_\Omega \mu (\nabla \cdot \underline{u})^2 d\Omega = 2(S(\underline{u}) - D(\underline{u}))$, so by (9), $\int_\Omega \mu (\nabla \cdot \underline{u})^2 d\Omega \leq 2D(\underline{u})$ and

$$\varphi(\underline{u}) = S(\underline{u}) + \int_\Omega \frac{\lambda}{2} (\underline{\nabla} \cdot \underline{u})^2 d\Omega \leq \left(2 + \max_\Omega \frac{\lambda}{\mu} \right) D(\underline{u}) \leq 2 \frac{1 - v_0}{1 - 2v_0} D(\underline{u}). \tag{13}$$

The Separate Displacement Ordering and Block Diagonal Preconditioners Based on Operator Splittings: The above spectral results can now be applied for the separate displacement ordering and block diagonal preconditioners for the stiffness matrix.

Consider the finite element variational formulation of (3),

$$a(\underline{u}, \underline{v}) = (\underline{g}, \underline{v}) \quad \text{for all } \underline{v} \in V_h^d \subset V, \tag{14}$$

where V_h is the chosen finite element subspace. The matrix vector formulation of (14) is $K\underline{u}_h = \underline{G}_h$. Using the separate displacement ordering, the corresponding stiffness matrix takes the form

$$K = \begin{bmatrix} K_{11} & K_{12} & K_{13} \\ K_{21} & K_{22} & K_{23} \\ K_{31} & K_{32} & K_{33} \end{bmatrix},$$

where the entries of K_{ij} equal $a(\varphi_k^{(j)}, \varphi_l^{(i)})$, where $\varphi_k^{(j)}, \varphi_l^{(i)}$ are the finite element basisfunctions for the node-points k and l, respectively. We have (for $i = 2$), $\varphi_l^{(i)} = (0, \hat{\varphi}_l^{(i)}, 0)$ where $\hat{\varphi}_l^{(i)}$ is the scalar basis function.

For the purpose of solving the system $K\underline{u}_h = \underline{G}_h$ by a preconditioned conjugate gradient method, we consider block diagonal preconditioners. An advantage with such preconditioners is that the corresponding systems for the diagonal blocks can be solved concurrently on d computer processors. Consider then first the preconditioner

$$K_D = \begin{bmatrix} K_{11} & 0 & 0 \\ 0 & K_{22} & 0 \\ 0 & 0 & K_{33} \end{bmatrix}. \tag{15}$$

It follows from (8) that the corresponding spectral condition number is bounded by

$$\kappa(K_D^{-1}K) \leq k_1 \frac{1 - \nu_0}{1 - 2\nu_0}.$$

Earlier derivations of somewhat less sharp or general bounds for this condition number can be found in [7] and [16].

In the preconditioner, each block can be solved by an "inner iteration" method such as an algebraic multilevel iteration method, used as a preconditioner for a conjugate gradient method. Thereby one can simply replace the matrices K_{ii} with such a method using a fixed number of iteration steps or use a variable number depending on a relative stopping criterion. In both cases the resulting preconditioner is variable, i.e. it is not the same for each outer iteration step, because the conjugate gradient method depends on an initial residual, which will change from one outer iteration to the next.

In the preconditioner (15), each block corresponds to a second order scalar problem with different ratio of anisotropy. This has the disadvantage that each block is different and the work-load is therefore not evenly distributed on the parallel computer when one processor is devoted to each block K_{ii}. Furthermore, as it turns out certain algebraic multilevel iteration methods (of red-black ordered type) do not have a robust performance with respect to anisotropy. Therefore, we consider a second version

$$\widehat{K}_D = \begin{bmatrix} \widehat{K}_{11} & 0 & 0 \\ 0 & \widehat{K}_{22} & 0 \\ 0 & 0 & \widehat{K}_{33} \end{bmatrix}.$$

Here $\widehat{K}_{ii}$ corresponds to the Laplacian operator for the ith displacement variable. The finite element space is the same for all variables but the boundary conditions can be different. The parts of $\widehat{K}_{ii}$ corresponding to the major part of the interior are identical. Therefore, when solving systems with $\widehat{K}_D$ concurrently, the load is nearly optimally distributed. Furthermore, the operators are not affected by anisotropy. Since we use a conforming finite element method, (8) holds also for subspaces of V, the following bounds of the spectral condition numbers can be derived for the above preconditioners.

Theorem 3.1. *Consider the stiffness matrix K for the Lamé–Navier's equation* (2) *and the preconditioners*

(*i*) $K_D = \text{blockdiag}(K_{11}, K_{22}, K_{33})$

(*ii*) $\widehat{K}_D = \text{blockdiag}(\widehat{K}_{11}, \widehat{K}_{22}, \widehat{K}_{33})$,

where $\widehat{K}_{ii}$ corresponds to the stiffness matrix for the Laplacian operator, using the same finite element space as for K. Then the spectral condition number satisfies

(*i*) $\kappa(K_D^{-1}K) \le k_1 \frac{1-\nu_0}{1-2\nu_0}$

(*ii*) $\kappa(\widehat{K}_D^{-1}K) \le k_1 \left(1 + d \frac{\nu_0}{1-2\nu_0}\right)$,

where k_1 is the constant in the Korn's inequality. In the special case when (11) *holds, we have the sharper bound,*

$$\kappa\left(\widehat{K}_D^{-1}K\right) \le 2\frac{1-\nu_0}{1-2\nu_0}.$$

Here $\nu_0 = \max \nu$, where ν is the Poisson contractivity ratio.

Proof: This follows from the previously given bounds for $\varphi(\underline{u})$. □

Note that with these preconditioners the condition number takes quite small values for common values of ν and they do not depend on the size (h) or order (p) of the finite element method used. In practice, each system $\widehat{K}_{ii}$ or K_{ii} is solved by inner iterations also and as has been shown in [4, 16, 27], one can then permit quite large inner iteration errors without influencing the outer iteration process much.

As mentioned previously, instead of using inner–outer iteration methods, one can use a fixed preconditioner

$$C = \begin{bmatrix} B_{11} & & 0 \\ & B_{22} & \\ 0 & & B_{33} \end{bmatrix}$$

for K, where each block B_{ii} is an approximation of K_{ii}. Here the inner iterations for K_{ii} have been replaced by a solution with matrix B_{ii}. In general, the number of outer iterations increases compared to the number of outer iterations when K_D or $\widehat{K}_D$ is used as preconditioner. Therefore there will be more matrix vector multiplications with K, which are costly as it involves d^2 matrix sub-blocks. If one is able to tune the inner iteration stopping criterion so that there will be sufficiently few inner iterations, then the inner–outer iteration method can be less costly. Furthermore, it requires less data-transport, which is important in avoiding data congestion since the available computer cache size is normally quite limited:

It follows from the above bounds that the condition number deteriorates for nearly incompressible materials where $\nu = \frac{1}{2}$. This can be handled by using a subspace projection method projecting off the null space (projection operator for $\nu = \frac{1}{2}$). Another method is to use a mixed variable method, including the pressure variable, see e.g. [13] for details and further references. This latter method allows working with the Schur complement which arises when the displacement variables are eliminated. The resulting equation is hence an equation for the pressure. As with all Schur complement systems it requires in general inner iterations. Thereby proper stopping criteria for the inner iterations are important. Such issues have been discussed in [14, 15], among others.

4. Element Based Preconditioning Methods

Standard (compatible) finite element methods are based on considering the variational formulation of a boundary value problem in a subspace of the function space of definition of the given operator. The subspace is spanned by a finite number of basic functions each with local support.

This can be used in proving spectral equivalence for operators having the same number of basisfunctions. If they are defined on the same element mesh, i.e. the same geometric partitioning of the domain of definition of the operator, then the analysis of the condition number can be done locally, element-wise. This important property tremendously simplifies both the construction and the analysis of element based preconditioners. Similarly, in connection with mesh refinements, one can use a partitioning of the basisfunctions in previous and added mesh points to form block diagonal or Schur complement preconditioners which can also be analyzed locally.

Such methods will be discussed in this section. We recall first the theoretical background and discuss some general techniques used in the construction of matrix preconditioners and their analysis.

4.1. Reduction Methods

Algebraic versions of operator splitting methods arise when some entries like positive off-diagonal entries in the matrix are deleted and moved to the diagonal entry in the same row, i.e. they are diagonally compensated for. If the reduction and movement takes place symmetrically, i.e. if an entry a_{ij} is deleted and added to a_{ii} and at the same time, a_{ji} is deleted and added to a_{jj}, then for symmetric matrices this corresponds to a perturbation of the given matrix with the matrix

$$a_{ij} \begin{bmatrix} 1 & & -1 \\ & \ddots & \\ -1 & & 1 \end{bmatrix} \begin{matrix} (i) \\ \\ (j) \end{matrix},$$

i.e. with a symmetric semidefinite matrix. This can be used in the analysis of modified incomplete factorization methods, see [2]. This type of method can be nicely analyzed in a local element context, see below. The arising matrix B is then normally monotone and various incomplete factorization methods of it can be used to give the final preconditioner for the original matrix A. See [2, 26] for further details.

A general important technique for finite element methods is to transform the arising integral over an arbitrary element to a standard reference element. In this section we consider the diffusion operator

$$\sum_{i,j=1}^{d} \frac{\partial}{\partial x_i}\left(a_{ij}\frac{\partial u}{\partial x_j}\right) + bu = f \text{ in } \Omega \tag{16}$$

with proper boundary conditions. Here $[a_{ij}]_{i,j=1}^{d}$ is assumed to be symmetric and positive definite and $b \geq 0$. For simplicity of the presentation we assume that they are constant on each element. The variational formulation of (16) is

$$a(u,v) = \int_\Omega \left(\sum_{i,j=1}^{d} a_{ij}\frac{\partial u}{\partial x_j}\frac{\partial v}{\partial x_i} + buv\right) dx\,dy \tag{17}$$

for all $u, v \in \mathring{H}^1(\Omega)$, satisfying the essential boundary conditions.

We illustrate the transformation method for a planar domain ($d = 2$) using a triangulation of it.

Transformation is equivalent to a change of the coefficients in the differential equation operator.

First we show that the analysis for an arbitrary triangle (e) with coordinates (x_i, y_i), $i = 1, 2, 3$ can be done on the reference triangle $(\tilde{e})$, with coordinates $(0,0)$, $(1,0)$, $(0,1)$. Transforming the finite element function between these triangles, the element bilinear form in a $2D$ problem where $b = 0$ (see e.g. [4]) becomes

$$\begin{aligned} a_e(u,v) &= a_{\tilde{e}}(\tilde{u},\tilde{v}) \\ &= \iint_{\tilde{e}} \left[\frac{\partial \tilde{u}}{\partial \tilde{x}}, \frac{\partial \tilde{u}}{\partial \tilde{y}}\right] \begin{bmatrix} (x_2 - x_1) & (y_2 - y_1) \\ (x_3 - x_1) & (y_3 - y_1) \end{bmatrix}^{-1} \begin{bmatrix} a_{11} & a_{12} \\ a_{21} & a_{22} \end{bmatrix} \\ &\qquad \times \begin{bmatrix} (x_2 - x_1) & (x_3 - x_1) \\ (y_2 - y_1) & (y_3 - y_1) \end{bmatrix}^{-1} \left[\frac{\partial \tilde{v}}{\partial \tilde{x}}, \frac{\partial \tilde{v}}{\partial \tilde{y}}\right] d\tilde{x}\,d\tilde{y} \end{aligned}$$

where $0 < \tilde{x}, \tilde{y} < 1$, i.e., it takes the form

$$\tilde{a}_e(\tilde{u},\tilde{v}) = \iint_{\tilde{e}} \sum_{i,j} \tilde{a}_{ij} \frac{\partial \tilde{u}}{\partial \tilde{x}_i}\frac{\partial \tilde{v}}{\partial \tilde{x}_j} d\tilde{x}\,d\tilde{y},$$

where $\tilde{x}_1 = \tilde{x}$, $\tilde{x}_2 = \tilde{y}$ and where the coefficients $\tilde{a}_{ij}$ depend on both the coordinates in e (or, equivalently, the angles in e) and the coefficients a_{ij} in the differential operator. A similar form holds in $3D$ problems and with $b > 0$.

We conclude that it suffices for the analysis of the finite element method to consider (17) for the reference triangle and arbitrary coefficients $[a_{ij}]$, or alternatively, for the operator $-\Delta$ (where $[a_{ij}] = \begin{bmatrix} 1 & 0 \\ 0 & 1 \end{bmatrix}$) and an arbitrary triangle e. The following Lemma will be useful for the analysis of the condition numbers when a reduction method has been used.

Lemma 4.1. *Let $\{A_i\}_{i=1}^n$ be symmetric positive semidefinite (s.p.s.d.) matrices and let $A_i = M_i - N_i$, $A = \sum_1^n A_i$, $M = \sum_{i=1}^n M_i$, $N = \sum_{i=1}^n N_i$, where M_i are s.p.s.d. Assume, for some positive α_i and $\beta_i \geq \alpha_i$, that*

$$\alpha_i M_i \leq N_i \leq \beta_i M_i,$$

where the inequalities are in a positive semidefinite sense. Then

$$(1 - \beta)M \leq A \leq (1 - \alpha)M,$$

where $\alpha = \min_i \alpha_i$, $\beta = \max_i \beta_i$.

Proof: Since $N = \sum_{i=1}^n N_i$, we have

$$N \leq \sum_{i=1}^{n} \beta_i M_i \leq \beta \sum_{i=1}^{n} M_i = \beta M$$

and similarly, $N \geq \alpha M$. Therefore,

$$\alpha M \leq N \leq \beta M$$

and, since $A = M - N$,

$$(1 - \beta)M \leq A \leq (1 - \alpha)M. \quad \square$$

The Lemma shows also that if all splittings $A_i = M_i - N_i$ are convergent, then the splitting $A = M - N$ is also convergent, and $\rho(M^{-1}N) \leq \max_i \rho(M_i^{-1}N_i)$, assuming here that M_i are nonsingular. Hence to construct preconditioners satisfying the bound $(1 - \alpha)/(1 - \beta)$ it suffices to construct local preconditioners satisfying the bounds $\alpha_i M_i \leq N_i \leq \beta_i N_i$ and, similarly, to construct a convergent splitting for the global matrix, it suffices to construct convergent splittings for the elementary matrices.

Example 4.1. Consider a triangulation of a planar domain of definition for a general diffusion operator (16), where for simplicity $b = 0$. As has been shown above, when analyzing spectral properties, we may equally well just consider the Laplacian operator for a general mesh. Each such triangular element contains the three vertex nodes and the three midedge nodes.

Let $K^{(1)}$ be the assembled piecewise linear finite element matrix for the four congruent triangles which arises when each pair of midedge points have been connected and let $K^{(2)}$ be the element matrix which arises when we use quadratic basisfunctions in all of the six nodepoints. Then (see e.g. [6]) ordering the midedge nodes first,

$$K^{(1)} = \frac{1}{2}\begin{bmatrix} 2d & -2c & -2b & 0 & -a & -a \\ -2c & 2d & -2a & -b & 0 & -b \\ -2b & -2a & 2d & -c & -c & 0 \\ 0 & -b & -c & b+c & 0 & 0 \\ -a & 0 & -c & 0 & a+c & 0 \\ -a & -b & 0 & 0 & 0 & a+b \end{bmatrix}$$

$$K^{(2)} = \frac{1}{6}\begin{bmatrix} 8d & -8c & -8b & 0 & -4a & -4a \\ -8c & 8d & -8a & -4b & 0 & -4b \\ -8b & -8a & 8d & -4c & -4c & 0 \\ 0 & -4b & -4c & 3(b+c) & c & b \\ -4a & 0 & -4c & c & 3(a+c) & a \\ -4a & -4b & 0 & b & a & 3(a+b) \end{bmatrix}$$

where $a = \cot\alpha$, $b = \cot\beta$, $c = \cot\gamma$, $d = a + b + c$ and α, β, γ are the angles in the given triangle.

We want to compute an upper bound for the condition number of $B^{-1}A$ when A and B are the assembled (global) finite element matrices for the piecewise quadratic and piecewise linear basisfunctions, respectively, for the whole mesh. As follows from Lemma 4.1, this can be done by computing the generalized condition number β/α of the local matrices, $K^{(1)}, K^{(2)}$ which are specially related by $\alpha K^{(1)} \le K^{(2)} \le \beta K^{(1)}$. The latter can be computed in the following way. We observe first that if we use *matrix reduction* and diagonal compensation of the off-diagonal entries in the matrix block $K_{11}^{(2)}$, then the reduced matrix is nothing but $\frac{4}{3}K^{(1)}$. Further, we have

$$K^{(2)} = \frac{4}{3}K^{(1)} - \frac{1}{6}\begin{bmatrix} 0 & 0 \\ 0 & \begin{matrix} (b+c) & -c & -b \\ -c & (a+c) & -a \\ -b & -a & (a+b) \end{matrix} \end{bmatrix} = \frac{4}{3}K^{(1)} - N \qquad (18)$$

and it remains to compute the spectral relation between $K^{(1)}$ and N. This can be done as follows.

Schur complement method (see [2]).

Let $K^{(1)} = B$, where $B = \begin{bmatrix} B_{11} & B_{12} \\ B_{21} & B_{22} \end{bmatrix}$ is partitioned consistently with the partitioning $N = \begin{bmatrix} 0 & 0 \\ 0 & N_{22} \end{bmatrix}$. Then $B = \begin{bmatrix} 0 & 0 \\ 0 & S \end{bmatrix} + QQ^T$ where $S = B_{22} - B_{21}B_{11}^{-1}B_{12}$ and $Q = \begin{bmatrix} B_{11}^{1/2} & 0 \\ B_{21}B_{11}^{-1/2} & I \end{bmatrix}$. Hence, $B \ge \begin{bmatrix} 0 & 0 \\ 0 & S \end{bmatrix}$ and it remains to find the spectral relation between S and N_{22}. This can be done algebraically. However, in the context of two-level finite element methods, it can be done in a more elegant way via the constant in the strengthened Cauchy–Bunyakowsky–Schwarz (CBS) inequality for the corresponding split subspaces.

Two-level finite element methods: To be specific we illustrate this method for triangular meshes. Two-level methods arise in mesh refinement methods. Given a 'coarse' triangle, it is subdivided in four congruent triangles by joining the midedge nodes. In the so arising six node points we can use either standard basisfunctions for the small triangles or hierarchical basisfunctions, where we keep the previous basisfunctions in the vertex node points of the coarse triangle and add new basisfunctions in the midedge nodes. The latter can be piecewise linear, with the support only on the adjacent three triangles (see Fig. 1), called the h-version.

Alternatively, we can use a piecewise quadratic basisfunction in the added node points (4, 5 and 6), called the p-version $(p = 2)$.

The following relation holds between the corresponding standard A and hierarchical ($\widehat{A} = \widehat{A}_h$ or $\widehat{A} = \widehat{A}^{(p)}$) basisfunction matrices. By assumption, A and $\widehat{A}$ are symmetric and positive definite.

Let $J = \begin{bmatrix} I_1 & J_{12} \\ 0 & I_2 \end{bmatrix}$ where I_1, I_2 are identity matrices corresponding to the coarse vertex node set and the midedge node set, respectively, and J_{12} is the interpolation matrix, with two non-zero entries $(=\frac{1}{2})$ in each row, corresponding to linear interpolation on the midedge node for each edge. The following relation holds then between functions (vectors) represented in the hierarchical and standard basis,

$$v_{SB} = J v_{HB}.$$

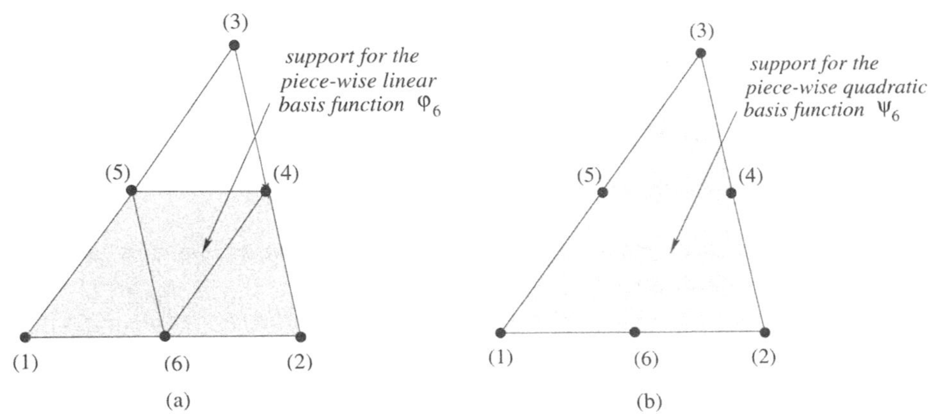

Figure 1. Support for the piecewise linear and quadratic basisfunctions

From this relation follows

$$\widehat{A} = J^T A J$$

and an elementary computation, using this shows the next relations between the corresponding matrix blocks,

$$\widehat{A}_{11} = A_{11}, \widehat{A}_{12} = A_{12} + A_{11}J_{12}, \widehat{A}_{21} = A_{21} + J_{12}^T A_{11},$$
$$\widehat{A}_{22} = A_{22} + J_{12}^T A_{12} + A_{21}J_{12} + J_{12}^T A_{11}J_{12}.$$

Further, $\widehat{A}_{22} = A_{2h}$, i.e. the standard basisfunction matrix for the coarse (unrefined) mesh and $\widehat{S} = S$, where $\widehat{S} = \widehat{A}_{22} - \widehat{A}_{21}A_{11}^{-1}\widehat{A}_{12}$ and $S = A_{22} - A_{21}A_{11}^{-1}A_{12}$.

Further the following spectral relations hold (see [2], [6], [8]):

$$(1-\gamma)\begin{bmatrix} A_{11} & 0 \\ 0 & \widehat{A}_{22} \end{bmatrix} \le \begin{bmatrix} A_{11} & \widehat{A}_{12} \\ \widehat{A}_{21} & \widehat{A}_{22} \end{bmatrix} \le (1+\gamma)\begin{bmatrix} A_{11} & 0 \\ 0 & \widehat{A}_{22} \end{bmatrix},$$

$$(1-\gamma^2)\widehat{A}_{22} \le S \le \widehat{A}_{22},$$

where $\gamma = \gamma_p = \left\{\rho\left(\widehat{A}_{22}^{-1/2} A_{21}A_{11}^{-1}A_{12}\widehat{A}_{22}^{-1/2}\right)\right\}^{1/2}$, and all inequalities are sharp.

Here γ, $0 \le \gamma < 1$, is identical to the constant in the strengthened CBS-inequality

$$u^T A v \le \gamma\left\{u^T A u \; v^T A v\right\}^{1/2},$$

which holds for all orthogonal vectors u, v, $u = [0,0,0,\alpha_1,\alpha_2,\alpha_3]^T$, $v = [\beta_1,\beta_2,\beta_3,0,0,0]^T$. We let γ_1, γ_2 denote the constants for the h-version (i.e. for $p = 1$) and the p-version (i.e. for $p = 2$), of hierarchical basisfunctions, respectively.

We return now to example 4.1. Here it holds (with a slight change in notations)

$$A_h^{(2)} = \frac{4}{3}A_{h/2}^{(1)} - \frac{1}{3}\begin{bmatrix} 0 & 0 \\ 0 & A_h^{(1)} \end{bmatrix}. \tag{19}$$

A relation between the CBS constants γ_1, γ_2.

Relation (19) implies the following relation between γ_1 and γ_2.

Theorem 4.1. *For any regularly refined finite element mesh into congruent elements, for which* (19) *is valid, one has*

$$\gamma_2^2 = \tfrac{4}{3}\gamma_1^2 \tag{20}$$

where γ_1, γ_2 are the CBS constants for piecewise linear and piecewise quadratic finite elements, respectively.

Proof: Taking Schur complements, (19) shows that

$$S^{(2)} = \tfrac{4}{3}S^{(1)} - \tfrac{1}{3}A_h^{(1)}.$$

Hence

$$\frac{x_2^T S^{(2)} x_2}{x_2^T A_h^{(2)} x_2} = \frac{4}{3}\frac{x_2^T S^{(1)} x_2}{x_2^T A_h^{(1)} x_2} - \frac{1}{3}$$

i.e.

$$\begin{aligned} 1 - \gamma_2^2 &= \min_{x_2} \frac{x_2^T S^{(2)} x_2}{x_2^T A_h^{(2)} x_2} = \frac{4}{3}\min_{x_2} \frac{x_2^T S^{(1)} x_2}{x_2^T A_h^{(1)} x_2} - \frac{1}{3} \\ &= \frac{4}{3}(1 - \gamma_1^2) - \frac{1}{3} = 1 - \frac{4}{3}\gamma_1^2. \quad \square \end{aligned}$$

The relation (20) has been shown previously in [23], [5] using a more involved derivation. The new derivation can be found in [6′].

As has been shown in [13], [11] local element methods can be used also to define preconditioners for the global matrix block A_{11}. This matrix is well-conditioned for model type problems but becomes increasingly ill-conditioned for problems where the mesh or the coefficients $[a_{ij}]$ deteriorate, such as for meshes with big aspect ratios or anisotropic coefficients. The above methods give preconditioners with a resulting condition number which remains bounded independent on the mesh or the coefficients in the differential operator.

Finally we mention that since all bounds have been derived locally, but do also hold for the global (assembled) matrices, the bounds do not depend on jumps in the coefficients or on varying element size. Hence, the preconditioning is robust.

References

[1] Ajiz, M. A., Jennings, A.: A robust incomplete Choleski conjugate gradient algorithm. Int. J. Num. Meth. Eng. *20*, 949–966 (1984).

[2] Axelsson, O.: Iterative solution methods. New York: Cambridge University Press, 1994.

[3] Axelsson, O.: Condition numbers for the study of the rate of convergence of the conjugate gradient method. In: Iterative Methods in Linear Algebra II (Margenov, S., Vassilevski, P. S., eds.), pp. 3–33. New Jersey: IMACS, 1996.

[4] Axelsson, O.: On iterative solvers in structural mechanics; separate displacement orderings and mixed variable methods. Math. Comput. Simul. *40*, 11–30 (1999).

[5] Axelsson, O.: Stabilization of algebraic multilevel iteration methods, additive methods. Numer. Alg. *21*, 23–47 (1999).

[6] Axelsson, O., Barker, V. A.: Finite element solution of boundary value problems, theory and computation. Orlando: Academic Press, 1984.

[6′] Axelsson, O., Blaheta R.: Two simple derivations of universal bounds for the C.B.S. inequality constant (in preparation).

[7] Axelsson, O., Gustafsson, I.: Iterative methods for the solution of the Navier's equations of elasticity. Comput. Meth. Appl. Mech. Eng. *15*, 241–258 (1978).

[8] Axelsson, O., Gustafsson, I.: Preconditioning and two-level multigrid methods of arbitrary degree of approximation. Math. Comp. *40*, 219–242 (1983).

[9] Axelsson, O., Kaporin, I.: Error norm estimation and stopping criteria in preconditioned conjugate gradient iterations. Num. Linear Algebra Appl. *8*, 265–286 (2001).

[10] Axelsson, O., Layton, W.: Defect correction methods for convection dominated, convection diffusion equations. RAIRO Modél. Math. Anal. Numér. *24*, 423–455 (1990).

[11] Axelsson, O., Margenov, S.: An optimal order multilevel preconditioner with respect to problem and discretization parameters, Report no. 0015 (2000), Department of Mathematics, University of Nijmegen, The Netherlands.

[12] Axelsson, O., Neytcheva, M. G., Polman, B.: The bordering method as a preconditioning method, Vestnik Mosc. Univ. Journal, Seria 15. Vicnisl. Math. Cybern. *1*, 3–25 (1995).

[13] Axelsson, O., Padiy, A.: On a robust and scalable linear elasticity solver based on a saddle point formulation. Int. J. Numer. Meth. Eng. *44*, 801–818 (1999).

[14] Axelsson, O., Vassilevski, P.: A black-box generalized conjugate gradient solver with inner iterations and variable-step preconditioning. SIAM J. Matrix Anal. Appl. *12*, 625–644 (1991).

[15] Axelsson, O., Vassilevski, P.: Construction of variable-step preconditioners for inner–outer iteration methods. In: Iterative methods in linear algebra (Beauwens, R., de Groen, P., eds.), pp. 1–14. Amsterdam: North-Holland, 1992.

[16] Blaheta, R.: Displacement decomposition-incomplete factorization preconditioning techniques for linear elasticity problems. Numer. Linear Algebra Appl. *1*, 107–128 (1994).

[17] Friedrichs, K. O.: On the boundary value problems of the theory of elasticity and Korn's inequality. Ann. Math. *48*, 441–471 (1947).

[18] Horgan, C. O.: Korn's inequalities and their applications in continuum mechanics. SIAM Rev. *37*, 491–511 (1995).

[19] Jennings, A., Malik, G. M.: Partial elimination. J. Inst. Math. Appl. *20*, 307–316 (1977).

[20] Jung, M., Langer, U., Semmler, U.: Two-level hierarchically preconditioned conjugate gradient methods for solving linear elasticity finite element equations. BIT *29*, 748–768 (1989).

[21] Kaporin, I. E.: High quality preconditioning of a general symmetric positive definite matrix based on its $U^TU + U^TR + R^TU$-decomposition. Numer. Linear Algebra Appl. *5*, 483–509 (1998).

[22] Korn, A.: Über einige Ungleichungen, welche in der Theorie der elastischen und elektrischen Schwingungen eine Rolle spielen. Bull. Int. Cracov. Acad. Umiejet, 706–724 (1908).

[23] Maitre, J. T., Musy, F.: The contraction number of a class of two level methods: an exact evaluation for some finite element subspaces and model problems. In: Multigrid methods (Hackbusch, W., Trottenberg, U., eds.), pp. 535–544. Lecture Notes in Mathematics, 960, Berlin Heidelberg New York: Springer, 1982.

[24] Ovtchinnikov, E. E., Xanthis, L. S.: A new Korn's type inequality for thin domains in application to iterative methods. Comput. Methods Appl. Mech. Eng. *138*, 299–315 (1996).

[25] Padiy, A., Axelsson, O., Polman, B.: Generalized augmented matrix preconditioning approach and its application to iterative solution of ill-conditioned algebraic systems. SIAM J. Matrix Anal. Appl. (to appear).

[26] Saint-George, P., Warzee, G., Notay, Y., Beauwens, R.: High-performance PCG solver for FEM structural analysis. Int. J. Numer. Meth. Eng. *39*, 1133–1160 (1996).

[27] Shlafman, S., Efrat, I.: Using Korn's inequality for an efficient iterative solution of structural analysis problems. In: Iterative Methods in Linear Algebra (Beauwens, R., de Groen, P., eds.), pp. 575–581. Amsterdam: North Holland 1992.

[28] Stüben, K.: Algebraic Multigrid (AMG): An introduction with applications. In: Multigrid (Trottenberg, U., Oosterlee, C. W., Schüller, A., eds.). New York: Academic Press, 2000.

O. Axelsson
Department of Mathematics
University of Nijmegen
Toernoidveld 1
625 ED Nijmegen
e-mail: axelsson@sci.kun.nl

Computing [Suppl] 15, 49–60 (2001)

Computing

A Box-Constrained Optimization Algorithm with Negative Curvature Directions and Spectral Projected Gradients

E. G. Birgin, USP, and **J. M. Martínez,** UNICAMP

Dedicated to Professor Tetsuro Yamamoto on the occasion of his 65th birthday

Abstract

A practical algorithm for box-constrained optimization is introduced. The algorithm combines an active-set strategy with spectral projected gradient iterations. In the interior of each face a strategy that deals efficiently with negative curvature is employed. Global convergence results are given. Numerical results are presented.

AMS Subject Classifications: 49M07, 49M10, 65K, 90C06, 90C20.

Key Words: Box constrained minimization, active set methods, spectral projected gradients, dogleg path methods.

1. Introduction

The problem considered in this paper is

$$\text{Minimize } f(x) \text{ subject to } x \in \Omega, \tag{1}$$

where

$$\Omega = \{x \in \mathbb{R}^n \mid \ell \leq x \leq u\}, \tag{2}$$

$\ell, u \in \mathbb{R}^n$, $\ell < u$ and $f : \mathbb{R}^n \to I\mathbb{R}$ has continuous second partial derivatives.

In a recent paper [9] a spectral projected gradient method (SPG) was introduced for solving optimization problems with convex constraints. This algorithm is easily implementable when the constraints are simple, as it is the case of the box (2). Numerical tests shown in [9] reveal that SPG can be quite efficient in very large and practical problems. See also [7, 8, 32].

In this research, SPG iterations are combined with an active set strategy that allows one to use negative curvature directions in the active face. Essentially, what we do here is to use SPG when an adequate test reveals that the current face must be abandoned and to use the algorithm recently introduced in [35] when the test recommends to stay in the current face.

The motivation for this research came from the observation of the behavior of some numerical algorithms. The algorithm described in [25] is a box-constrained method for large-scale problems that uses trust regions and, at each iteration, solves a box-constrained quadratic subproblem using a specific quadratic solver. The quadratic solver was introduced in [22] and updated in [6, 14, 24, 25]. A related algorithm was introduced in [16] where, also, interesting physical interpretations were given. See, also, [17–20]. This box-constraint quadratic algorithm uses an active-set strategy by means of which one stays in a face or abandons it according to the relation between components of the gradient. The quadratic method solved successfully many applied problems [17–19] and was shown to be more efficient than other large-scale quadratic solvers in numerical studies [14, 15]. This motivated us to apply the principles of the quadratic solver directly to the original bound constrained problem (1), instead of using these principles on a subproblem. In fact, the philosophy of [6, 14, 22, 24, 25] seems to be independent of the fact of dealing with a quadratic or not.

The present research represents our first attempt of constructing an algorithm that solves (1) using the principles of [22]. For this implementation we used two recently introduced tools for unconstrained optimization and convex-constrained optimization. The unconstrained optimization algorithm used inside the faces is the dogleg method defined in [35], which is able to deal efficiently with negative curvatures. The algorithm used for leaving the faces is the spectral projected gradient introduced in [9] (see also [23, 33, 34]) which uses first order information in a very efficient and economical way. Although our final objective is a method for large-scale problems (we have in mind its association with augmented Lagrangian algorithms and other algorithms for general constrained optimization like the ones in [11, 12, 29–31]), we found it useful to begin our numerical research with an implementation that uses full factorization of matrices.

In practical numerical analysis, the efficiency of many algorithms for a class of problems is strongly linked to the efficiency of algorithms for solving simpler problems. The reason is that simpler problems play the role of subproblems that must be solved (perhaps many times) by the main algorithm. For example, the efficiency of trust region methods like the one described in [25] is related to the development of algorithms for minimizing box-constrained quadratics. On the other hand, the efficiency of active set methods is related to the development of unconstrained minimization algorithms. The development of new attractive unconstrained optimization methods like the one in [35] naturally stimulates its application to active-set strategies in constrained minimization. This was an independent motivation for our present research.

The organization of this paper is as follows. The algorithm is described in Section 2. In Section 3, basic global convergence theorems are proved. In Section 4 we present numerical experiments. Conclusions are given in Section 5.

2. The Algorithm

As in [22], let us divide the feasible set Ω into disjoint open faces, as follows. For all $I \subset \{1, 2, \ldots, n, n+1, n+2, \ldots, 2n\}$, we define

$$F_I = \{x \in \Omega \mid x_i = \ell_i \text{ if } i \in I, x_i = u_i \text{ if } n+i \in I, \ell_i < x_i < u_i \text{ otherwise}\}.$$

We also define V_I the smallest affine subspace that contains F_I and S_I the parallel linear subspace to V_I.

Throughout this paper, $\|\cdot\|$ will be the Euclidean norm, although many times it can be replaced by an arbitrary norm on $\mathbb{R}^n$. For all $y \in \mathbb{R}^n$, the orthogonal projection of y onto a convex set S will be denoted $P_S(y)$. The (continuous) projected gradient at $x \in \Omega$ is defined as

$$g_P(x) = P_\Omega(x - \nabla f(x)) - x.$$

For all $x \in F_I$, we define

$$g_I(x) = P_{S_I}[g_P(x)].$$

The main algorithm considered in this paper is described below.

Algorithm 2.1. Assume that $x^0 \in \Omega$ is an arbitrary initial point, $\eta \in (0,1)$ and $0 < \sigma_{\min} \leq \sigma_{\max} < \infty$. Let F_I be the face that contains the current iterate x^k. Assume that $g_P(x^k) \neq 0$ (otherwise the algorithm terminates). At the main iteration of the algorithm we perform the test:

$$\|g_I(x^k)\| / \|g_P(x^k)\| \geq \eta. \tag{3}$$

If (3) takes place, we judge that it is convenient that the new iterate belongs to $\bar{F}_I$ (the closure of F_I) and, so, we compute x^{k+1} using Algorithm 2.2 below. If (3) does not hold, we decide that some constraints should be abandoned and, so, the new iterate x^{k+1} is computed using the SPG iteration described by Algorithm 2.3. In this case, before the computation of x^{k+1} we compute the spectral gradient coefficient σ_k in the following way. If $k > 0$ then

$$\sigma'_k = \frac{(x^k - x^{k-1})^T(\nabla f(x^k) - \nabla f(x^{k-1}))}{(x^k - x^{k-1})^T(x^k - x^{k-1})}.$$

Otherwise, define $\kappa = \frac{\nabla f(x^k)^T \nabla f(x^k)}{\nabla f(x^k)^T H(x^k) \nabla f(x^k)}$. If $\kappa > 0$ set $\sigma'_k = \kappa$, else define $\sigma'_k = 1/\|g_P(x^k)\|$. Finally,

$$\sigma_k = \begin{cases} \sigma_{\max}, & \text{if } \sigma'_k < 0; \\ \min\{\sigma_{\max}, \max\{\sigma_{\min}, \sigma'_k\}\}, & \text{otherwise.} \end{cases}$$

The algorithm used inside the faces is the following:

Algorithm 2.2. Assume, without loss of generality, that the first m variables are free ($\ell_j < x_j^k < u_j \ \forall i = 1, \dots, m$) at face F_I and $x_j = x_j^k$ for all $x \in F_I$. Define

$$\varphi(x_1, \dots, x_m) = f(x_1, \dots, x_m, x_{m+1}^k, \dots, x_n^k).$$

So, $\varphi : \mathbb{R}^m \to \mathbb{R}$. Let Γ be the piecewise linear path defined by the Zhang-Xu algorithm [35] for the minimization of φ, using $(x_1^k, \dots, x_m^k)$ as current point. If $\Gamma \subset \Omega$, compute $y \in \mathbb{R}^m$ as the "next iterate" given by the algorithm [35] and define $x^{k+1} = (y_1, \dots, y_m, x_{m+1}^k, \dots, x_n^k)$. Otherwise, proceed in one of the following ways:

(i) Let y be the trial point defined by the algorithm [35]. If

$$f(P_\Omega(y_1, \dots, y_m, x_{m+1}^k, \dots, x_n^k)) < f(x^k),$$

define $x^{k+1} = P_\Omega(y_1, \dots, y_m, x_{m+1}^k, \dots, x_n^k)$. Otherwise, proceed as in (ii).

(ii) Replace Γ by the connected component of $\Gamma \cap \Omega$ that contains x^k and compute y as the next iterate of the dogleg algorithm [35], taking the point of Γ that belongs to the boundary of F_I as first trial point. Define

$$x^{k+1} = (y_1, \dots, y_m, x_{m+1}^k, \dots, x_n^k).$$

Algorithm 2.3 is the algorithm used when it is necessary to leave the current face, according to the test (3).

Algorithm 2.3. Compute x^{k+1} as the next iterate of a monotone SPG2 iteration [9] with the spectral step σ_k.

Remark. Observe that $x^{k+1} \notin \bar{F}_I$ if $x^k \in F_I$ and x^{k+1} is computed by Algorithm 2.3. In this case, (3) does not hold, so $\|g_P(x^k)\| > \|g_I(x^k)\|$. Since the components corresponding to the free variables of $g_I(x^k)$ and $g_P(x^k)$ are the same, this means that $g_P(x^k)$ has nonnull components corresponding fixed variables. Therefore, $x^k + \alpha g_P(x^k) \notin \bar{F}_I$ for all $\alpha > 0$. So, $P_\Omega(x^k + \alpha g_P(x^k)) \notin \bar{F}_I$ for all $\alpha > 0$. But, according to the SPG2 iteration,

$$x^{k+1} = x^k + \alpha'[P_\Omega(x^k + \alpha g_P(x^k)) - x^k]$$

for some $\alpha > 0, \alpha' > 0$. This implies that $x^{k+1} \notin \bar{F}_I$.

3. Convergence

Theorem 3.1. *Algorithm 2.1 is well defined.*

Proof: This is a trivial consequence of the fact that the algorithms defined in [35] and [9] are well defined. □

Theorem 3.2. *Suppose that there exists $\bar{k} \in \{0, 1, 2, \ldots\}$ such that $x^k \in F_I$ for all $k \geq \bar{k}$. Then, every limit point of $\{x^k\}$ is first-order stationary and*, moreover, *the Hessian submatrix corresponding to the free variables is positive semidefinite at each limit point.*

Proof: In this case, x^{k+1} is computed by Algorithm 2.2 for all $k \geq \bar{k}$. This corresponds, essentially, to use the algorithm [35] for the unconstrained minimization of φ. Therefore, with slight modifications, the convergence theorems of [35] can be applied. Thus, the gradient $\nabla\varphi(x^k)$ tends to zero and the Hessian of φ is positive semidefinite at the limit point. But, by a straightforward projection argument, it follows that $\|g_I(x^k)\| \leq \|\nabla\varphi(x^k)\|$, therefore $\|g_I(x^k)\| \to 0$. Since (3) holds, this implies that $\|g_P(x^k)\| \to 0$. So, every limit point is first-order stationary. □

Theorem 3.3. *Suppose that for all $k \in \{0, 1, 2, \ldots\}, x^k \in F_I$, there exists $k' > k$ such that $x^{k'} \notin F_I$. Then, there exists a limit point of $\{x^k\}$ that is first-order stationary.*

Proof: If $x^{k+1} \notin F_I$, then

$$x^{k+1} \in \bar{F}_I - F_I \tag{4}$$

or

$$x^{k+1} \notin \bar{F}_I. \tag{5}$$

But, since the number of constraints is finite, (4) cannot take place infinitely many times unless (5) also holds infinitely many times. Therefore, the hypothesis of the theorem implies that (5) holds for infinitely many iterations. This implies that there exists a subsequence $\{x^k, k \in K\}$ such that x^{k+1} is computed by Algorithm 2.3 for all $k \in K$. Therefore, the whole sequence can be considered as monotone decreasing f-sequence ($f(x^{k+1}) \leq f(x^k) \forall k$) with infinitely many spectral gradient iterations. Therefore, updating the proof of convergence of SPG2 given in [9], we conclude that all the limit points of $\{x^k, k \in K\}$ are stationary. This completes the proof.

Theorem 3.4. *Suppose that all the stationary points of* (1) *are nondegenerate.* ($\frac{\partial f}{\partial x_i}(x) = 0$ *only if* $\ell_i < x_i < u_i$.) *Then, the hypothesis of Theorem* 3.2 (*and, hence, its thesis*) *must hold.*

Proof: Suppose that the thesis is not true. Therefore, the hypothesis of Theorem 3.3 holds. Since the number of faces is finite, there exists a face F_I and an infinite set K such that $x^k \in F_I$ and $x^{k+1} \notin \bar{F}_I$ for all $k \in K$. Let x^* be a limit point of $\{x^k, k \in K\}$. By Theorem 3.3, this point is first-order stationary. Without loss of generality, assume that in F_I the first m variables are free and the remaining $n - m$ are on its lower bound. Therefore,

$$\frac{\partial f}{\partial x_j}(x^*) > 0 \quad \forall j = m+1, \ldots, n.$$

By continuity this implies that, for k large enough, $k \in K$,

$$\frac{\partial f}{\partial x_j}(x^k) > 0 \quad \forall\, j = m+1, \ldots, n.$$

Therefore, for all $k \in K$, k large enough, the nonnull components of $g_P(x^k)$ are the nonnull components of $g_I(x^k)$. So, $g_P(x^k) = g_I(x^k)$ and, hence, the test (3) is satisfied. So, $x^{k+1} \in \bar{F}_I$, which is a contradiction. □

4. Numerical Experiments

In order to assess the reliability of the new algorithm, we tested it against the well known package LANCELOT [11] using a set of bound constrained problems with more than 10 variables and less than 500 variables from the CUTE [10] collection (version of May 28th, 1998). We present the results in two tables. The first corresponds to non-quadratic problems with less than 100 variables and the second to problems with more than 100 variables. As a whole, we have 60 problems. In both cases, our method used $\eta = 0.1$. In the small-dimensional set Algorithm 2.2 used the strategy (ii). The strategy (i) was used in the large-dimensional tests.

In our numerical experiments we used all except one of the defalut options for LANCELOT, we chose a full-matrix preconditioner (bandsolver with band $= n$). This is the adequate option that must be tested against an algorithm that uses full factorization of matrices, as the one presented in this paper. Therefore, at a LANCELOT iteration, the sequence of procedures to obtain the trial point are:

(i) Obtaining the "Cauchy point" [11]. If the Cauchy point is a reasonable approximate solution of the quadratic model, then it is chosen as the trial point. Otherwise, the step (ii) is executed.

(ii) Perform preconditioned conjugate gradient iterations in the face determined by the Cauchy point. If a face of lower dimension is encountered, preconditioned conjugate gradient iterations continue in the smaller face. This finishes when the norm of the internal gradient to the face is small enough. No constraints are abandoned in this process.

If the Cauchy point is interior, the Hessian is positive definite, and the first conjugate gradient iteration is interior too, then the trial point is a full Newton step and coincides with the first conjugate gradient iteration. More conjugate gradient iterations can be necessary if the conjugate-gradient iterate falls outside the trust region or if the Hessian is not positive definite.

For deciding when to stop the execution of the algorithms declaring convergence we used the criterion $\|g_P(x_k)\|_\infty \leq 10^{-5}$. We also stopped the execution when 1000 function evaluations were completed without achieving convergence.

All the experiments were run in a SPARCstation Sun Ultra 1, with an UltraSPARC 64 bits processor, 167-MHz clock and 128-MBytes of RAM memory. All codes are in Fortran and were compiled with f 77 compiler (SC 4.0 Fortran 77 4.0) using the -O optimization compiler option.

In the tables, GE means "gradient evaluations", FE means "function evaluations" and TIME is CPU time in seconds. For the new method, we report HE (Hessian evaluations) and in LANCELOT we report CG (conjugate gradient iterations). The number of Hessian evaluations in LANCELOT is GE-1.

Looking at Tables 1 and 2 we observe that:

1. In 9 problems the algorithms clearly arrived to different solutions. In 6 cases (SINEALI(20), S368(50), LINVERSE(19), S368(100), HADAMALS(400) and NCVXBQP3(100)) the solution obtained by LANCELOT was better than the one obtained by the new method.
2. In 8 problems, it is not clear whether different functional values at the final point correspond to different solutions or to different final precisions. The new method was better than LANCELOT in 5 of these cases.
3. Concerning the 43 problems where both methods arrived to the same solution, we observed that our method used less computer time than LANCELOT in 35 problems. However, it must be warned that in 19 of these problems the CPU time used by LANCELOT cannot be justified in terms of GE, FE or CG. Considering that the same compilation options have been used for both methods, these time differences must be due to implementation details. It is worth noting that, to compute the objective function as well as its gradient and Hessian, LANCELOT exploits structure in a more efficient way than the one provided by the interface tools of the CUTE collection [10].
4. The number of SPG iterations is, in our method, equal to the difference between GE-1 and HE. It is perhaps surprising that, in most cases, this

Table 1. Non-quadratic bound constrained problems with dimension between 10 and 99

Problem (n)	New method					LANCELOT				
	GE	FE	HE	TIME	f	GE	FE	CG	TIME	f
CHEBYQAD (50)	138	186	137	26.24	5.386D − 03	70	88	176	6.85	5.386D − 03
DECONVB (61)	51	60	46	1.08	8.638D − 03	16	20	108	0.31	5.720D − 03
EXPLIN (12)	22	22	19	0.01	−6.850D + 03	11	13	11	0.03	−6.850D + 03
EXPLIN2 (12)	22	22	19	0.01	−7.092D + 03	12	13	15	0.03	−7.092D + 03
EXPQUAD (12)	18	18	16	0.01	−4.201D + 03	12	14	20	0.04	−4.201D + 03
HADAMALS (36)	40	41	38	0.19	1.303D + 01	17	18	129	0.15	1.308D + 01
HATFLDC (25)	5	5	4	0.01	8.264D − 14	5	5	3	0.03	7.770D − 19
HS110 (50)	3	3	2	0.02	−9.990D + 09	2	2	0	0.01	−9.990D + 09
LINVERSE (19)	7	7	6	0.01	7.000D + 00	15	18	29	0.06	6.000D + 00
MCCORMCK (50)	6	6	5	0.02	−4.613D + 01	6	7	5	0.05	−4.613D + 01
NONSCOMP (50)	10	10	9	0.02	1.180D − 18	9	9	8	0.06	8.207D − 17
PROBPENL (50)	174	211	172	0.99	−2.331D + 05	2	3	0	0.03	3.918D − 06
QR3DLS (40)	31	39	30	0.12	6.707D − 13	28	33	32	0.17	8.376D − 13
QRTQUAD (12)	20	31	17	0.01	−3.608D + 03	36	44	42	0.08	−3.608D + 03
S368 (50)	20	20	19	0.92	−3.063D + 01	7	8	9	0.46	−3.400D + 01
SCON1LS (52)	629	882	628	2.17	1.086D − 16	339	397	1011	1.95	2.169D − 10
SINEALI (20)	9	9	8	0.01	−1.896D + 03	872	1000	1211	1.81	−1.901D + 03

Table 2. Bound constrained problems with dimension between 100 and 500

Problem (n)	New method					LANCELOT				
	GE	FE	HE	TIME	f	GE	FE	CG	TIME	f
BDEXP (100)	13	13	12	0.09	3.471D − 05	11	11	10	0.22	3.918D − 05
BIGGSB1 (100)	103	287	52	0.31	1.500D − 02	52	51	50	0.35	1.500D − 02
CHENHARK (100)	6	6	5	0.04	−2.000D + 00	25	24	61	0.25	−2.000D + 00
CVXBQP1 (100)	6	6	5	0.04	2.273D + 02	2	1	0	0.07	2.273D + 02
EXPLIN (120)	40	40	37	0.07	−7.238D + 05	14	13	50	0.11	−7.238D + 05
EXPLIN2 (120)	41	42	38	0.07	−7.245D + 05	12	11	24	0.10	−7.245D + 05
EXPQUAD (120)	24	28	21	0.58	−3.626D + 06	16	18	46	0.53	−3.626D + 06
GRIDGENA (170)	4	4	3	0.08	4.800D + 02	4	3	3	0.14	4.800D + 02
HADAMALS (400)	93	114	84	101.71	2.098D + 02	29	28	1989	35.79	1.121D + 02
HARKERP2 (100)	37	37	36	14.74	−5.000D − 01	2	1	2	0.14	−5.000D − 01
JNLBRNG1 (100)	5	6	3	0.03	−1.790D − 01	2	1	1	0.06	−1.790D − 01
JNLBRNG2 (100)	5	5	3	0.03	−3.953D + 00	3	2	2	0.07	−3.953D + 00
JNLBRNGA (100)	6	6	2	0.02	−3.612D − 01	3	2	2	0.07	−3.612D − 01
JNLBRNGB (100)	6	6	3	0.02	−7.255D + 00	4	3	3	0.07	−7.255D + 00
LINVERSE (199)	12	16	11	0.28	6.800D + 01	25	30	473	5.49	6.800D + 01
MCCORMCK (100)	6	6	5	0.05	−9.179D + 01	6	7	5	0.13	−9.179D + 01
NCVXBQP1 (100)	52	52	51	0.34	−1.996D + 06	2	1	0	0.07	−1.996D + 06
NCVXBQP2 (100)	72	72	70	0.49	−1.333D + 06	3	2	4	0.07	−1.333D + 06
NCVXBQP3 (100)	56	56	52	0.40	−6.601D + 05	4	3	4	0.08	−6.708D + 05
NOBNDTOR (484)	14	26	7	1.14	−4.980D − 01	7	6	6	2.58	−4.980D − 01
NONSCOMP (100)	10	10	9	0.06	3.652D − 18	9	9	8	0.18	2.387D − 16

OBSTCLAE (100)	6	6	4	0.04	1.398D + 00	3	2	29	0.12	1.398D + 00
OBSTCLAL (100)	8	9	3	0.02	1.398D + 00	4	3	3	0.07	1.398D + 00
OBSTCLBL (100)	7	7	3	0.02	2.875D + 00	3	2	6	0.07	2.875D + 00
OBSTCLBM (100)	3	3	2	0.02	2.875D + 00	2	1	3	0.07	2.875D + 00
OBSTCLBU (100)	4	4	1	0.01	2.875D + 00	2	1	1	0.06	2.875D + 00
PROBPENL (500)	701	1000	700	2438.96	−9.215D + 04	2	1	0	4.44	3.992D − 07
QR3DLS (155)	163	215	162	17.44	7.084D − 12	112	142	286	10.36	2.719D − 11
QRTQUAD (120)	21	31	18	0.49	−3.625D + 06	100	126	160	2.26	−3.625D + 06
S368 (100)	19	20	18	4.39	−1.260D + 02	7	7	8	2.21	−1.337D + 02
SCON1LS (102)	568	772	567	7.26	6.906D − 16	684	820	755	12.40	1.945D − 16
TORSION1 (484)	16	28	7	0.76	−4.561D − 01	8	7	7	1.09	−4.561D − 01
TORSION2 (484)	10	13	6	1.02	−4.561D − 01	5	4	112	8.58	−4.561D − 01
TORSION3 (484)	8	9	3	0.28	−1.242D + 00	4	3	3	0.33	−1.242D + 00
TORSION4 (484)	11	11	7	0.75	−1.242D + 00	5	4	115	2.49	−1.242D + 00
TORSION5 (484)	4	7	1	0.09	−2.885D + 00	2	1	1	0.28	−2.885D + 00
TORSION6 (484)	7	7	5	0.61	−2.885D + 00	4	3	24	0.52	−2.885D + 00
TORSIONA (484)	14	23	6	0.68	−4.161D − 01	7	6	6	1.03	−4.161D − 01
TORSIONB (484)	10	15	6	1.08	−4.161D − 01	5	4	114	10.39	−4.161D − 01
TORSIONC (484)	8	10	3	0.24	−1.199D + 00	4	3	3	0.34	−1.199D + 00
TORSIOND (484)	11	11	7	0.78	−1.199D + 00	6	5	166	6.97	−1.199D + 00
TORSIONE (484)	4	7	1	0.09	−2.841D + 00	2	1	1	0.28	−2.841D + 00
TORSIONF (484)	7	7	5	0.63	−2.841D + 00	4	3	34	0.62	−2.841D + 00

number is very small. This means that, in many cases, the new method works on faces of decreasing dimensions. This may be due to a characteristic of the initial points of the test set [10]. Therefore, the internal algorithm [35] have more influence in the overall behavior of the method than the strategy for leaving the faces.

5. Conclusions

We have introduced a reliable algorithm for solving small to medium dimensional box constrained problems. The algorithm is of active-set type and inside the faces it uses the method introduced in [35]. Since the algorithm [35] is able to deal with negative curvature regions and, in fact, cannot converge to a stationary point where the Hessian is indefinite, we expected that its combination with SPG could have the property of finding better minimizers than alternative first-order methods. Numerical experiments did not confirm this conjecture. Whereas in some cases we obtained lower functional values than LANCELOT, the opposite situation occurred in at least as many cases. In the cases in which our method and LANCELOT found the same solution, the computer time of the new method tends to be smaller, but in many cases this seems to be due to implementation features of LANCELOT that are not easy to detect.

In these set of problems it has been quite remarkable the monotone decrease property of the dimension of the current face. We do not think that such a property will be maintained in large-scale problems (the kind of situation where the quadratic algorithm [22] and its updatings showed to be effective). Therefore, the objective of finding an efficient algorithm for general box constraints that uses the philosophy of [22] remains valid. The present research shows that, perhaps, adapting the algorithm [35] to large-scale optimization by means of sparse Cholesky and Bunch–Parlett factorizations is not as promising as we expected. However, much research should be expected along these lines. In any case, we think that the work in [35] opened the possibility of dealing very cheaply with trust-region paths. Refinements of the main ideas of [35] will possibly give rise of even more efficient implementations of the unconstrained algorithm and can motivate further adaptations to active-set strategies in constrained optimization. The dialog between unconstrained methods and active-set strategies will continue, as far as interesting unconstrained research continues to be done.

With respect to small and medium-scale problems, it is probably interesting to test the algorithm introduced in [13] as internal-face solver. Its comparison with [35] in the unconstrained case is, of course, of independent interest. However, the large-scale situation is the most challenging one to be addressed in the near future. Large-scale bound constrained (or even unconstrained) problems lie in the nucleus of many efficient algorithms for complementarity problems and variational inequalities (see [1–5, 21, 26–28] and many others) which are largely used in modern Engineering.

Acknowledgement

We are indebted to an anonymous referee whose comments helped us to improve the paper. E. G. Birgin was supported by PRONEX-Optimization 76.79.1008-00 and FAPESP (Grant 99/08029-9) J. M. Martínez was supported by PRONEX-Optimization 76.79.1008-00, FAPESP (Grant 90-3724-6), CNPq and FAEP-UNICAMP.

References

[1] Andreani, R., Friedlander, A., Martínez, J. M.: On the solution of finite-dimensional variational inequalities using smooth optimization with simple bounds. J. Optim. Theory Appl. *94*, 635–657 (1997).

[2] Andreani, R., Martínez, J. M.: On the solution of the extended linear complementarity problem. Linear Alg. Appl. *281*, 247–257 (1998).

[3] Andreani, R., Martínez, J. M.: Solving complementarity problems by means of a new smooth constrained nonlinear solver. In: Reformulation – Nonsmooth, Piecewise Smooth, Semismooth and Smoothing Methods (Fukushima, M., Qi, L., eds.), pp. 1–24. New York: Kluwer, 1999.

[4] Andreani, R., Martínez, J. M.: On the reformulation of nonlinear complementarity problems using the Fischer-Burmeister function. Appl. Math. Lett. *12*, 7–12 (1999).

[5] Andreani, R., Martínez, J. M.: Reformulation of variational inequalities on a simplex and compactification of complementarity problems. SIAM J. Optim. *10*, 878–895 (2000).

[6] Bielschowsky, R. H., Friedlander, A., Gomes, F. M., Martínez, J. M., Raydan, M.: An adaptive algorithm for bound constrained quadratic minimization. Invest. Operat. *7*, 67–102 (1998).

[7] Birgin, E. G., Biloti, R., Tygel, M., Santos, L. T.: Restricted optimization: a clue to a fast and accurate implementation of the common reflection surface stack method. J. Appl. Geophys. *42*, 143–155 (1999).

[8] Birgin, E. G., Chambouleyron, I., Martínez, J. M.: Estimation of the optical constants and the thickness of thin films using unconstrained optimization. J. Comput. Phys. *151*, 862–880 (1999).

[9] Birgin, E. G., Martínez, J. M., Raydan, M.: Nonmonotone spectral projected gradient methods on convex sets. SIAM J. Optim. *10*, 1196–1211 (2000).

[10] Bongartz, I., Conn, A. R., Gould, N. I. M., Toint, Ph. L.: CUTE: constrained and unconstrained testing environment. ACM Trans. Math. Software *21*, 123–160 (1995).

[11] Conn, A. R., Gould, N. I. M., Toint, Ph. L.: Global convergence of a class of trust region algorithms for optimization with simple bounds. SIAM J. Numer. Anal. *25*, 433–460 (1988).

[12] Conn, A. R., Gould, N. I. M., Toint, Ph. L.: A globally convergent augmented Lagrangean algorithm for optimization with general constraints and simple bounds. SIAM J. Numer. Anal. *28*, 545–572 (1991).

[13] Dennis, J. E., Echebest, N., Guardarucci, M., Martínez, J. M., Scolnik, H., Vacchino, C.: A curvilinear search using tridiagonal secant updates for unconstrained optimization. SIAM J. Optim. *1*, 352–372 (1991).

[14] Diniz-Ehrhardt, M. A., Dostál, Z., Gomes-Ruggiero, M. A., Martínez, J. M., Santos, S. A.: Nonmonotone strategy for minimization of quadratics with simple constraints. Appl. Math. (in press).

[15] Diniz-Ehrhardt, M. A., Gomes-Ruggiero, M. A., Santos, S. A.: Comparing the numerical performance of two trust-region algorithms for large-scale bound-constrained minimization. In: International Workshop of Numerical Linear Algebra and Optimization (Sampaio, R. J. B., Yuan, J. Y., eds.), pp. 23–24. Department of Mathematics, Universidade Federal do Paraná, Brazil, (1997).

[16] Dostál, Z.: Box constrained quadratic programming with proportioning and projections. SIAM J. Optim. *7*, 871–887 (1997).

[17] Dostál, Z., Friedlander, A., Santos, S. A.: Solution of coercive and semicoercive contact problems by FETI domain decomposition. Cont. Math. *218*, 82–93 (1998).

[18] Dostál, Z., Friedlander, A., Santos, S. A.: Augmented Lagrangians with adaptive precision control for quadratic programming with equality constraints. Comput. Optim. Appl. *14*, 1–17 (1999).

[19] Dostál, Z., Friedlander, A., Santos, S. A.: Adaptive precision control in quadratic programming with simple bounds and/or equalities. In: High Performance Algorithms and Software in Nonlinear Optimization: Applied Kluwer (De Leone, R., Murli, A., Pardalos, P., Toraldo, G., eds.), New York: Kluwer (in press).

[20] Dostál, Z., Gomes, F. A. M., Santos, S. A.: Solution of contact problems by FETI domain decomposition with natural coarse space projections. Comput. Methods Appl. Mech. Eng. (in press).
[21] Fernandes, L., Friedlander, A., Guedes, M., Júdice, J.: Solution of General Linear Complementarity Problem using smooth optimization and its application to bilinear programming and LCP. Appl. Math. Optim. *43*, 1–19 (2001).
[22] Friedlander, A., Martínez, J. M.: On the maximization of a concave quadratic function with box constraints. SIAM J. Optim. *4*, 177–192 (1994).
[23] Friedlander, A., Martínez, J. M., Molina, B., Raydan, M.: Gradient methods with retards and generalizations. SIAM J. Numer. Anal. *36*, 275–289 (1999).
[24] Friedlander, A., Martínez, J. M., Raydan, M.: A new method for large-scale box constrained convex quadratic minimization problems. Optim. Methods Software *5*, 57–74 (1995).
[25] Friedlander, A., Martínez, J. M., Santos, S. A.: A new trust region algorithm for bound constrained minimization. Appl. Math. Optim. *30*, 235–266 (1994).
[26] Friedlander, A., Martínez, J. M., Santos, S. A.: On the resolution of large scale linearly constrained convex minimization problems. SIAM J. Optim. *4*, 331–339 (1994).
[27] Friedlander, A., Martínez, J. M., Santos, S. A.: Solution of linear complementarity problems using minimization with simple bounds. J. Global Optim. *6*, 1–15 (1995).
[28] Friedlander, A., Martínez, J. M., Santos, S. A.: A new strategy for solving variational inequalities on bounded polytopes. Numer. Funct. Anal. Optim. *16*, 653–668 (1995).
[29] Gomes, F. M., Maciel, M. C., Martínez, J. M.: Nonlinear programming algorithms using trust regions and augmented Lagrangians with nonmonotone penalty parameters. Math. Program. *84*, 161–200 (1999).
[30] Krejić, N., Martínez, J. M., Mello, M. P., Pilotta, E. A.: Validation of an augmented Lagrangian algorithm with a Gauss-Newton Hessian approximation using a set of hard-spheres problems. Comput. Optim. Appl. *16*, 247–263 (2000).
[31] Martínez, J. M.: BOX-QUACAN and the implementation of Augmented Lagrangian algorithms for minimization with inequality constraints. Comput. Appl. Math. *19*, 31–56 (2000).
[32] Mulato, M., Chambouleyron, I., Birgin, E. G., Martínez, J. M.: Determination of thickness and optical constants of a-Si:H films from transmittance data. Appl. Phys. Lett. *77*, 2133–2135 (2000).
[33] Raydan, M.: On the Barzilai and Borwein choice of steplength for the gradient method. IMA J. Numer. Anal. *13*, 321–326 (1993).
[34] Raydan, M.: The Barzilai and Borwein gradient method for the large scale unconstrained minimization problem. SIAM J. Optim. *7*, 26–33 (1997).
[35] Zhang, J., Xu, C.: A class of indefinite dogleg path methods for unconstrained minimization. SIAM J. Optim. *9*, 646–667 (1999).

E. G. Birgin
Department of Computer Science IME-USP
University of São Paulo
Rua do Matão 1010, Cidade Universitária
05508-900, São Paulo SP
Brazil
e-mail: egbirgin@ime.usp.br

J. M. Martínez
Department of Applied Mathematics
IMECC-UNICAMP
University of Campinas
CP 6065, 13081-970 Campinas SP
Brazil
e-mail: martinez@ime.unicamp.br

Computing [Suppl] 15, 61–77 (2001)

Computing

Inclusions and Existence Proofs for Solutions of a Nonlinear Boundary Value Problem by Spectral Numerical Methods

B. Breuer and **M. Plum,** Karlsruhe, and **P. J. McKenna,** Storrs

Dedicated to Professor Tetsuro Yamamoto on the occasion of his 65th birthday

Abstract

For a specific elliptic boundary value problem with quadratic nonlinearity, we give a partial positive answer to an old conjecture concerning the number of solutions. This result is obtained via an existence and enclosure method. For computing the highly accurate solutions needed for this method, a spectral two-grid procedure (combined with a numerical Mountain-Pass algorithm and a Newton iteration) is proposed. Furthermore, Emden's equation is shown to admit completely spurious approximate solutions which nevertheless have "small" defects – a powerful argument for rigorous enclosure methods.

AMS Subject Classifications: 35J25, 65N15, 65N35.

Key Words: Existence and inclusion method, nonlinear elliptic boundary value problem, spectral methods.

1. Introduction

In this article the boundary value problem

$$\begin{aligned} -\Delta u = u^2 - \lambda h \quad &\text{in } \Omega_\zeta := (0,\zeta) \times \left(0, \frac{1}{\zeta}\right), \\ u = 0 \quad &\text{on } \partial\Omega_\zeta, \end{aligned} \tag{1}$$

with $h(x_1, x_2) := \sin(\frac{\pi}{\zeta}x_1)\sin(\pi\zeta x_2)$ is considered. We are looking for solutions $u \in H_2^B(\Omega_\zeta)$ (denoting the closure of $\{u \in C_2(\overline{\Omega_\zeta}) : u = 0 \text{ on } \partial\Omega_\zeta\}$ in $H_2(\Omega_\zeta)$).

This problem has its roots in work of Ambrosetti and Prodi, Kazdan and Warner, Lazer and McKenna, Hofer, Solimini, and Dancer (see [1, 6, 9, 10, 12, 19]), who were concerned with the more general problem where Ω is a bounded, simply connected Lipschitz domain in $\mathbb{R}^n$, u^2 is replaced by $f(u)$, and h by the first eigenfunction of $-\Delta$ with Dirichlet boundary conditions. This work gave rise to the conjecture that, in the case $f'(+\infty) = +\infty$, the problem has at least four solutions if λ is sufficiently large. While in the one-dimensional case even the unboundedness of the number of solutions (as $\lambda \to +\infty$) has been proved, the conjecture is open since twenty years for problems which cannot be treated by ordinary differential equation techniques (as our problem (1)).

Here we will show, using a computer aided approach (existence and inclusion method), that for $\zeta = 1$ and $\lambda = 800$ there exist at least four essentially different solutions (i.e. none of these four solutions is an elementary symmetry transform of another one). This gives at least a partial positive answer to the conjecture.

Moreover we will investigate the occurrence of approximate solutions for $\lambda = 0$ (Emden's equation) and $\zeta > 2.5$ which are not symmetric with respect to the axis $\{x = 1\zeta/2\}$.

Since we intended to obtain high precision numerical solutions, we started with the numerical Mountain-Pass algorithm developed in [5, pp. 419], in order to calculate an approximate solution $u_0 \in H_2^B(\Omega)$. Details of this procedure, as well as a more detailed historical classification of problem (1), will be presented in a forthcoming paper.

Starting from this approximate solution we used the Newton algorithm:

$$\begin{aligned} -\Delta p_n - 2u_n p_n &= \Delta u_n + u_n^2 - \lambda h =: d_n \quad \left(p_n \in H_2^B(\Omega)\right) \\ u_{n+1} &:= u_n + p_n \end{aligned} \qquad (n = 0, 1, \ldots, n_0), \qquad (2)$$

with n_0 denoting the index at which the iteration is stopped, either because u_{n_0} is sufficiently accurate or because the maximal iteration number has been exceeded.

All iteration steps were implemented in a space of spectral basis functions. As a result, the high smoothness of the approximate solutions $\omega \in H_2^B(\Omega)$ of (1) provides very low defect bounds, which are necessary for inclusions of solutions, as will be seen in the following.

2. Theoretical Foundations

2.1. An Existence and Inclusion Theorem

We consider the boundary value problem

$$-\Delta u + F(x, u) = 0 \text{ in } \Omega, \quad u = 0 \text{ on } \partial\Omega, \qquad (3)$$

where the boundary of the bounded domain $\Omega \subset \mathbb{R}^n$ (with $n \in \{2, 3\}$) is globally Lipschitz continuous and piecwise C_2-smooth and "regular", in the sense that the mapping $-\Delta : H_2^B(\Omega) \longrightarrow L_2(\Omega)$ is onto.

The nonlinearity F and its derivative $\partial F/\partial u$ are assumed to be continuous on $\bar{\Omega} \times \mathbb{R}$. For existence and inclusion results we need the following quantities:

- a verified upper bound δ for the defect of a numerical approximate solution $\omega \in H_2^B(\Omega)$ of (3):

$$\| - \Delta\omega + F(\cdot, \omega)\|_2 \leq \delta \qquad (4)$$

• a constant K such that

$$\|u\|_\infty \leq K\|L[u]\|_2 \quad \forall u \in H_2^B(\Omega), \tag{5}$$

where

$$L : H_2^B(\Omega) \longrightarrow L_2(\Omega),\ u \quad \mapsto \quad -\Delta u + cu, \ \text{ with } c(x) := \frac{\partial F}{\partial u}(x, \omega(x))\ (x \in \bar{\Omega}), \tag{6}$$

moreover a monotonically nondecreasing function $G : [0,\infty) \longrightarrow [0,\infty)$, such that with c given in (6), the following holds true:

• $|F(x, \omega(x) + y) - F(x, \omega(x)) - c(x)y| \leq G(|y|) \quad (x \in \Omega, y \in \mathbb{R})$

• $G(t) = o(t) \quad \text{for } t \to 0$

Theorem 2.1. (*see* [15, *p.* 59]) *If there is a constant* $\alpha \geq 0$ *satisfying*

$$\delta \leq \frac{\alpha}{K} - \sqrt{|\Omega|} \cdot G(\alpha), \tag{7}$$

then there exists a solution $u \in H_2^B(\Omega)$ *of* (3) *such that*

$$\|u - \omega\|_\infty \leq \alpha.$$

Remarks:

- As $G(t) = o(t)$ for $t \to 0$, the derivative of the right-hand side of (7) (as a function of α) equals $1/K$ for $\alpha = 0$, so at least for small α the right-hand side will be positive. Hence (7) holds for some "small" α, if the defect bound δ is sufficiently small, i.e. the approximate solution ω is calculated with sufficient accuracy; the required precision of ω is determined by (7).
- The above existence and inclusion result does not require monotonicity properties of the operator L, i.e. existence and inclusion results can be obtained for boundary value problems where the linearisation L is indefinite (this is the case for several approximate solutions of (1); here existence and inclusion results cannot be obtained with the "classical" methods of upper and lower solutions).
- The theory can be generalized in various ways, for example to more general nonlinearities (F depends on ∇u, see [17]), or to weak solutions, see [16]. Likewise, the boundary condition in (3) can be formulated more generally:

  $$B[u] = s \quad \text{on } \partial\Omega$$

  where B is an operator of Dirichlet, Neumann or mixed type; $s : \partial\Omega \longrightarrow \mathbb{R}$ is a function with $s = B[\bar{s}]$ (in the trace sense) for some $\bar{s} \in H_2(\Omega)$.
- An alternative method for providing existence and inclusion statements for nonlinear elliptic problems, which however we will not report on in more detail here, has been developed by M.T. Nakao (see, e.g. [14]).

2.2. Calculation of K

We define $\|u_{xx}\|_2^2 := \sum_{i,j=1}^{n} \|\frac{\partial^2 u}{\partial x_i \partial x_j}\|_2^2$ and choose constants $\underline{c}$ and $\bar{c}$ such that

$$\underline{c} \le c(x) \le \bar{c} \quad (x \in \bar{\Omega}). \tag{8}$$

A constant K satisfying (5) can be calculated by determining constants C_i and K_i $(i = 1, 2, 3)$ with

$$\|u\|_\infty \le C_0\|u\|_2 + C_1\|\nabla u\|_2 + C_2\|u_{xx}\|_2 \quad \forall\, u \in H_2(\Omega) \tag{9}$$

and

$$\|u\|_2 \le K_0\|L[u]\|_2, \quad \|\nabla u\|_2 \le K_1\|L[u]\|_2, \quad \|u_{xx}\|_2 \le K_2\|L[u]\|_2 \quad \forall\, u \in H_2^B(\Omega). \tag{10}$$

Then, (5) holds with $K := C_0K_0 + C_1K_1 + C_2K_2$.

In the case of a rectangle $\Omega = \prod_{i=1}^{n}(0, l_i)$ (with $n \in \{2, 3\}$), such constants C_i are given by:

$$C_0 = \gamma_0(\beta)\frac{1}{\sqrt{t_1 \cdot t_2(\cdot t_3)}}, \quad C_1 = \gamma_1(\beta)\sqrt{\frac{t_1^2 + t_2^2(+t_3^2)}{3 \cdot t_1 \cdot t_2(\cdot t_3)}},$$

$$C_2 = \gamma_2(\beta)\frac{1}{3}\sqrt{\frac{[t_1^2 + t_2^2(+t_3^2)]^2 + \frac{4}{5}[t_1^4 + t_2^4(+t_3^4)]}{t_1 \cdot t_2(\cdot t_3)}}$$

with

$$\gamma_0(\beta) = \frac{1}{2}\beta\sqrt{\frac{\beta - 1}{2\beta - 3}}, \quad \gamma_1(\beta) = \sqrt{\frac{2\beta}{2\beta - 1}}, \quad \gamma_2(\beta) = \frac{1}{2}\sqrt{\frac{3}{(2\beta + 1)(\beta + 1)}} \quad \text{if } n = 2,$$

$$\gamma_0(\beta) = \frac{\beta(\beta - 1)}{\sqrt{3 \cdot (2\beta - 3)}}, \quad \gamma_1(\beta) = \frac{2}{3}\beta\sqrt{\frac{5}{2\beta - 1}}, \quad \gamma_2(\beta) = \frac{1}{3}\sqrt{\frac{7}{2\beta + 1}} \quad \text{if } n = 3,$$

where $\beta > 3/2$ and $0 < t_i \le l_i$ are arbitrary, see [15, p. 41] and [15, p. 43].

As t_1, t_2 and β can be chosen arbitrarily in a wide range, the constant K can sometimes be reduced essentially (depending on the values of K_0, K_1 and K_2), so in view of (7) the requirements for a numerical solution can be weakened significantly in some cases.

The values K_i depend on the respective numerical solution via the definition of L. By eigenfunction series expansion we see that the first inequality in (10) holds (see [15, p. 43]) with $K_0 = \sigma^{-1}$ where

$$0 < \sigma \leq \min\{|\lambda| : \lambda \text{ eigenvalue of } L \text{ on } H_2^B(\Omega)\}.$$

Hence, for the calculation of K_0, bounds for eigenvalues are required. For this purpose, many methods have been developed (e.g. by Rayleigh-Ritz, Kato, Weinstein, Lehmann, Goerisch; see [18] and the references therein), which allow to reduce the calculation of the required eigenvalue bounds to matrix eigenvalue problems, for which effective algorithms exist.

Partial integration yields that for K_1 we can choose (see [15, p. 44]):

$$K_1 = \begin{cases} (K_0(1 - \underline{c}K_0))^{\frac{1}{2}} & \text{if } \underline{c}K_0 \leq \frac{1}{2} \\ \frac{1}{2\sqrt{\underline{c}}} & \text{otherwise} \end{cases}.$$

For convex domains Ω, an admissible K_2 is given by (see [15, p. 46]):

$$K_2 = 1 + K_0 \cdot \max\{-\underline{c}, \tfrac{1}{2}(\bar{c} - \underline{c})\}.$$

If Ω is not convex, then for the calculation of K_2 the (piecewise) curvature of the boundary $\partial\Omega$ has to be taken into account (see [15, pp. 46]).

Remarks:

- When calculating the constant K, the main work has to be put into the determination of a verified upper bound, as small as possible, for K_0. For this purpose it has to be guaranteed that there is no further eigenvalue of L between zero and the smallest modulus of all eigenvalues of L which were enclosed by numerical means. This can be carried out using an eigenvalue homotopy, see Section 5.
- For problem (1), $|\Omega| = 1$, and we can choose $G(t) = t^2$. So we can find a constant α satisfying (7) if $4\delta K^2 \leq 1$, and for α we get:

$$\alpha = \frac{2\delta K}{1 + \sqrt{1 - 4\delta K^2}}. \tag{11}$$

3. A Spectral Multigrid Algorithm

3.1. Discretisation of the Newton step (2)

In the following let $\Omega = (0,1) \times (0,1)$. For the calculation of precise approximate solutions (starting from a rough approximation u_0), we choose the spectral ansatz space ($N \in \mathbb{N}$):

$$V_N := \text{span}(\{\sin(i\pi x_1)\sin(j\pi x_2) \mid 1 \leq i, j \leq N - 1\}),$$

which is the linear hull of some eigenfunctions of the Laplacian on $(0,1) \times (0,1)$.

For computing an approximate solution $p_n \in V_N$ of (2) we use a collocation method, i.e. we require (where $c_n := -2u_n$ in the following):

$$-\Delta p_n\left(\frac{k}{N},\frac{l}{N}\right)+c_n\left(\frac{k}{N},\frac{l}{N}\right)p_n\left(\frac{k}{N},\frac{l}{N}\right)=d_n\left(\frac{k}{N},\frac{l}{N}\right)\quad (1\le k,l\le N-1). \quad (12)$$

For reading (12) as a system of linear equations, we set:

Definition 3.1. *For* $1 \le i,j,k,l \le N-1$ *let* $B, C_n, E \in \mathbb{R}^{(N-1)^2\times(N-1)^2}$, $R_n \in \mathbb{R}^{(N-1)^2}$ *be given by*

- $B_{(k-1)(N-1)+l,(i-1)(N-1)+j} := \sin\left(k\frac{i}{N}\pi\right)\sin\left(l\frac{j}{N}\pi\right)$
- $(C_n)_{(k-1)(N-1)+l,(i-1)(N-1)+j} := \begin{cases} c_n\left(\frac{k}{N},\frac{l}{N}\right) & \text{if } i=k \text{ and } j=l \\ 0 & \text{otherwise} \end{cases}$
- $E_{(k-1)(N-1)+l,(i-1)(N-1)+j} := \begin{cases} \pi^2(i^2+j^2) & \text{if } i=k \text{ and } j=l \\ 0 & \text{otherwise} \end{cases}$
- $(R_n)_{(k-1)(N-1)+l} := d_n\left(\frac{k}{N},\frac{l}{N}\right)$

So, the vector $\underline{p_n}$ of the Fourier coefficients of p_n is determined by the linear system

$$(BE + C_nB)\underline{p_n} = R_n. \quad (13)$$

As the matrix on the left-hand side of (13) is dense and as the operator L turns out to be indefinite for all upcoming interesting cases we will need a multigrid method for solving (13) in convenient time and to avoid storage problems.

3.2. A Two-Grid Method

First we perform a good preconditioning of the linear system (13) corresponding to the discretisation of the Newton step (2), so that by means of some appropriate iteration steps applied to the preconditioned system, the remaining defect can be displayed conveniently on a rough grid (see e.g. [8]).

Definition 3.2. *Let* $V \in \mathbb{R}^{(N-1)^2\times(N-1)^2}$,

$$V_{(k-1)(N-1)+l,(i-1)(N-1)+j} := \begin{cases} \pi\sqrt{i^2+j^2} & \text{if } i=k \text{ and } j=l \\ 0 & \text{otherwise} \end{cases}$$

With the substitution

$$\underline{p_n} = V^{-1}\underline{q_n} \quad \left(\underline{q_n} \in \mathbb{R}^{(N-1)^2}\right) \quad (14)$$

and using $B^2 = \frac{N^2}{4}I$ we get the equivalent linear system

$$\left(I + \frac{4}{N^2} V^{-1} B C_n B V^{-1}\right) \underline{q_n} = \frac{4}{N^2} V^{-1} B R_n. \tag{15}$$

It can be shown that most of the eigenvalues of the matrix $A := I + \frac{4}{N^2} V^{-1} B C_n B V^{-1}$ are very close to 1 (see [2, pp. 6]), while the eigenvalues of the unpreconditioned matrix in (13) are spread within $[-240000, 240000]$ (if $N = 128$), so that (13) is difficult to access by iterative methods directly.

Remarks:

- In order to obtain sufficiently precise approximate solutions for (1), we choose $N = 128$. In all four treated cases about 99% of the 16129 eigenvalues of A are in $[0.9, 1.1]$, and about 94% in $[0.99, 1.01]$.
- The eigenpairs of A are collocation approximations for the eigenvalue problem

$$\begin{aligned} -\Delta u + cu &= -\lambda \Delta u \quad \text{in } \Omega, \\ u &= 0 \quad \text{on } \partial\Omega. \end{aligned} \tag{16}$$

For a proof, see [2, p. 6]. Hence, eigenfunctions corresponding to eigenvalues of (15) which are "far away" from 1, will be of "low" frequency, i.e. they can be displayed well on a rough grid.

In order to treat equation (15) by iteration methods we introduce a relaxation parameter ϖ. In view of the above results, most of the eigenvalues of the iteration matrix T_ϖ,

$$T_\varpi := I - \varpi A,$$

will be close to $1 - \varpi$; for a convenient setup of the two-grid method, ϖ should be chosen close to 1.

These considerations yield a SOR method with a relaxation parameter ϖ for calculating an approximate solution of (15):

$$\begin{aligned} \underline{q_n}^{(1)} &:= \varpi \frac{4}{N^2} V^{-1} B R_n, \\ \underline{q_n}^{(m+1)} &= T_\varpi \underline{q_n}^{(m)} + \varpi \frac{4}{N^2} V^{-1} B R_n. \end{aligned} \tag{17}$$

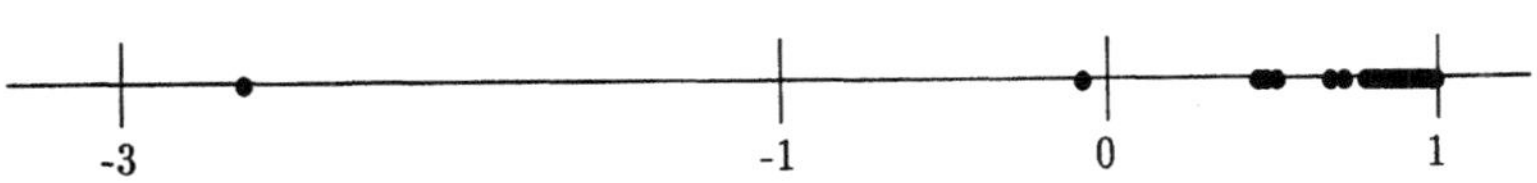

Figure 3.1. Spectrum of A for an approximate solution of (1)

Note that in practice the iteration steps can be carried out very quickly by means of the Fast Fourier Transformation (multiplying a vector v with the matrix B corresponds to the calculation of its Fourier transform!).

We have the following

Theorem 3.1. *Let $\underline{q_n}$ be the exact solution of* (15) *and $\underline{q_n}^{(m+1)}$ be an approximate solution of* (15) *calculated according to* (17). *Let $\{\lambda_i \mid i = 1, \dots, (N-1)^2\}$ be the set of all eigenvalues of A and $\Pi_{E_{\lambda_i}}$ the orthogonal projection onto the eigenspace E_{λ_i}. Then, if A is nonsingular,*

$$\underline{q_n} - \underline{q_n}^{(m+1)} = \frac{4}{N^2} \sum_{i=1}^{(N-1)^2} \frac{(1 - \varpi\lambda_i)^{m+1}}{\lambda_i} \Pi_{E_{\lambda_i}}(V^{-1}BR_n).$$

With $I_1 := \{1 \leq i \leq (N-1)^2 \mid |1 - \varpi\lambda_i| \leq \rho\}$ (for some fixed $0 < \rho \ll 1$) and $I_0 := \{1, \dots, (N-1)^2\} \setminus I_1$ we therefore have:

$$\underline{q_n} - \underline{q_n}^{(m+1)} = \frac{4}{N^2} \sum_{i \in I_0} \frac{(1 - \varpi\lambda_i)^{m+1}}{\lambda_i} \Pi_{E_{\lambda_i}}(V^{-1}BR_n) + O(\rho^{m+1}) \text{ as } m \longrightarrow \infty. \quad (18)$$

For the proof, see [2, p. 9]. This theorem yields that the difference $\underline{q_n} - \underline{q_n}^{(m+1)}$ consists essentially only of a linear combination of eigenvectors of A with corresponding eigenvalues "far away" from 1. As the corresponding eigenfunctions are of low frequency, we can calculate an approximate solution $p_n \in V_N$ of (13) on the grid $\{(\frac{k}{N}, \frac{l}{N}) \mid 1 \leq k, l \leq N-1\}$, by resubstitution of $\underline{q_n}^{(m+1)}$ according to (14) and by adding the coarse grid correction r (let $(\underline{q_n}^{(m+1)})_{ij}$ denote the $[(i-1)(N-1)+j]$-th component of $\underline{q_n}^{(m+1)}$):

$$p_n\left(\frac{k}{N}, \frac{l}{N}\right) = \sum_{i=1}^{N-1} \sum_{j=1}^{N-1} \frac{1}{\sqrt{(i^2 + j^2)\pi^2}} \left(\underline{q_n}^{(m+1)}\right)_{ij} \sin\left(i\frac{k}{N}\pi\right) \sin\left(j\frac{l}{N}\pi\right) + r\left(\frac{k}{N}, \frac{l}{N}\right),$$

where the Fourier coefficients $\underline{r}$ of r are calculated via

$$(B^g E^g + C_n^g B^g)\underline{r} = R_n^g \quad (19)$$

with B^g, E^g, C_n^g, R_n^g as in Definition 3.1, but with $N^g := N/k$, for some $k \in \mathbb{N}$. In our examples we chose $k = 4$, respectively $k = 8$, so that we could solve (19) by means of a simple LU decomposition. By this approach we can carry out the Newton step in V_N even for large N sufficiently well (the Newton algorithm converges) and in convenient time. Hence, the iteration method (2) leads to an approximate solution $u_{n_0} \in V_N$ with very small defect.

Remark: If we look for solutions which are symmetric with respect to the axis $\{x_1 = 1/2\}$ or/and $\{x_2 = 1/2\}$, we can reduce the dimension of the linear system (13) to one half respectively to one quarter by symmetrisation of the ansatz space V_N (see [2, p. 11]).

4. Calculation of the Defect

In the last paragraph we presented a method for calculating the Fourier coefficients $\hat{u}_{ij}(1 \le i,j \le N-1)$ of $u_{n_0} \in V_N$ such that the discrete defect (let F denote the nonlinearity in (3))

$$d^2 := \frac{1}{N^2}\sum_{k=1}^{N-1}\sum_{l=1}^{N-1}\left(\Delta u_{n_0}\left(\frac{k}{N},\frac{l}{N}\right) - F\left(\left(\frac{k}{N},\frac{l}{N}\right), u_{n_0}\left(\frac{k}{N},\frac{l}{N}\right)\right)\right)^2$$

becomes smaller than a given tolerance ε if the Newton algorithm converges.

For our special case $F((x_1,x_2),u) := -u^2 + \lambda \sin(\pi x_1)\sin(\pi x_2)$, the continuous defect

$$\delta^2 := \int_\Omega \left(\Delta u_{n_0}(x_1,x_2) + u_{n_0}(x_1,x_2)^2 - \lambda \sin(\pi x_1)\sin(\pi x_2)\right)^2 dx$$

(needed according to (4)) can be expressed exactly by finite sums involving the Fourier coefficients of u_{n_0}. For this purpose we write $u_{n_0}^2$ as a sum of cos products (instead of evaluating $u_{n_0}^4$ directly, which would lead to a run time behaviour of $O(N^8)$); this yields, in the end, a run time behaviour of $O(N^4)$ for calculating the value of the defect integral. In practice, interval arithmetic has to be used here in order to obtain verified bounds.

The upper bounds for the defects of the four approximate solutions ω_i, $i = 1,2,3,4$ computed via (2) are given in Table 1.

We close this section making some remarks on the realisation of the proposed methods: We use an IBM RS 6000 Workstation, Model 43P-140 with a 332 MHZ PowerPC Processor. All data are computed in double precision according to IEEE standard. The interval arithmetic needed for computing the defect bound δ (see the description above) and for other verified computations (e.g. the eigenvalue bounds needed for obtaining K_0; see next section) is provided by the Profil Interval Library, Version 2.1 (see [11]) (note that the above discussion shows that no evaluations of the *sin* function are needed).

We choose the parameter value $\varpi = 0.8$ for all four approximate solutions. Two steps of the SOR method (17) and the mentioned coarse grid correction are sufficient to provide, in all cases, approximate solutions to the linear problems

Table 1

	$N = 32$	$N = 64$	$N = 128$
ω_1	$\delta^{(1)} = 0.0699$	$\delta^{(1)} = 0.0105$	$\delta^{(1)} = 0.0023$
ω_2	$\delta^{(2)} = 0.0814$	$\delta^{(2)} = 0.0143$	$\delta^{(2)} = 0.0041$
ω_3	$\delta^{(3)} = 0.3476$	$\delta^{(3)} = 0.0618$	$\delta^{(3)} = 0.0059$[1]
ω_4	$\delta^{(4)} = 0.4688$	$\delta^{(4)} = 0.0834$	$\delta^{(4)} = 0.0151$

[1] On a 200×200 grid

arising in the Newton steps which are good enough to diminish the discrete defect d quickly in the course of the Newton iteration. With a tolerance $\varepsilon = 10^{-4}$, up to five Newton steps are necessary to achieve the required accuracy.

On a 128×128 grid, one step of the SOR method (17) takes about 22.3 minutes (**without** using a FFT method; so further improvement is reasonable here). For a convenient implementation we use ten auxiliary vectors of size $127^2 = 16129$; hence, we have a memory demand of about 10.3 MB (in double precision IEEE standard one variable of `DOUBLE` format is represented in 64 Bit = 8 Bytes). The LU decomposition on a 32×32 grid for the coarse grid correction takes about 66.7 seconds, so that in total one Newton step on a 128×128 grid costs about 24 minutes (without symmetry properties).

As the complexity of the LU decomposition is $O(N^6)$ (with N denoting the number of grid points in one direction), each doubling of N will extend the time for the coarse grid correction by a factor of $2^6 = 64$. An attempt to solve the system (13) directly by a LU decomposition would hence take about 75.8 hours, and for storing the respective dense matrix of (13) (which is of dimension $127^2 = 16129$) we would need about 17.6 GB memory.

5. Bounds for Eigenvalues

5.1. Calculation of Approximate Eigenpairs of L

In order to enclose solutions we need a rigorous bound for the norm of L^{-1} (see subsection 2.2). This requires sufficiently accurate bounds for the smallest modulus of all eigenvalues of

$$L : \left\{ \begin{array}{ccc} H_2^B(\Omega) & \longrightarrow & L_2(\Omega) \\ u & \mapsto & -\Delta u + cu \end{array} \right\}, \quad \text{with } c(x) := \frac{\partial F}{\partial u}(x, \omega(x)) \ \ (x \in \bar{\Omega}).$$

In our situation we have $\omega = u_{n_0}$ and $c = -2u_{n_0}$. Again a collocation ansatz yields approximate eigenpairs. An approximate eigenvalue $\tilde{\lambda}$ and an approximate eigenfunction $\tilde{\Phi} \in V_N$ are determined by:

$$-\Delta\tilde{\Phi}\left(\frac{k}{N}, \frac{l}{N}\right) + c\left(\frac{k}{N}, \frac{l}{N}\right)\tilde{\Phi}\left(\frac{k}{N}, \frac{l}{N}\right) = \tilde{\lambda}\tilde{\Phi}\left(\frac{k}{N}, \frac{l}{N}\right) \quad (1 \le k, l \le N-1).$$

With the matrix notations used in paragraph 3.1 we get the following matrix eigenvalue problem for the vector $\underline{\tilde{\Phi}}$ of the Fourier coefficients of $\tilde{\Phi}$ (with C defined by $c = -2u_{n_0}$, analogously to Definition 3.1):

$$(BE + CB)\underline{\tilde{\Phi}} = \tilde{\lambda}B\underline{\tilde{\Phi}}. \tag{20}$$

By the substitution $\underline{\Psi} := B\underline{\tilde{\Phi}}$ and use of $B^2 = \frac{N^2}{4}I$ we obtain a symmetric and simple matrix eigenvalue problem:

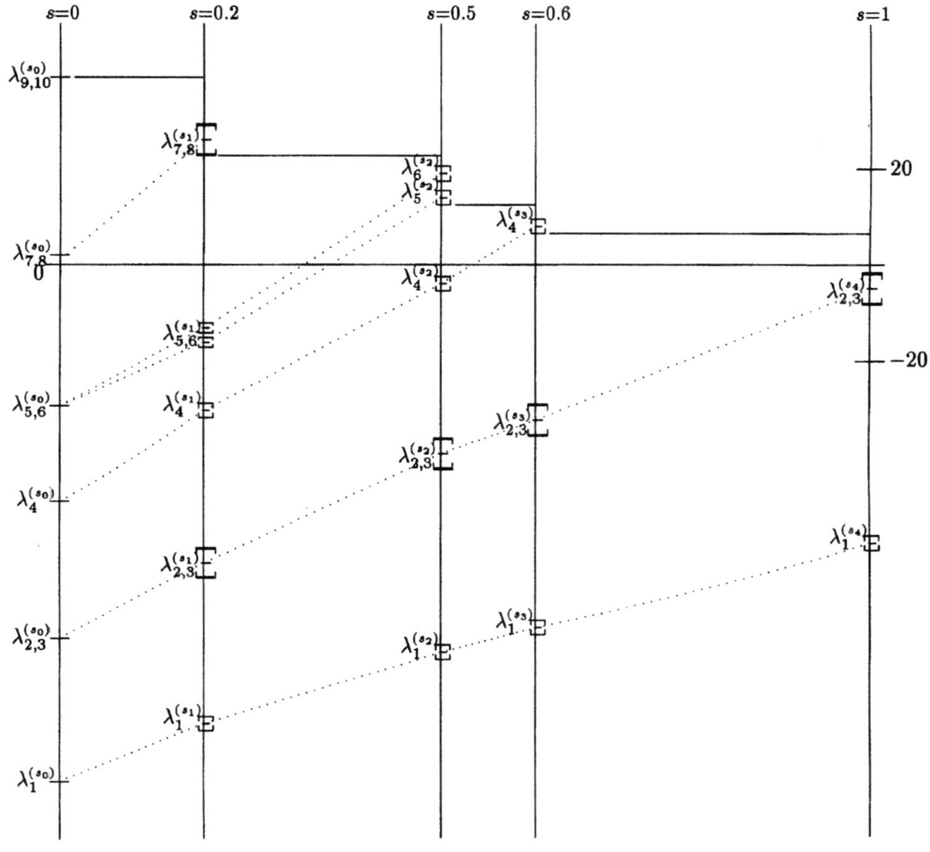

Figure 5.1. Eigenvalue homotopy for the second approximate solution of (1). Here we need a verified lower bound for the fourth eigenvalue of L. As we had to "drop" four eigenvalues, we started with eight eigenvalues in the beginning

$$\left(\frac{4}{N^2}BEB + C\right)\underline{\Psi} = \tilde{\lambda}\underline{\Psi}. \qquad (21)$$

The matrix product BEB is of block structure and each coefficient is given by simple explicit formulas. So this eigenvalue problem can be set up in $O(N^3)$ and, due to its structure, be solved easily by a `MATLAB` routine (for the simple eigenvalue problem (21) we used the `Eig()` routine based essentially on block-oriented QR decomposition). With $N = 32$ we obtained very good approximate eigenpairs (i.e. the corresponding Weinstein defects are very small, see below).

Note that the usual variational methods (like the Rayleigh-Ritz method), in a spectral ansatz space, lead to substantial problems when setting up the corresponding matrix eigenvalue problem (here the run time behaviour is $O(N^6)$!).

As L is symmetric with respect to the standard scalar product in $L_2(\Omega)$, the theorem by D. Weinstein (see [3, Cor. 6.20]) respectively a "clustered" version in the case of approximate eigenvalues close to each other (see [18, Theorem 1, Theorem 2]) yields small intervals which contain eigenvalues of L. But as already mentioned, the determination of K_0 requires that between zero and the calculated eigenvalue closest to zero (respectively the calculated Weinstein interval) there is no further eigenvalue of L. For this purpose we need an eigenvalue homotopy for calculating a rough but verified lower bound for some higher eigenvalue of L (with index information).

Here we set up such a homotopy method by transforming an operator L_0 with explicitly known spectrum into L (by using $L_s := (1-s)L_0 + sL$ with $0 \leq s \leq 1$), such that the eigenvalues $\lambda_i^{(s)}$ of L_s, as functions of s, are monotonically increasing, i.e., we choose $L_0[u] := -\Delta u + \underline{c}u$, with $\underline{c}$ satisfying (8). We refer to [18] for the details of the homotopy method; here, we restrict ourselves to the graphical display of homotopy results obtained in our concrete context of problem (1); see Fig. 5.1.

5.2. *Results*

For the needed eigenvalues of $L = -\Delta - 2\omega$ for the four different approximate solutions ω_i $(i = 1,2,3,4)$ we obtained enclosure intervals shown in Table 2, by D. Weinstein's theorem combined with the homotopy method (see subsection 5.1):

In all cases the lower bound ρ_4 for the fourth eigenvalue is larger than the computed upper bound of the third eigenvalue. All required bounds for the calculation of K_0 were computed in a verified way.

The knowledge of K_0 and therefore of K (see paragraph 2.2) and an upper bound for the defect δ yields an enclosure constant α according to (7) resp. (11). The decisive condition (7) is satisfied for all four approximate solutions, so that in all four cases we obtain the desired enclosure and existence statement for respective solutions $u_1, \ldots u_4$, by Theorem 2.1 (for $\lambda = 800$ and $h(x_1,x_2) = \sin(\pi x_1)\sin(\pi x_2)$). We obtained error bounds α_i for $\|u_i - \omega_i\|_\infty$ $(i = 1,2,3,4)$ as shown in Table 3.

The first two approximate solutions ω_1 and ω_2 are both symmetric with respect to the axes $\{x_1 = 1/2\}$ and $\{x_2 = 1/2\}$, whereas the third solution only has one

Table 2

	λ_1	λ_2	λ_3	ρ_4	K_0
ω_1	50.8^{5453}_{2560}	74.8^{8843}_{5752}	74.8^{8843}_{5752}	91.34	0.019676
ω_2	-58.9^{7504}_{6498}	-3.8^{909}_{554}	-3.8^{909}_{554}	7.01	0.25701
ω_3	-62.3^{267}_{733}	-3.6^{369}_{801}	[6.9301,7.0082]	27.65	0.144299
ω_4	$-65.^{2667}_{1336}$	$6.^{7745}_{6269}$	$6.^{7745}_{6269}$	12.38	0.150901

Table 3

Approximate solution	α_i	Approximate solution	α_i
ω_1	$5.8222 \cdot 10^{-4}$	ω_2	0.0180
ω_3	0.0222	ω_4	0.0586

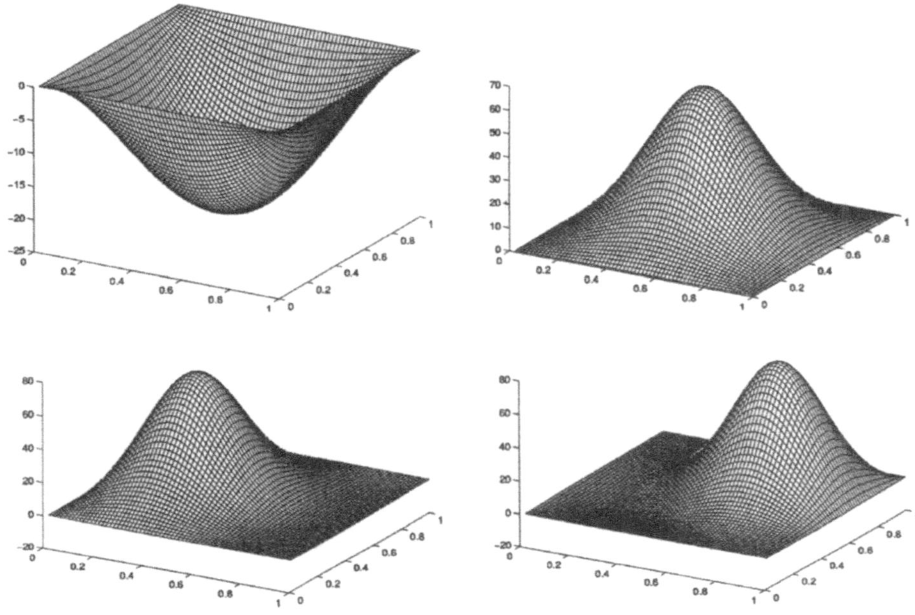

Figure 5.2. Four essentially different approximate solutions of (1) for $\lambda = 800$ and $\zeta = 1.0$

symmetry property with respect to $\{x_2 = 1/2\}$ and the fourth solution is neither symmetric to the first nor to the second axis (indeed it is symmetric with respect to the axis $\{x = y\}$, a fact which could not be exploited for reducing the effort when setting up the linear system, see paragraph 3.1).

The decisive error bound α is obtained in a verified way due to the calculation of verified intervals for δ and K by interval arithmetic. In particular $u_1, \ldots, u_4$ are pairwise different, as simple calculations show $\|\omega_i - \omega_j\|_\infty > \alpha_i + \alpha_j$ for $1 \le i,j \le 4$, $i \ne j$.

6. Emden's Equation

Emden's equation on the rectangle $\Omega_\zeta := (0, \zeta) \times (0, 1/\zeta)$ reads

$$\begin{aligned} -\Delta u &= u^2 \quad \text{in } \Omega_\zeta, \\ u &= 0 \quad \text{on } \partial\Omega_\zeta. \end{aligned} \tag{22}$$

We consider (22) for different values of $\zeta \ge 1$. For $\zeta = 1$ we computed, by a path following method (see [17, p. 39]), an approximate solution for (22) (with $\zeta = 1$). Proceeding from $\zeta = 1$ to larger ζ in a second path following, we obtained a (rough) approximation for (22). Starting with this function we performed the same steps as in paragraph 3 to obtain precise approximate solutions. In practice we transformed (22) to the unit square, so that all steps mentioned in Section 3 (collocation method, ansatz space V_N, preconditioning) could be carried out on this domain.

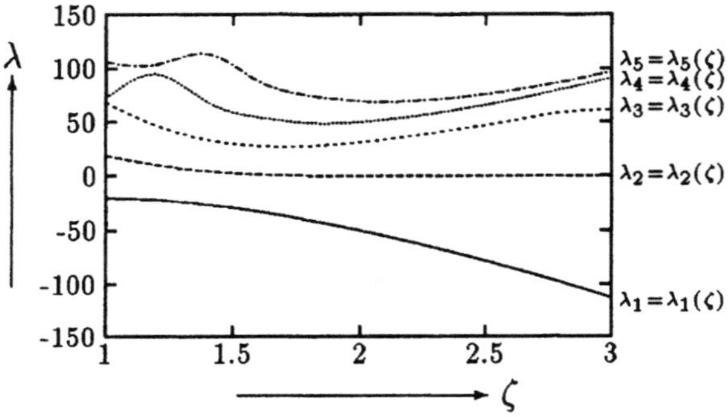

Figure 6.1. Spectrum of L_ζ

Table 4. Location of the maxima of appr. solutions of (22) for $\zeta = 2.9$

$u_{\max}$ at	defect
(0.500,0.5)	0.03364
(0.436,0.5)	0.03374
(0.317,0.5)	0.03544
(0.283,0.5)	1.15823

An investigation of the spectrum of $L_\zeta[u] =: -\frac{1}{\zeta^2}\frac{\partial^2 u}{\partial x^2} - \zeta^2 \frac{\partial^2 u}{\partial y^2} - 2\omega u$ on Ω_1 yields for increasing ζ in the diagram shown in Fig. 6.1[2].

Here we observe an interesting phenomenon: Besides the fully symmetric approximate solution shown in Fig. 6.2, we find approximations which are no longer symmetric with respect to the axis $\{x = \frac{1}{2}\zeta\}$ (see Fig. 6.3). Within a wide range, the defects of these approximate solutions are all in the same order of magnitude (see Table 4).

However, according to Gidas, Ni and Nirenberg (see [7, Remark 2, p. 230]), (22) only admits fully symmetric solutions. So, for large ζ we have approximate solutions with small defect, but an essential error.

Note that, in agreement with this situation, obviously our enclosure and existence theory cannot be applied here as the second eigenvalue of L_ζ cannot be separated from zero for $\zeta > 2.5$, hence L_ζ is almost singular.

While the asymmetric approximate solution (shown in Fig. 6.3) is therefore spurious, the fully symmetric solution shown in Fig. 6.2 is a "true" one, as can be shown by applying a slight modification of our existence and inclusion method within the class of fully symmetric functions (or, what amounts to the same, by applying it on one quarter of Ω_ζ, with Neumann boundary conditions on the new boundary parts); the full symmetry removes the critical eigenvalue λ_2, so that moderate constants K_0 and K are obtained, and (7) is satisfied.

[2] Only the "interesting" eigenvalues of L_ζ were considered, i.e. only the ones where the corresponding eigenfunctions are symmetric with respect to $\{y = 1/(2\zeta)\}$.

This example is a clear warning that even approximate solutions of high "accuracy", i.e. with very low defects, can have huge errors! The quality of an approximate solution cannot be judged only on the basis of its defect.

But what is the reason for the occurrence of these spurious asymmetric solutions, and what do we see in Figs. 6.2 and 6.3?

The idea can best be understood by a simple example. Consider the equation

$$y'' - y + y^2 = 0.$$

If studied on an infinite interval $(-\infty, \infty)$ this equation can be shown to have one solution $y(x)$ which is positive, and goes exponentially to zero as $|x| \to \infty$. This can be seen by studying the phase-plane, where this solution lies on the trajectory $(y')^2 - y^2 + \frac{2}{3}y^3 = 0$. Of course, any translate (in x-direction) of this solution is still a solution which goes to zero at infinity.

On the other hand, on the interval $(-R, R)$ with Dirichlet boundary conditions, this equation admits only a unique positive solution. It is an exercise in uniqueness of the initial value problem to show that this solution is symmetric about the origin.

The problem is that if R is large enough, many translates of the exponentially decaying solution (on $(-\infty, \infty)$) will appear to solve (approximately) the Dirichlet problem on the interval $(-R, R)$. What we see in Figs. 6.2 and 6.3 is a two-dimensional version of that idea. It is a relatively straightforward exercise, using the techniques of [13] to prove the following Theorem. The equation

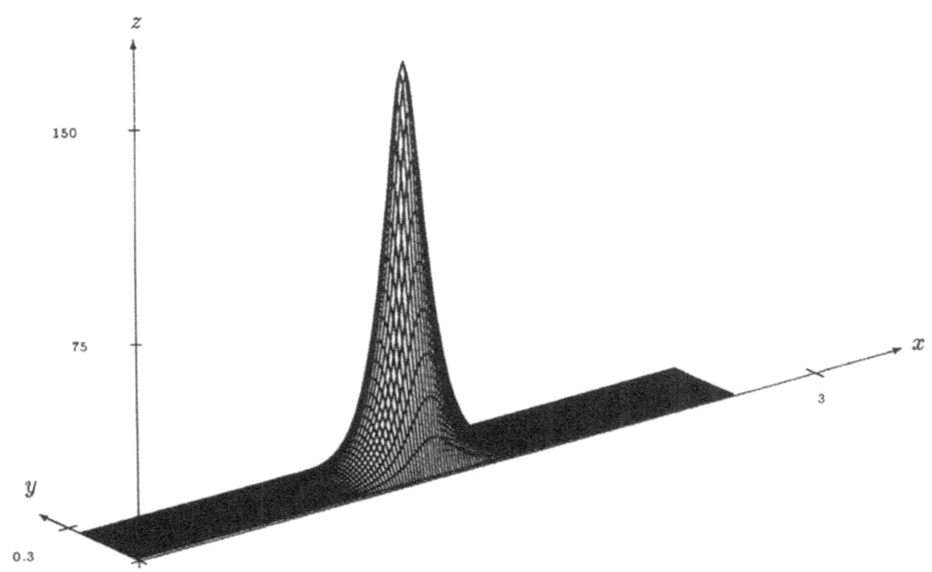

Figure 6.2. Symmetric approximate solution of (22) for $\zeta = 2.9$

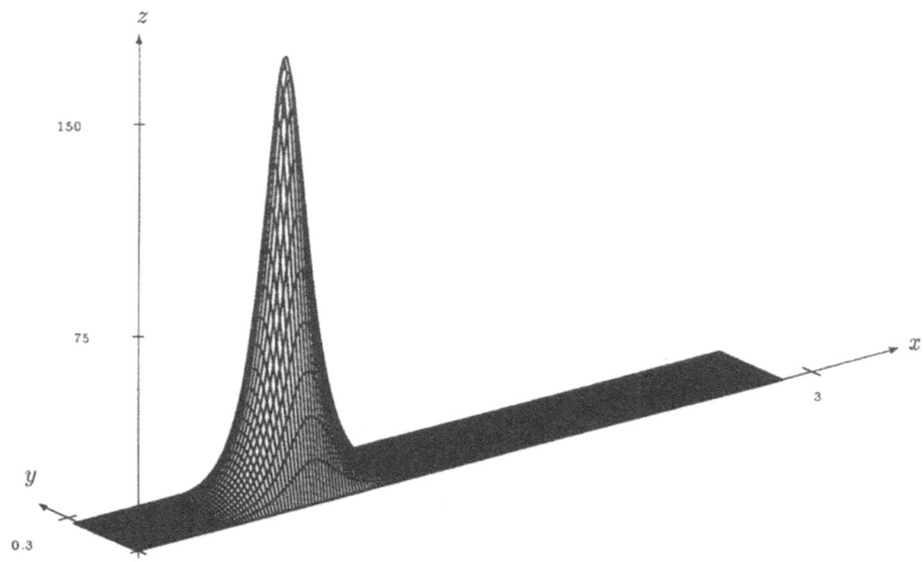

Figure 6.3. "Spurious" approximate solution of (22) for $\zeta = 2.9$

$$\Delta u + u^2 = 0$$

has a positive solution on the infinite strip $(-\infty, \infty) \times (0, \frac{1}{\zeta})$ satisfying the boundary conditions $u(x, 0) = u(x, \frac{1}{\zeta}) = 0$ for all x and $u(x, y) \to 0$ as $|x| \to \infty$.

Proving the existence of solutions on infinite domains such as a strip poses special problems. Standard methods on finite domains usually rely heavily on a priori estimates which establish compactness of an approximating sequence of near-solutions. On infinite domains, this cannot work because the solution-set is a continuum, since any translate of the solution will still be a solution. Thus, one must find an approximating solution sequence which is localised at some region, so that one can establish convergence to a non-zero solution. This is usually called the method of concentrated compactness, and a description can be found in [13].

To prove the above theorem, one follows this general outline. First, one finds an approximating sequence of "almost-mountain-pass" type. Then, one shows that elements of this sequence can be translated so that they remain bounded away from zero at, say, the origin. One can use the fact that the bounded sequence of solutions converges weakly to a solution. The last piece is to show that this solution is not the zero solution, since it is the limit of functions which are in some sense bounded away from zero. An example of this type of argument can be found in [4].

We strongly suspect, although are not able to prove, that these solutions decay exponentially as $|x| \to \infty$. Then, although the solution on the domain $(0, \zeta) \times (0, \frac{1}{\zeta})$ with Dirichlet boundary conditions must be symmetric about the axis $\{x = \frac{1}{2}\zeta\}$, if ζ is sufficiently large, a solution of the problem on the infinite strip will appear numerically to be a solution on the rectangle, thus allowing many translates of the solution on the strip to appear to be solutions on the rectangle. Again, we suspect but are not able to prove that except for translation, the solution on the infinite strip is unique (and therefore symmetric about the y-axis).

Acknowledgement

The authors are grateful to the anonymous referee for very helpful comments and suggestions.

References

[1] Ambrosetti, A., Prodi, G.: On the Inversion of some Differentiable Mappings with Singularities between Banach Spaces. Ann. Math. Pura Appl. *93*, 231–247 (1973).
[2] Breuer, B., Plum, M.: Lsungseinschließungen bei einem nichtlinearen Randwertproblem mittels eines Fourierreihenansatzes. Internal Report, Preprint Nr. 98/9, IWRMM Karlsruhe, 1998.
[3] Chatelin, F.: Spectral approximations of linear operators. New York: Academic Press 1983.
[4] Chen, Y., McKenna, P. J.: Traveling waves in a nonlinearly suspended beam: theoretical results and numerical observations. J. Differ. Eq. *136*, 325–355 (1997).
[5] Choi, Y. S., McKenna, P. J.: A mountain pass method for the numerical solution of semilinear elliptic problems. Nonlinear Anal. Theory Meth. Appl. *20*, 417–437 (1993).
[6] Dancer, E. N.: A counterexample to the Lazer-McKenna conjecture. Nonlinear Anal. T.M.A. *13*, 19–22 (1982).
[7] Gidas, B., Ni, W. M., Nirenberg, L.: Symmetry and related properties via the maximum principle. Commun. Math. Phys. *68*, 209–243 (1979).
[8] Hackbusch, W.: Multi-grid methods and applications. Springer Series in Computational Mathematics, Vol. 4. Berlin Heidelberg New York: Springer 1985.
[9] Hofer, H.: Variational and topological methods in partially ordered Hilbert spaces. Math. Ann. *261*, 493–514 (1982).
[10] Kazdan, J. L., Warner, F.: Remarks on quasilinear elliptic equations. Comm. Pure Appl. Math. *28*, 567–597 (1975).
[11] Knüppel, O.: Programmer's Runtime Optimized Fast Interval Library. Bericht 93.4, Technische Universität Hamburg-Harburg, 1993.
[12] Lazer, A. C., McKenna, P. J.: On the number of solutions of a nonlinear Dirichlet problem. J. Math. Anal. Appl. *84*, 282–294 (1981).
[13] Lions, P.-L.: On positive solutions of semilinear elliptic equations in unbounded domains. Nonlinear diffusion equations and their equilibrium states, II, Berkeley, CA, 85–122 (1986). Math. Sci. Res. Inst. Publ., 13. Berlin Heidelberg New York Tokyo: Springer 1988.
[14] Nakao, M. T.: Solving nonlinear elliptic problems with result verification using an H^{-1} type residual iteration. Computing [Suppl] *9*, 161–173 (1993).
[15] Plum, M.: Explicit H_2-estimates and pointwise bounds for solutions of second-order elliptic boundary value problems. J. Math. Anal. Appl. *165*, 36–61 (1992).
[16] Plum, M.: Enclosures for weak solutions of nonlinear elliptic boundary value problems, pp. 505–521. Singapore: World Scientific 1994.
[17] Plum, M.: Numerical existence proofs and explicit bounds for solutions of nonlinear elliptic boundary value problems. Computing *49*, 25–44 (1992).
[18] Plum, M.: Guaranteed numerical bounds for eigenvalues. In: Spectral theory and computational methods of sturm-liouville problems (Hinton, D., Schaefer, W., eds.), Lecture Notes in Pure and Applied Mathematics, vol. 191. New York, Basel: Marcel Dekker 1997.
[19] Solimini, S.: Some remarks on the number of solutions of some nonlinear elliptic equations. Anal. Nonlin., I.H.P. *2*, 143–156 (1985).

B. Breuer
M. Plum
Mathematisches Institut
I Universität Karlsruhe
Englerstrasse 2
D-76128 Karlsruhe
Germany
e-mail: bodo@ma1euler.mathematik.uni-karlsruhe.de

P. J. McKenna
Department of Mathematics
U9, University of Connecticut
196 Auditorium Road
Storrs, CT 06269-3009
U.S.A.

Computing [Suppl] 15, 79–90 (2001)

Computing

A Superlinearly and Globally Convergent Method for Reaction and Diffusion Problems with a Non-Lipschitzian Operator

X. Chen, Matsue

Dedicated to Professor Tetsuro Yamamoto on the occasion of his 65th birthday

Abstract

This paper proposes a superlinearly and globally convergent method for reaction and diffusion problems with a non-Lipschitz operator. We reformulate the problem as a system of equations with locally Lipschitzian functions. Then the system is solved by using a smoothing Newton method which converges superlinearly and globally.

AMS Subject Classification: 65H10.

Key Words: Non-Lipschitzian operator, smoothing Newton method.

1. Introduction

We consider the following system of nonlinear equations

$$F(x) := Ax + Cf(x) - b = 0 \tag{1.1}$$

where A is an $n \times n$ symmetric positive definite matrix, C is an $n \times n$ diagonal matrix with positive diagonal entries c_i, $i = 1, 2, \dots, n$,

$$f_i(x) = f_i(x_i) = \begin{cases} x_i^p, & x_i \geq 0 \\ 0, & x_i < 0, \end{cases}$$

and b is a vector in R^n. Here $p \in (0, 1)$ is a constant. System (1.1) arises from finite element approximations or finite difference approximations for reaction-diffusion problems. A typical problem is as follows [1, 2]. Let Ω be a bounded domain in R^2 with a Lipschitz boundary $\partial\Omega$. Given a positive number λ, find u such that

$$\begin{aligned} -\Delta u + \lambda\xi(u) &= 0 \text{ in } \Omega \\ u &= 1 \text{ on } \partial\Omega, \end{aligned}$$

where

$$\xi(u) = \begin{cases} u^p, & u \geq 0 \\ 0, & u < 0. \end{cases}$$

The difficulty to solve (1.1) is that F is not local Lipschitz. In the last decade, many superlinearly and globally convergent algorithms for nonsmooth equations defined by a locally Lipschitzian operator have been developed [6, 7, 9]. The Rademacher theorem, the Clarke generalized Jacobian and the semismoothness play key roles in convergence analysis of Newton type methods for nonsmooth equations with locally Lipschitzian operators. The Rademacher theorem states that a locally Lipschitzian operator is almost everywhere differentiable. According to the Rademacher theorem, if F is local Lipschitz, the Clarke generalized Jacobian can be defined by [6]

$$\partial F(x) = \mathrm{co}\{\lim_{\substack{x^k \to x \\ x^k \in D_F}} F'(x^k)\},$$

where D_F denotes the set of points at which F is differentiable and co denotes the covex hull. A locally Lipschitzian function is called semismooth at x^* if the limit

$$\lim_{\substack{V \in \partial F(x^* + th') \\ h' \to h, t \downarrow 0}} \{Vh'\}$$

exists for any $h \in R^n$ [9]. Unfortunately, F defined in (1.1) is not local Lipschitz. These powerful tools are not applicable for solving (1.1).

Globally convergent methods for solving (1.1) have been studied in [1, 2]. However, the convergence rate of these methods is not expected faster than linear.

The aim of this paper is to present a superlinearly and globally convergent method for solving (1.1). In Section 2, we reformulate the system of equations (1.1) as a system of equations defined by a locally Lipschitzian function, and study the Clarke generalized Jacobian and the semismoothness of the new function. In Section 3, we give a smooth approximation function of the locally Lipschitzian function. In Section 4, we study a smoothing Newton method for solving (1.1) and show that the method is superlinearly and globally convergent. Moreover, we illustrate the reformulation and the method by a numerical example.

The set of all positive real numbers is denoted by $R_{++} = \{t \mid t > 0, t \in R\}$. The index set is denoted by $N = \{1, 2, \ldots, n\}$. We use $\|\cdot\|$ to denote the Euclidean norm.

2. Lipschitz Reformulation

Let $\omega : R \to R$ be defined by

$$\omega(t) := \begin{cases} t^{1/p}, & t \geq 0 \\ t, & t < 0. \end{cases}$$

The function ω is strictly monotonically increasing. Hence the inverse of ω exists and has the form

$$\omega^{-1}(s) = \begin{cases} s^p, & s \geq 0 \\ s, & s < 0. \end{cases}$$

Moreover $\omega(t) \geq 0$ if and only if $t \geq 0$.

Let

$$g(y) = (\omega(y_1), \omega(y_2), \ldots, \omega(y_n))^T.$$

Then by definitions of f and g, we have

$$\begin{aligned} f_i(g_i(y_i)) &= \begin{cases} g_i(y_i)^p, & y_i \geq 0 \\ 0, & y_i < 0 \end{cases} \\ &= \begin{cases} y_i, & y_i \geq 0 \\ 0, & y_i < 0 \end{cases} \\ &= \max(0, y_i), \quad i \in N. \end{aligned}$$

Now we define a Lipschitz function $H : R^{2n} \to R^{2n}$ as

$$H(x,y) = \begin{pmatrix} Ax + C\max(0,y) - b \\ x - g(y) \end{pmatrix},$$

where "max" denotes the componentwise maximum.

It is easy to see that if (x,y) is a solution of

$$H(x,y) = 0 \tag{2.1}$$

then x is a solution of (1.1). Conversely, if x is a solution of (1.1) then $(x, g^{-1}(x))$ is a solution of (2.1).

The function H is not differentiable only at points (x,y) where $y_i = 0$ for some $i \in N$. In other words, the set of points at which H is differentiable is

$$D_H = \{(x,y) \mid y_i \neq 0, \quad \text{for all } i \in N\}.$$

Since H is local Lipschitz, we can define the Clarke generalized Jacobian of H. Let $r : R \to R$ be defined by

$$r(t) = \begin{cases} 1 + t^{(1-p)/p}/p & t > 0 \\ 1 & t \leq 0. \end{cases}$$

Let I be the $n \times n$ identity matrix and

$$R(y) = \text{diag}(r(y_1), r(y_2), \ldots, r(y_n)).$$

Theorem 2.1. *The Clarke generalized Jacobian of H at* (x, y) *is equal to the set of matrices*

$$\begin{pmatrix} A & CQ_y \\ I & -R(y) + \max(0, Q_y) \end{pmatrix},$$

where

$$Q_y = \operatorname{diag}(q_1, q_2, \ldots, q_n)$$

and

$$q_i \in \partial \max(0, y_i) = \begin{cases} \{1\} & t > 0 \\ [0, 1] & t = 0 \\ \{0\} & t < 0 \end{cases} \qquad i \in N.$$

Proof: It is easy to verify that for $(x, y) \in D_H$,

$$H'(x, y) = \begin{pmatrix} A & CQ_y \\ I & -R(y) + \max(0, Q_y) \end{pmatrix}.$$

By Corollary 2 of Proposition 2.3.3 in [6],

$$\begin{aligned} \partial H(x, y) &= \partial \begin{pmatrix} Ax \\ x \end{pmatrix} + \begin{pmatrix} C & 0 \\ 0 & -I \end{pmatrix} \partial \begin{pmatrix} \max(0, y) \\ g(y) \end{pmatrix} \\ &= \begin{pmatrix} A & 0 \\ I & 0 \end{pmatrix} + \operatorname{co} \left\{ \lim_{\substack{y^k \to y \\ y^k \in D_H}} \begin{pmatrix} 0 & CQ_{y^k} \\ 0 & -R(y^k) + \max(0, Q_{y^k}) \end{pmatrix} \right\}. \end{aligned}$$

Since both Q_y and $-R(y) + \max(0, Q_y)$ are diagonal, we can consider each y_i, for $i \in N$ independently. If $y_i = 0$, then

$$\begin{aligned} &\operatorname{co} \left\{ \lim_{\substack{y_i^k \to y_i \\ y_i^k \neq 0}} \begin{pmatrix} c_i (Q_{y^k})_{ii} \\ -r(y_i^k) + \max(0, (Q_{y^k})_{ii}) \end{pmatrix} \right\} \\ &= \operatorname{co} \left\{ \begin{pmatrix} c_i \\ 0 \end{pmatrix}, \begin{pmatrix} 0 \\ -1 \end{pmatrix} \right\} \\ &= \left\{ \begin{pmatrix} c_i \alpha \\ -1 + \alpha \end{pmatrix}, 0 \le \alpha \le 1 \right\} \\ &= \left\{ \begin{pmatrix} c_i q_i \\ -r(0) + q_i \end{pmatrix}, q_i \in \partial \max(0, y_i) \right\}. \end{aligned}$$

This completes the proof. □

Theorem 2.2. *At every point* $(x,y) \in R^{2n}$, *all elements of* $\partial H(x,y)$ *are nonsingular.*

Proof: For fixed $(x,y) \in R^{2n}$ and elements $q_i \in \partial \max(y_i, 0)$, $i \in N$, let $Q = Q_y$ and

$$L = R(y) - \max(0, Q).$$

It is known that $\begin{pmatrix} A & CQ \\ I & -L \end{pmatrix}$ is nonsingular if and only if the Schur complement $-L - A^{-1}CQ$ is nonsingular. Since A, C, Q and L are symmetric, the Schur complement is nonsingular if and only if $LA + QC$ is nonsingular.

To show the nonsingularity of $LA + QC$, we let E be a diagonal matrix whose diagonal entries are

$$E_{ii} = \begin{cases} 1/L_{ii}, & L_{ii} \neq 0 \\ 1, & L_{ii} = 0. \end{cases}$$

The matrix E is nonsingular and EL is a diagonal matrix whose diagonal entries are either 0 or 1. Moreover, $L_{ii} = 0$ implies that $Q_{ii} = 1$. Now

$$\begin{aligned} LA + QC &= E^{-1}E(LAC^{-1} + Q)C \\ &= E^{-1}(I + EL(AC^{-1} - I) + E(L + Q) - I)C. \end{aligned} \tag{2.2}$$

Since A is symmetric positive definite and C is a diagonal matrix with $c_{ii} > 0$, $C^{-1}AC^{-1}$ is symmetric positive definite and so it is a P-matrix. Hence $AC^{-1} = CC^{-1}AC^{-1}$ is a P-matrix. By Theorem 4.4 in [8], together with that EL is a diagonal matrix whose entries are either 0 or 1,

$$I + EL(AC^{-1} - I)$$

is nonsingular, and so a P-matrix. Moreover, all diagonal entries of the diagonal matrix $E(L + Q) - I$ are nonnegative. Therefore

$$I + EL(AC^{-1} - I) + E(L + Q) - I$$

is a P-matrix. By (2.2), $AL + CQ$ is nonsingular. □

Remark 2.1. The function H is a piecewise continuously differentiable function. According to Theorem 4.1 in [9], H is semismooth in R^{2n}.

3. Smoothing Function of *H*

In this section, we study smoothing functions of H. The nonsmoothness of H appears in two terms: $\max(0, y)$ and $g(y)$. To define a smoothing function of H, we set

$$\theta(t) = \begin{cases} t^{1/p} & t \geq 0 \\ 0 & t < 0. \end{cases}$$

It is easy to see that θ is continuously differentiable in R. Moreover,

$$\omega(t) = \min(\theta(t), t) = t - \max(t - \theta(t), 0), \quad \text{for } t \leq 1.$$

Now for $\max(0, t)$, we use the following smoothing function

$$\phi(t, \epsilon) = \begin{cases} \max(0, t) & |t| \geq \epsilon \\ \frac{1}{4\epsilon}(t + \epsilon)^2 & |t| < \epsilon. \end{cases}$$

Let $\alpha \in (0, p^{p/(1-p)}]$. We define the following smoothing function for ω.

$$\psi(t, \epsilon) = \begin{cases} t - \frac{1}{4\epsilon}(t - \theta(t) + \epsilon)^2, & t \leq \alpha \text{ and } |t - \theta(t)| < \epsilon \\ \omega(t), & \text{otherwise.} \end{cases}$$

Proposition 3.1. *Functions* ϕ, $\psi : R \times R_{++} \to R$ *satisfy the following properties.*

1. *For every fixed* ϵ, $\phi(t, \epsilon)$ *and* $\psi(t, \epsilon)$ *are continuously differentiable with respect to* t *in* R.
2. *For every* $t \in R$,

$$|\phi(t, \epsilon) - \max(0, t)| \leq \epsilon.$$

$$|\psi(t, \epsilon) - \omega(t)| \leq \epsilon.$$

3. *For every fixed* $t \in R$,

$$\lim_{\epsilon \downarrow 0} \frac{\partial \phi(t, \epsilon)}{\partial t} = \phi^o(t) \in \partial \max(0, t)$$

and

$$\lim_{\epsilon \downarrow 0} \frac{\partial \psi(t, \epsilon)}{\partial t} = \psi^o(t) \in \partial \omega(t).$$

Proof: The proof for the smoothing function ϕ can be found in [4]. Now we show the part for ψ.

1. For $t > \alpha$, ω is continuously differentiable and so ψ. Let $t \leq \alpha$. Then $t - \theta(t) > 0$ if only only if $t > 0$. Suppose that $\bar{t} - \theta(\bar{t}) = \epsilon$. Then

$$\bar{t} - \frac{1}{4\epsilon}(\bar{t} - \theta(\bar{t}) + \epsilon)^2 = \bar{t} - \epsilon = \theta(\bar{t}) = \min(\theta(\bar{t}), \bar{t}) = \omega(\bar{t})$$

and

$$1-\frac{1}{2\epsilon}(\bar{t}-\theta(\bar{t})+\epsilon)(1-\theta'(\bar{t}))=\theta'(\bar{t})=\omega'(\bar{t}).$$

Moreover if $\bar{t}-\theta(\bar{t})=-\epsilon$, then

$$\bar{t}-\frac{1}{4\epsilon}(\bar{t}-\theta(\bar{t})+\epsilon)^2=\bar{t}=\min(\theta(\bar{t}),\bar{t})=\omega(\bar{t})$$

and

$$1-\frac{1}{2\epsilon}(\bar{t}-\theta(\bar{t})+\epsilon)(1-\theta'(\bar{t}))=1=\omega'(\bar{t}).$$

Hence ψ is continuously differentiable.

2. If $t>\alpha$ or $t\leq\alpha$ and $|t-\theta(t)|\geq\epsilon$, we have

$$|\psi(t,\epsilon)-\omega(t)|=0\leq\epsilon.$$

If $t\in[0,\alpha]$ and $|t-\theta(t)|<\epsilon$, then $t\geq\theta(t)=t^{1/p}=\omega(t)$,

$$\psi(t,\epsilon)-\omega(t)\geq t-\frac{1}{4\epsilon}(2\epsilon)^2-\omega(t)=t-\epsilon-t^{1/p}\geq-\epsilon$$

and

$$\psi(t,\epsilon)-\omega(t)\leq t-\omega(t)\leq\epsilon+\theta(t)-\omega(t)=\epsilon.$$

Hence $|\psi(t,\epsilon)-\omega(t)|\leq\epsilon$.

If $t<0$ and $|t-\theta(t)|<\epsilon$, we have $\theta(t)=0$ and $\omega(t)=t\geq-\epsilon$. Thus

$$|\psi(t,\epsilon)-\omega(t)|=\frac{1}{4\epsilon}(t-\theta(t)+\epsilon)^2\leq\epsilon.$$

3. For $t>0$, we have $\lim_{\epsilon\downarrow 0}\psi'(t,\epsilon)=\omega'(t)$. For $t<0$, we have $\lim_{\epsilon\downarrow 0}\psi'(t,\epsilon)=1=\omega'(t)$. For $t=0$, we have $\lim_{\epsilon\downarrow 0}\psi'(0,\epsilon)=1-\frac{1}{2}(1-\theta'(0))=\frac{1}{2}\in\partial\omega(0)=[0,1]$. □

Using the functions ϕ and ψ, we can define a smoothing function of H. Let

$$\Phi(y,\epsilon)=(\phi(y_1,\epsilon),\ldots,\phi(y_n,\epsilon))$$

$$\Psi(y,\epsilon)=(\psi(y_1,\epsilon),\ldots,\psi(y_n,\epsilon))$$

and

$$\mathscr{H}(x,y,\epsilon) = \begin{pmatrix} Ax + C\Phi(y,\epsilon) - b \\ x - \Psi(y,\epsilon) \end{pmatrix}.$$

For brevity we let $z = (x, y)$.

According to Proposition 3.1, we have the following theorem.

Theorem 3.1. *Function $\mathscr{H} : R^{2n} \times R_{++} \to R^{2n}$ satisfies the following properties:*

1. *For every fixed $\epsilon > 0$, $\mathscr{H}$ is continuously differentiable with respect to z in R^{2n}.*
2. *For every $z \in R^{2n}$,*

$$\|\mathscr{H}(z,\epsilon) - H(z)\| \le \epsilon\sqrt{n(\|C\|^2 + 1)}.$$

3. *For every fixed $z \in R^{2n}$,*

$$\lim_{\epsilon\downarrow 0} \mathscr{H}_z(z,\epsilon) =: \mathscr{H}^o(z) \in \partial H(z).$$

Theorem 3.2. *$\mathscr{H}_z(z,\epsilon)$ is nonsingular at every point $(z,\epsilon) \in R^{2n} \times R_{++}$.*

Proof: By the definition of θ, we have $\theta'(t) \in [0,1]$ for $t \le \alpha \le p^{p/(1-p)}$. This implies that $\psi'(t,\epsilon) > 0$ for all $t \in R$, and $\epsilon \in R_{++}$. Hence for a fixed $(x,y,\epsilon) \in R^{2n} \times R_{++}$, the diagonal entries of $\Psi'(y,\epsilon)$ are positive. Suppose that there is $(u,v) \in R^{2n}$ such that

$$\mathscr{H}_z(z,\epsilon)\begin{pmatrix} u \\ v \end{pmatrix} = 0.$$

Then $Au + C\Phi'(y,\epsilon)v = 0$ and $u = \Psi'(y,\epsilon)v$. Note that A is a symmetric positive definite matrix and $C\Phi'(y,\epsilon) \ge 0$ is a diagonal matrix. We have

$$0 = (u, Au + C\Phi'(y,\epsilon)v) = (u, Au + C\Phi'(y,\epsilon)\Psi'(y,\epsilon)^{-1}u) \ge (u, Au) \ge 0.$$

Therefore $u = 0$ and so $v = \Psi'(y,\epsilon)^{-1}u = 0$. Since (u,v) is arbitrarily chosen, $\mathscr{H}_z(z,\epsilon)$ is nonsingular. □

4. An Algorithm and an Example

In this section we study an algorithm which is an application of Algorithm 3.1 in [5] to the system of equations (2.1).

Algorithm 4.1. *Given $\rho, \tau, \eta \in (0,1)$, and a starting point $z^0 \in R^{2n}$. Choose a scalar $\sigma \in (0, 1-\tau)$. Let $\nu = \tau/(2\sqrt{2n}\max\{1, \|C\|\})$. Let $\beta_0 = \|H(z^0)\|$ and $\epsilon_0 = \nu\beta_0$.*

For $k \geq 0$:

1. *Find a solution $\hat{d}^k$ of the system of linear equations*

$$H(z^k) + \mathscr{H}^o(z^k)d = 0.$$

If $\|H(z^k + \hat{d}^k)\| \leq \eta\beta_k$, let $z^{k+1} = z^k + \hat{d}^k$ and perform Step 3. Otherwise perform Step 2.

2. *Find a solution d^k of the system of linear equations*

$$H(z^k) + \mathscr{H}_z(z^k, \epsilon_k)d = 0.$$

Let m_k be the smallest nonnegative integer m such that

$$\|\mathscr{H}(z^k + \rho^m d^k, \epsilon_k)\|^2 - \|\mathscr{H}(z^k, \epsilon)\|^2 \leq -\sigma\rho^m \|H(z^k)\|^2.$$

Set $t_k = \rho^{m_k}$ and $z^{k+1} = z^k + t_k d^k$.

3. 3.1 *If $\|H(z^{k+1})\| = 0$, terminate.*

3.2 *If*

$$0 < \|H(z^{k+1})\| \leq \max\{\eta\beta_k, \tau^{-1}\|H(z^{k+1}) - \mathscr{H}(z^{k+1}, \epsilon_k)\|\},$$

let

$$\beta_{k+1} = \|H(z^{k+1})\| \text{ and } \epsilon_{k+1} = \min\left\{\nu\beta_{k+1}, \frac{\epsilon_k}{2}\right\}.$$

3.3 *Otherwise, let $\beta_{k+1} = \beta_k$ and $\epsilon_{k+1} = \epsilon_k$.*

Theorem 4.1. *The system of equations* (2.1) *has a unique solution.*

Proof: The system (1.1) is equivalent to the following minimization problem

$$\min_{x \in R^n} \frac{1}{2} x^T A x + \sum_{i=1}^{n} c_i \int_0^{x_i} f(t)dt - b^T x. \tag{4.1}$$

Moreover we have

$$\begin{aligned}
&(Ax + f(x) - Ay - f(y))^T (x - y) \\
&\quad = (x - y)^T A(x - y) + (f(x) - f(y))^T (x - y) \\
&\quad \geq (x - y)^T A(x - y) \\
&\quad \geq \lambda_1 \|x - y\|^2
\end{aligned}$$

and

$$\frac{1}{2}x^T Ax + \sum_{i=1}^{n} \int_0^{x_i} f(t)dt - b^T x \geq \frac{1}{2}x^T Ax - b^T x \geq \frac{\lambda_1}{2} x^T x - b^T x,$$

where $\lambda_1 > 0$ is the smallest eigenvalue of A. Hence (1.1) is a strictly convex minimization problem with bounded level set. It turns out that (1.1) has a unique solution x^*, and so (2.1) has a unique solution $(x^*, g^{-1}(x^*))$. □

Theorem 4.2. *For any* $\gamma > 0$, *the set*

$$S_\gamma = \{z \mid \|H(z)\|^2 \leq \gamma\}$$

is nonempty and bounded.

Proof: From Theorem 4.1, S_γ contains the solution of (2.1), and so it is nonempty.

Moreover for every $(x, y) \in S_\gamma$,

$$|x_i - g_i(y_i)| \leq \gamma, \quad i \in N.$$

This implies that there is $u \in R^n$ with $\|u\|_\infty \leq \gamma$ such that

$$y_i = g_i^{-1}(x_i + u_i) = (x_i + u_i)^p, \quad i \in N.$$

Hence if $\|(x, y)\| \to \infty$, then $\|x\| \to \infty$. Moreover,

$$\begin{aligned}
\|H(z)\|^2 &= \|Ax + C\max(0, y) - b\|^2 + \|x - g(y)\|^2 \\
&\geq \|Ax + C\max(0, y)\|^2 - 2(Ax + C\max(0, y))^T b \\
&= \|Ax + C\max(0, (x+u)^p)\| - 2(Ax + C\max(0, (x+u)^p))^T b \\
&= \|Ax + D(x+u)\| - 2(Ax + D(x+u))^T b \\
&\geq \|Ax + D(x+u)\|^2 - 2\|Ax + D(x+u)\|\|b\|
\end{aligned}$$

where $D = \text{diag}(d_i)$ whose diagonal elements are

$$d_i = \begin{cases} c_i(x_i + u_i)^{p-1} & x_i + u_i > 0 \\ 0 & x_i + u_i \leq 0. \end{cases}$$

Since A is symmetric positive definite, $D \geq 0$ and $\|u\| \leq \gamma$, this implies that if $\|x\| \to \infty$, then $\|Ax + D(x+u)\| \to \infty$ and so $\|H(z)\| \to \infty$. Hence S_γ is bounded. □

Theorem 4.3. *For any starting point* $z^0 \in R^{2n}$, *Algorithm* 4.1 *is well defined and the generated sequence* $\{z^k\}$ *remains in the level set* $S_{(1+\tau)\|H(z^0)\|}$ *and converges to the unique solution* z^* *of* (2.1). *Moreover, the convergence rate is superlinear.*

Proof: By Theorem 3.1 and Theorem 4.2, Assumptions A1 and A2 in [5] holds. Hence it follows from Theorem 3.1 in [5] that Algorithm 4.1 is well defined and the generated sequence $\{z^k\}$ remains in $S_{(1+\tau)\|H(z^0)\|}$ and satisfies

$$\lim_{k\to\infty} \|H(z^k)\| = 0.$$

By Theorem 4.1, $H(z)$ has a unique zero z^*. Therefore, $\{z^k\}$ converges to z^*.

Proposition 4 in [3] states that all elements of $\partial H(z)$ are nonsingular if and only if $\|H^o(z)^{-1}\| \le \Gamma < \infty$ in a neighborhood of z. Hence from Theorem 2.2, all assumptions of Theorem 3.2 in [5] hold, which ensures the superlinear convergence property. □

To illustrate the smoothing Newton method, we consider the following example [2].

Example 4.1.

$$-\Delta u + \frac{9}{(1-p)^2}\max(0, u^p) = \frac{9}{r(1-p)^2}\left(\frac{3r-1}{2}\right)^{\frac{2p}{1-p}} h\left(r - \frac{1}{3}\right), \quad \text{in } \Omega = (0,1)\times(0,1)$$

$$u(r) = \left(\frac{3r-1}{2}\right)^{\frac{2}{1-p}} h\left(r - \frac{1}{3}\right), \quad \text{on } \partial\Omega,$$

where $r^2 = x^2 + y^2$ and h is the Heaviside function.

This problem has the solution

$$u(r) = \left(\frac{3r-1}{2}\right)^{\frac{2}{1-p}} h\left(r - \frac{1}{3}\right).$$

Application of the five-point difference scheme with mesh size $1/(\sqrt{n}+1)$ to this problem gives a system of equations $F(x) = 0$. We rewrite the system as a system of equations $H(z) = 0$.

We used Algorithm 4.1 to solve $H(z) = 0$. In our numerical experiment, parameters of Algorithm 4.1 were chosen as

$$\rho = 0.8, \quad \tau = 0.6, \quad \eta = 0.96, \quad \sigma = 0.3.$$

We took the starting point $z^0 = 0$ and stoped the algorithm when $\|H(z^k)\| \le 10^{-10}$.

Numerical results are obtained by using Matlab on a IBM PC. Numerical results show that the Lipschitz reformulation is stable and Algorithm 4.1 is superlinearly

Table 1. Numerical result of Example 4.1

p	n	k	$\|H(z^{k-2})\|$	$\|H(z^{k-1})\|$	$\|H(z^k)\|$
0.1	225	21	2.0E−5	1.2E−8	1.3E−14
0.1	625	26	2.4E−6	1.8E−10	2.2E−14
0.3	225	9	5.8E−4	2.5E−6	5.0E−11
0.3	625	12	5.2E−5	8.8E−8	1.7E−12
0.5	225	6	2.1E−3	8.5E−7	1.0E−12
0.5	625	7	7.8E−5	3.7E−8	5.3E−13

and globally convergent. In Table 1, we present $\|H(z^k)\|$ in the last three iterations for different n and p.

Acknowledgement

This work was supported in part by the Scientific Research Grant C11640119 from the Japan Society of the Promotion of Science.

References

[1] Aziz, A. K., Stephens, A. B., Suri, M.: Numerical methods for reaction-diffusion problems with non-differentiable kinetics. Numer. Math. *53*, 1–11 (1988).
[2] Barrett, J. W., Shanahan, R. M.: Finite element approximation of a model reaction-diffusion problem with a non-Lipschitz nonlinearity. Numer. Math. *59*, 217–242 (1991).
[3] Chen, B., Chen, X.: A global linear and local quadratic continuation smoothing method for variational inequalities with box constraints. Comput. Optim. Appl. *17*, 131–158 (2000).
[4] Chen, X.: Smoothing methods for complementarity problems and their applications: a survey. J. Oper. Res. Soc. Jpn. *43*, 32–47 (2000).
[5] Chen, X., Ye, Y.: On homotopy-smoothing methods for box-constrained variational inequalites, SIAM J. Control. Optim. *37*, 589–616 (1999).
[6] Clarke, F. H.: Optimization and nonsmooth analysis. New York: Wiley, 1983.
[7] Fukushima, M., Qi, L. (eds).: Reformulation: nonsmooth, piecewise smooth, semismooth and smoothing methods. New York: Kluwer, 1999.
[8] Gabriel, S. A., Moré, J. J.: Smoothing of mixed complementarity problems. In: 'Complementarity and Variational Problems: State of the Art,' (Ferris, M. C., Pang, J. S., eds.), pp. 105–116. Philadelphia: SIAM, 1997.
[9] Qi, L., Sun, J.: A nonsmooth version of Newton's method. Math. Program. *58*, 353–367 (1993).

X. Chen
Department of Mathematics and Computer Science
Shimane University
Matsue 690-8504
Japan
e-mail: chen@math.shimane-u.ac.jp

Computing [Suppl] 15, 91–104 (2001)

Computing

On Linear Asynchronous Iterations when the Spectral Radius of the Modulus Matrix is One

A. Frommer, Wuppertal, and **P. Spiteri,** Toulouse

Dedicated to Professor Tetsuro Yamamoto on the occasion of his 65th birthday

Abstract

A classical result on linear asynchronous iterations states that convergence occurs if and only if the spectral radius of the modulus matrix is less than 1. The present paper shows that if one introduces very mild restrictions on the admissible asynchronous processes, one gets convergence for a larger class of matrices for which the spectral radius of the modulus matrix is allowed to be equal to 1. The mild restrictions are virtually always fulfilled in practical implementations. In this manner, our result contributes to the better understanding of the different hypotheses underlying mathematical models for asynchronous iterations.

Key Words: Linear systems, iterative methods, parallel computing.

1. Introduction

In this paper we consider genuine parallel iterative methods for solving a linear system

$$Ax = b, \quad \text{where } A \in \mathbb{C}^{n\times n}, \; x, b \in \mathbb{C}^n. \tag{1}$$

The iterative methods we deal with rely on a fixed point formulation for (1) of the form

$$x = Hx + c, \quad \text{where } H = (h_{ij}) \in \mathbb{C}^{n\times n} \text{ and } c \in \mathbb{C}^n. \tag{2}$$

We assume that H and c are constructed such that x is a solution of the linear system (1) if and only if x is a solution of the fixed point equation (2). For example, H may arise from an appropriate splitting of A, see, for example, [20].

As is well known, the total step method

$$x^{k+1} = Hx^k + c, \quad k = 0, 1, \ldots, \; x^0 \text{ given}, \tag{3}$$

converges to a unique solution of (2) for any starting vector x^0 if and only if $\rho(H) < 1$, see [20] for example. On a parallel computer, in order to avoid idle times due to imbalanced load or communication delays, efficient implementations of the total step method actually lead to new iterative schemes whose behaviour can only partly be determined *a priori*. These iterations belong to the so-called class of asynchronous methods, and the purpose of the present paper is to further contribute to the understanding of sufficient conditions guaranteeing the convergence of such methods.

Since the pioneering work by Chazan and Miranker [6], the condition $\rho(|H|) < 1$ is known as a necessary and sufficient condition for the convergence of asynchronous iterations. However, the precise mathematical model for the asynchronous iterations underlying the Chazan–Miranker result (see [6] and [3] for a slight modification) is very general. Virtually any practical situation can in fact be modelled mathematically by assuming slightly more restricitive hypotheses than in the Chazan–Miranker model, and this has indeed been done repeatedly in the literature on asynchronous iterations. Examples include iterations for singular systems [1, 13, 17] (see also [2] for cases with a varying iteration operator), the flexible communication model of [7, 8, 15], results on monotone convergence [10, 14] or iterations relying on interval arithmetic [11]. However, the linear case (3) with $\rho(H) < 1$ has never been reconsidered under such natural, but slightly more restrictive assumptions. We do so in the present paper, and we will find that asynchronous iterations converge for a larger class of matrices H than those with $\rho(|H|) < 1$. We believe that, besides establishing a new convergence theorem on linear asynchronous iterations, the main interest of our result is that it contributes to clarify the role of the various hypotheses in the mathematical models on asychronous iterations.

The paper is organized as follows. We introduce our notation and some basic results in Section 2. Section 3 presents and discusses the mathematical model for linear asynchronous iterations. This section also reviews the various hypotheses imposed on the model and states known convergence results. The main part of the paper is Section 4 where we prove our new convergence theorem. We end with some conclusions in Section 5.

2. Auxiliary Results

In this section we briefly introduce our notation and state some auxiliary results.

For a given matrix $H \in \mathbb{C}^{n\times n}$, its (i,j) entry will usually be denoted by h_{ij} but sometimes, we will find it convenient to write $(H)_{ij}$ instead. The matrix H is termed *nonnegative*, $H \geq 0$, if all its entries are nonnegative. It will be termed *positive*, $H > 0$, if all entries are positive. We adopt a similar notation and terminology for vectors. The *modulus matrix* $|H| \in \mathbb{R}^{n\times n}$ of H is defined entrywise through

$$(|H|)_{ij} = |h_{ij}|, \quad i,j = 1,\ldots,n.$$

By $\rho(H)$ we denote the spectral radius of H, i.e. the largest modulus of the eigenvalues of H. The following result can be found in [16], e.g.

Lemma 1. *For any matrix $H \in \mathbb{C}^{n\times n}$ we have*

$$\rho(H) \leq \rho(|H|).$$

The non-zero pattern of a matrix $H \in \mathbb{C}^{n\times n}$ defines a directed graph $G(H) = (V, E)$ with vertices V and edges E via

$$V = \{1, \ldots, n\}, \quad E = \{(i,j) \mid h_{ij} \neq 0\} \subseteq V \times V.$$

A (directed) *path of length q* from vertex i to vertex j in G is a finite sequence of vertices $(i = i_0, i_1, \ldots, i_{q-1}, i_q = j)$ such that $(i_{l-1}, i_l) \in E$ for $l = 1, \ldots, q$. The matrix H is termed *irreducible* if in $G(H)$ there is a path from any vertex i to any other vertex $j \neq i$.

For a given vertex i, we define its (reverse) level sets $L_\mu(i)$ recursively as follows:

$$L_0(i) = \{i\}$$

$$L_\mu(i) = \{j \in V \mid \text{there exists } l \in L_{\mu-1}(i) \text{ such that } (j,l) \in E\} - \bigcup_{\nu=0}^{\mu-1} L_\nu,$$

$$\text{for } \mu = 1, \ldots, n.$$

Clearly, the level sets $L_\mu(i)$ bin those vertices j for which there is a path from j to i by the length of the shortest such path. A shortest path has always length less than n. Consequently, $L_n(i) = \emptyset$ for all i, and if H is irreducible, then $\cup_{\mu=0}^{n-1} L_\mu(i) = \{1, \ldots, n\}$ for all i.

A *signature* matrix is a diagonal matrix all diagonal elements of which have modulus 1. Similarity transformations with signature matrices play a major role when taking a closer look on when there can be equality in Lemma 1:

Lemma 2. *Assume that H is irreducible. Then $\rho(H) = \rho(|H|)$ if and only if there exists a signature matrix S and a complex number γ with modulus* 1 *such that* $\gamma S^{-1}HS \geq 0$.

The proof can be found, for example, in [4, Th. 2.2.14].

We end this section with a purely technical result which will be used in Section 4 for the proof of Theorem 3. Here, as in the sequel, $\arg(a)$ denotes the argument of a complex number $a \neq 0$, and this argument is unique mod 2π.

Lemma 3. *Let be given n sequences of complex numbers $\{a_j^k\}_{k=0}^\infty$, an additional such sequence $\{b^k\}_{k=0}^\infty$ and n complex numbers $h_j, j = 1, \ldots, n$ such that*

$$b^k = \sum_{j=1}^n h_j a_j^k, \quad k = 0, 1, \ldots \quad \textit{and} \quad \sum_{j=1}^n |h_j| = 1.$$

Moreover, assume that there exists $c > 0$ *such that*

$$\lim_{k\to\infty} |a_j^k| = c \text{ and } \lim_{k\to\infty} b^k = c.$$

Let $\alpha_j^k = \arg(a_j^k)$, $\beta^k = \arg(b^k)$ *and* $\eta_j = \arg(h_j)$. *Then for all* j *for which* $h_j \neq 0$ *we have*

$$\lim_{k\to\infty} \cos(\beta^k - \alpha_j^k - \eta_j) = 1.$$

Proof: If $h_j \neq 0$ for just one j, the result is trivial. So assume that $h_j \neq 0$ for at least two indices j. Then $|h_j| < 1$ for all j, which will prevent us from dividing by zero in the algebraic manipulations to come. Fix any j for which $h_j \neq 0$. We have

$$|b^k - h_j a_j^k| = \left| \sum_{l=1, l\neq j}^{n} h_l a_l^k \right|. \tag{4}$$

For the left part of (4) we get, as for any difference of two complex numbers,

$$\left|b^k - h_j a_j^k\right| = \left| |b^k| - |h_j a_j^k| \right| \cdot \sqrt{1 + p_j^k (1 - \cos(\beta^k - \alpha_j^k - \eta_j))}$$

$$\text{with } p_j^k = 2 \cdot \frac{\left|b^k\right| \cdot \left|h_j a_j^k\right|}{\left|b^k - h_j a_j^k\right|^2}. \tag{5}$$

Using the triangle inequality on the right hand part of (4), we get

$$|b^k - h_j a_j^k| \leq \sum_{l=1, l\neq j}^{n} |h_l| \cdot |a_l^k|. \tag{6}$$

For the right hand side of (6) we get

$$\lim_{k\to\infty} \sum_{l=1, l\neq j}^{n} |h_l| \cdot |a_l^k| = \sum_{l=1, l\neq j}^{n} |h_l| c = (1 - |h_j|)c.$$

On the other hand, for the right hand side of (5), we have

$$\lim_{k\to\infty} \left| |b^k| - |h_j a_j^k| \right| = (1 - |h_j|)c \text{ and } \lim_{k\to\infty} p_j^k = 2|h_j|/(1 - |h_j|)^2 > 0.$$

So the square root in (5) tends to 1, and since p_j^k tends to a positive limit we have $\lim_{k\to\infty}(1 - \cos((\beta^k - \alpha_j^k - \eta_j)) = 0.$ □

3. Asynchronous Linear Iterations

Assume that we are given a linear system in the fixed point form (2). Asynchronous iterations (2) may be regarded as a whole class of iterative methods derived from the total step method (3). One now allows that only certain components of the iterate are updated at a given time step and that more than just the previous iterate may be used in the updating process. The precise definition is as follows, see [3, 5].

Definition 1. *For* $k = 1, 2, \ldots$ *let* $J_k \subseteq \{1, \ldots, n\}$ *and* $(s_1(k), \ldots, s_n(k)) \in \mathbb{N}_0^n$ *be such that*

$$s_i(k) \leq k - 1 \quad \textit{for } i = 1, \ldots, n, \;\; k = 1, 2, \ldots, \tag{7}$$

$$\lim_{k \to \infty} s_i(k) = +\infty \quad \textit{for } i = 1, \ldots, n, \tag{8}$$

$$\textit{for every } i \in \{1, \ldots, n\} \textit{ the set } \{k \mid i \in J_k\} \textit{ is unbounded.} \tag{9}$$

Then the iteration

$$x_i^{k+1} = \begin{cases} \sum_{j=1}^{n} h_{ij} x_j^{s_j(k)} + c_i & \textit{if } j \in J_k \\ x_i^k & \textit{if } j \notin J_k \end{cases} \tag{10}$$

is termed an asynchronous iteration.

We will call the sets J_k the *active components*. The differences $k - s_i(k)$ represent the delay between the current iteration step k and the step $s_i(k)$ in which we calculated the components actually used in the current step. For the sake of brevity, but somehow abusively, we will in the sequel reserve the word *delay* to just denote $s_i(k)$.

Asynchronous iterations arise naturally on parallel computers if one eliminates synchronization points in order to avoid idle times on the individual processors. Detailed discussions can be found in the surveys in [5, 12]. Typically, then, the components of the vector to compute are distributed in (non-overlapping) blocks to the processors. The iteration counter k in (10) has then to be interpreted as counting every step in which any of the processors has finished its next update. The delays $s_i(k)$ account for the fact that processors will usually need different times to accomplish their individual updates and that the most recent data may not be available due to communication delays. The active components J_k represent the block of components assigned to the processor which does the k-th update.

In such a situation, the following more restrictive hypotheses than (7)–(9) will be satisfied:

$$s_i(k) = k - 1 \quad \text{if } i \in J_k \tag{11}$$

$$\text{there exists } d \in \mathbb{N} \text{ such that } k - d \leq s_i(k) \leq k - 1 \quad \text{for all } i \text{ and } k \tag{12}$$

$$\text{there exists } r \in \mathbb{N} \text{ such that } \bigcup_{l=k}^{k+r} J_l = \{1, \ldots, n\} \quad \text{for all } k \tag{13}$$

Here, (11) holds because a processor has always the latest information of 'its own' block available, and (12) as well as (13) simply reflect the fact that each communication and each update on the processors will neither become arbitrarily slow nor arbitrarily fast. We refer to [5, 12] for additional details. It is worth noting, though, that (11) will most likely *not* be fulfilled in the (less natural) situation where at least one component of the iteration vector is assigned to several processors, i.e. when we distribute *overlapping* blocks of components onto the processors.

Under the assumptions (7), (8) and (9), the convergence behaviour of the asynchronous iteration (10) has been studied in [3, 6, 19]. In particular, the first and second part of the following theorem date back to [3] and [6], respectively.

Theorem 1. *Assume that* (7) *to* (9) *are fulfilled.*

(*i*) If $\rho(|H|) < 1$ *the asynchronous iteration* (10) *converges to* x^*, *the unique fixed point of* (2), *i.e. for every starting vector* x^0 *we have* $\lim_{k\to\infty} x^k = x^*$.

(*ii*) *If* $\rho(|H|) \geq 1$ *and if* (2) *has at least one fixed point* x^*, *then there exists a starting vector* x^0, *a sequence of delays* $s_i(k)$ *and a sequence of active components* J_k, *satisfying* (7) *to* (9) *and even* (12) *and* (13) *such that the iterates* x^k *of* (10) *do not converge to* x^*.

The proof for part (ii), as given in [6] (see also [9]), constructs appropriate delays and active components as well as a starting vector such that the iterates do not converge to x^*. This construction, while satisfying (12) and (13) crucially relies on *not* fulfilling (11).

Theorem 1 is complemented by the following result due to Lubachevsky and Mitra [13] in the context of asynchronous iterations for Markov chains.

Theorem 2. *Assume* $H \geq 0$ *is irreducible,* $\rho(H) = 1$ *and let* $c = 0$ *in* (10). *Moreover, let the starting vector* x^0 *be nonnegative and let there exist at least one index i such that* $h_{ii} > 0$, $x_i^0 > 0$ *and* $s_i(k) = k - 1$ *whenever* $i \in J_k$. *Finally, assume that the asynchronous iteration* (10) *satisfies the more restrictive hypotheses* (12) *and* (13). *Then*

$$\lim_{k\to\infty} x^k = x^*$$

where $x^* > 0$ *is a positive vector satisfying* $Hx^* = x^*$.

As a first comment, let us note that [13] actually deals with an asynchronous model which looks more general than ours allowing for the delays $s_j(k)$ to also depend on the index $i \in J_k$. However, this is only a formal generalization, since one can assume all sets of active components J_k to be singletons as can be seen as follows: step k in (10) can alternatively be rewritten as a sequence of $|J_k|$ new steps, where in each new step we have just one active component. The delays have then to be redefined, but the new delays together with the new sets (singletons) of active components will still satisfy (7) to (9) if the old ones did; and similarly for (11) to (13) (with redefinition of d and r).

As another comment, note that Theorem 2 is by no means in conflict with part (ii) of Theorem 1: Under the assumptions of Theorem 2, the equation $x^* = Hx^*$ has a one-dimensional subspace as its solution set, and the theorem just says that the asynchronous iteration converges to one (non-zero) point of that subspace. As was shown in [13], this point will actually depend on the delays and the active components of the asynchronous iteration. Note also that Theorem 2 assumes a particular starting vector and (11) to be fulfilled for at least one particular component i. For further discussion, see also [18].

In his paper [17], Pott proved a generalization of Theorem 2. He assumes that H is reducible and then requires conditions similar to those of Theorem 2 for each diagonal block in the reducible normal form of H which has spectral radius 1.

Theorems 1 and 2 set the stage for the work to be presented in this paper. Since we want to prove convergence results for the case $\rho(|H|) = 1$, we certainly should look at matrices which are not nonnegative (Theorem 2) and we will have to assume more restrictive hypotheses than (7)–(9) (Theorem 1).

Our main result in the next section will identify a class of such matrices for which (11)–(13) will turn out to be sufficient for convergence. This demonstrates that condition (11) crucially affects the convergence of asynchronous iterations, whereas replacing (7)–(9) by (12) and (13) does not (Theorem 1 (ii)).

Let us end this section by introducing another useful piece of notation. For a given time step k and a component i, we will use $r_i(k)$ to denote the latest time step before k at which the i-th component was modified, i.e.

$$r_i(k) = \max\{l \mid i \in J_l \text{ and } l \leq k\}. \tag{14}$$

We thus have

$$x_i^k = x_i^{k-1} = \cdots = x_i^{r_i(k)} = \sum_{j=1}^{n} h_{ij} x_j^{s_j(r_i(k))} + c_i,$$

and, if (13) is satisfied,

$$k - r_i(k) \leq r \quad \text{for all } i. \tag{15}$$

4. A New Convergence Result

We start with a note on the scaling invariance of asynchronous iterations. Clearly, the synchronous total step method (3) is invariant under similarity transformations. By this, we mean that if we take any non-singular matrix $T \in \mathbb{R}^{n\times n}$, the iterates x^k of (3) are related to the iterates $\widehat{x}^k$ of

$$\widehat{x}^k = (T^{-1}HT)\widehat{x}^{k-1} + T^{-1}c, \quad k = 1, 2, \ldots,$$

through $T\widehat{x}^k = x^k, k = 0, 1, \ldots$ An analogous property holds for asynchronous iterations only if we assume T to be diagonal. In the convergence analysis to come, we can therefore without loss of generality assume an appropriate diagonal scaling $H \leftarrow D^{-1}HD$ with a non-singular diagonal matrix D. We will use such a diagonal scaling based on the following observation: If H is irreducible and $\rho(|H|) = 1$, the Perron-Frobenius theorem (see, e.g., [4, Th. 1.4]) shows that there exists a positive vector $u \in \mathbb{R}^n$ such that $|H|u = u$. This means that we have

$$\frac{1}{u_i}\sum_{j=1}^{1} |h_{ij}|u_j = 1, \quad i = 1, \ldots, n.$$

Taking $D = \text{diag}(u)$ and scaling $H \leftarrow D^{-1}HD$ we may therefore assume that, in addition,

$$\sum_{j=1}^{n} |h_{ij}| = 1, \quad i = 1, \ldots, n, \tag{16}$$

while still $\rho(|H|) = 1$.

Theorem 3. *Assume that the matrix H is irreducible, that $\rho(H) < \rho(|H|) = 1$ and, without loss of generality, that H satisfies* (16). *In addition, let all diagonal elements h_{ii} of H be real and positive. Moreover, assume that the asynchronous iteration* (10) *satisfies* (11) *to* (13). *Then the iterates x^k converge to x^*, the unique fixed point of $x^* = Hx^* + c$.*

Proof: Clearly, since $\rho(H) < 1$ the fixed point x^* exists and is unique. We will show that the error vectors $e^k = x^k - x^*$ converge to 0. The errors e^k satisfy

$$e_i^{k+1} = \begin{cases} \sum_{j=1}^{n} h_{ij} e_j^{s_j(k)} & \text{if } j \in J_k \\ e_i^k & \text{if } j \notin J_k \end{cases}. \tag{17}$$

Let us first define quantities ϵ_k as

$$\epsilon_k = \max_{i=0}^{d-1} \|e^{k-i}\|_\infty, \quad k = 0, 1, \ldots,$$

where we set $e^{-d+1} = e^{-d+2} = \cdots = e^0$. Note that for given $k > 0$, we have

$$\|e^k\|_\infty \leq \epsilon_{k-1}. \tag{18}$$

Indeed, any component e_i^k of e^k either satisfies $e_i^k = e_i^{k-1}$ which implies $|e_i^k| = |e_i^{k-1}| \leq \epsilon_{k-1}$ or it satisfies $e_i^k = \sum_{j=1}^{n} h_{ij} e_j^{s_j(k)}$ with $k - d \leq s_j(k) \leq k - 1$. This yields

$$|e_i^k| \leq \sum_{j=1}^{n} |h_{ij}| \left| e_j^{s_j(k)} \right| \leq \sum_{j=1}^{n} |h_{ij}| \epsilon_{k-1} = \epsilon_{k-1},$$

thus proving (18). By definition of ϵ_k this gives

$$\epsilon_k \leq \epsilon_{k-1} \quad \text{for } k = 1, 2, \ldots,$$

which shows that the sequence $\{\epsilon_k\}$ is convergent,

$$\lim_{k\to\infty} \epsilon_k = c \geq 0. \tag{19}$$

Our ultimate goal is to show $c = 0$. To this purpose, we from now on assume that, on the contrary, $c > 0$ and the whole rest of this proof is concerned with obtaining a contradiction. This will be achieved by establishing the following four major conclusions:

(i) For all i we have $\lim_{k\to\infty} |e_i^k| = c$.

(ii) Denoting $\theta_i^k = \arg(e_i^k)$ and $\eta_{ij} = \arg(h_{ij})$, we have for all k

$$\lim_{k\to\infty} \cos\left(\theta_i^k - \theta_j^{s_j(r_i(k))} - \eta_{ij}\right) = 1 \quad \text{for all } i \text{ and } j \text{ for which } h_{ij} \neq 0.$$

(See (14) for the definition of $r_i(k)$.)

(iii) For all i and j the limit $\lim_{k\to\infty} e_i^k / e_j^k$ exists.

(iv) There exists a signature matrix S such that $S^{-1}HS \geq 0$.

Note that by Lemma 2 assertion (iv) is a contradiction to $\rho(H) < \rho(|H|)$, showing that our initial assumption $c > 0$ was wrong, thus proving the theorem.

To show (i), let us first introduce the notation σ for the modulus of the smallest non-zero entry of $H = (h_{ij})$:

$$\sigma = \min\{|h_{ij}| \mid h_{ij} \neq 0\}.$$

Since $\lim_{k\to\infty} \epsilon_k = c$, we have $\limsup_{k\to\infty} |e_i^k| \leq c$ for all i. We therefore must only show that $\liminf_{k\to\infty} |e_i^k| \geq c$. Assume that, on the contrary, this is not the case for some component i. This means that there exists $\delta > 0$ such that

$$|e_i^k| \leq c - \delta \quad \text{for infinitely many } k. \tag{20}$$

The idea is to show that this implies $\epsilon^k < c$ for some k which is a contradiction to (19), since the ϵ_k converge monotonically from above. Unfortunately, establishing this is rather technical. To start, let us fix a constant α satisfying

$$0 < \alpha < \frac{\sigma^{(r+d+1)n}}{1 - \sigma^{(r+d+1)n}} \cdot \delta,$$

where d and r are defined in (12) and (13). Note first that then

$$\alpha < \frac{\sigma^\nu}{1-\sigma^\nu} \cdot \delta \quad \text{for } \nu = 1, \ldots, (r+d+1)n. \tag{21}$$

By (19) there exists $k(\alpha)$ such that

$$|e_j^k| \le c + \alpha \quad \text{for all } j = 1, \ldots, n \text{ and all } k \ge k(\alpha). \tag{22}$$

Let $k_0 \ge k(\alpha) + d$ be such that $|e_i^{k_0}| \le c - \delta$. Such k_0 exists by virtue of (20). We will show the following: If m is a component belonging to the level set $L_\mu(i)$, then

$$\begin{aligned} |e_m^k| \le c - (\sigma^{(r+d)n+\mu}\delta - (1-\sigma^{(r+d)n+\mu})\alpha) < c \\ \text{for } k = k_0 + (r+d)(\mu-1), \ldots, k_0 + (r+d)n. \end{aligned} \tag{23}$$

Clearly, since $\cup_{\mu=0}^{n-1} L_\mu(i) = \{1, \ldots, n\}$, we then get $\epsilon^{k_0+(r+d)n} < c$, the sought-after contradiction to (19). To show (23), note first that the second inequality is immediate from (21).

To show the first inequality we start with $\mu = 0$, so $L_\mu(i) = L_0(i) = \{i\}$. Let $k_1, \ldots, k_l$ denote all time steps in $\{k_0 + 1, \ldots, k_0 + (r+d)n\}$ at which the i-th component is updated, i.e. $k_l = r_i(k_0 + (r+d)n)$ and $k_{\nu-1} = r_i(k_\nu)$, $\nu = 1, \ldots, l$, $r_i(k)$ having been defined in (14). Trivially, $l \le (r+d)n$. We show

$$|e_i^{k_\nu}| \le c - (\sigma^\nu\delta - (1-\sigma^\nu)\alpha), \quad \nu = 0, \ldots, l. \tag{24}$$

For $\nu = 0$ this is true by the choice of k_0. For $\nu \ge 1$ observe that $s_j(k_\nu) \ge k(\alpha)$ for all j and $e_i^{k_\nu - 1} = e_i^{k_{\nu-1}}$. Therefore, we get

$$\begin{aligned} |e_i^{k_\nu}| &= \left| h_{ii} e_i^{k_{\nu-1}} + \sum_{j=1, j\ne i}^{n} h_{ij} e_j^{s_j(k_\nu)} \right| \\ &\le |h_{ii}| |e_i^{k_{\nu-1}}| + \sum_{j=1, j\ne i}^{n} |h_{ij}| \cdot \left| e_j^{s_j(k_\nu)} \right| \\ &\le |h_{ii}| \, |e_i^{k_{\nu-1}}| + (1 - |h_{ii}|) \cdot (c + \alpha). \end{aligned}$$

Here, the last inequality is due to (16) and (22). Using the induction hypothesis for $\nu - 1$ we further obtain

$$\begin{aligned} |e_i^{k_\nu}| &\le |h_{ii}|(c - (\sigma^{\nu-1}\delta - (1-\sigma^{\nu-1})\alpha)) + (1 - |h_{ii}|) \cdot (c + \alpha) \\ &= c - |h_{ii}| \cdot (\sigma^{\nu-1}\delta - (1-\sigma^{\nu-1})\alpha) + (1 - |h_{ii}|) \cdot \alpha \\ &\le c - \sigma \cdot (\sigma^{\nu-1}\delta - (1-\sigma^{\nu-1})\alpha) + (1-\sigma) \cdot \alpha \\ &= c - (\sigma^\nu\delta - (1-\sigma^\nu)\alpha). \end{aligned}$$

This terminates the inductive proof for (24). It also establishes (23) for $\mu = 0$, since $\sigma^\nu\delta - (1-\sigma^\nu)\alpha \ge \sigma^{(r+d)n}\delta - \left(1 - \sigma^{(r+d)n}\alpha\right) > 0$ for $\nu = 0, \ldots, l$.

As a next step, consider the level 1 set $L_1(i)$ of i, i.e. all components m for which $h_{mi} \neq 0$. Whenever such component m is updated at time step $k \in \{k_0 + (r+d), \dots, k_0 + (r+d)n\}$ (it is actually updated at least $n-1$ times in this time interval), we get

$$\begin{aligned} |e_m^k| &= \left| h_{mi} e_i^{k-1} + \sum_{j=1, j\neq i}^{n} h_{mj} e_j^{s_j(k)} \right| \\ &\leq |h_{mi}| \cdot \left| e_i^{s_i(k)} \right| + (1 - |h_{ii}|) \cdot (c + \alpha). \end{aligned}$$

Here, $s_i(k) \geq k - d \geq k_0$, so that we can use (23) for $\mu = 0$ to get

$$\begin{aligned} |e_m^k| &\leq |h_{mi}| \cdot (c - (\sigma^{(r+d)n}\delta - (1 - \sigma^{(r+d)n})\alpha)) + (1 - |h_{mi}|) \cdot (c + \alpha) \\ &= c - |h_{mi}| \cdot (\sigma^{(r+d)n}\delta - (1 - \sigma^{(r+d)n})\alpha) + (1 - |h_{mi}|)\alpha \\ &\leq c - \sigma \cdot (\sigma^{(r+d)n}\delta - (1 - \sigma^{(r+d)n})\alpha) + (1 - \sigma)\alpha \\ &= c - (\sigma^{(r+d)n+1}\delta - (1 - \sigma^{(r+d)n+1})\alpha). \end{aligned}$$

This proves (23) for the level set $L_1(i)$. It should now be apparent that we can use induction on the level sets $L_\mu(i)$ to establish (23) for the remaining level sets of i. Since the proof proceeds in a completely analogous manner as for $\mu = 1$ we refrain from reproducing the details here and consider (i) as settled.

The proof for part (ii) is much shorter. Because of (i) (with $c > 0$), we can assume without loss of generality that all e_i^k are nonzero. In particular, all arguments θ_i^k are then well defined. Given (i), for any fixed i we can now apply Lemma 3 (with $b^k = e_i^k$ and $a_i^k = e_j^{s_j(r_i(k))}$) and we obtain (ii).

We now turn to prove (iii). We start with an important observation arising from (ii) in the special case $j = i$ where we have $s_i(r_i(k)) = r_i(k) - 1$. Since we assumed $h_{ii} > 0$ we have $\eta_{ii} = 0$ so that (ii) reads $\lim_{k\to\infty} \theta_i^k - \theta_i^{r_i(k)-1} = 0 \bmod 2\pi$. For every l with $r_i(k) < l \leq k$ we trivially have $\theta_i^l = \theta_i^{l-1} \bmod 2\pi$, since $e_i^l = e_i^{l-1}$. Altogether we therefore have

$$\lim_{k\to\infty} \theta_i^k - \theta_i^{k-1} = 0 \bmod 2\pi \quad \text{for all } i. \tag{25}$$

Now, to establish (iii), because of (i), we only have to show that $\lim_{k\to\infty}(\theta_i^k - \theta_j^k)$ exists mod 2π for all i and j. We will first show that in fact

$$\lim_{k\to\infty} \cos(\theta_i^k - \theta_j^k - \eta_{ij}) = 1 \quad \text{whenever } h_{ij} \neq 0. \tag{26}$$

To this purpose, let us write

$$\theta_i^k - \theta_j^k = \theta_i^k - \theta_j^{s_j(r_i(k))} + \sum_{l=s_j(r_i(k))+1}^{k} (\theta_j^{l-1} - \theta_j^l).$$

Because of (12), (13) and (15) the sum contains at most $r + d$ summands. By (25) each of the summands $(\theta_j^{l-1} - \theta_j^l)$ approaches 0 mod 2π as $k \to \infty$. Therefore, (26) follows by using (ii) for the first term $\theta_i^k - \theta_j^{s_j(r_i(k))}$ and we have proved (iii) for all i and j with $h_{ij} \neq 0$. If $h_{ij} = 0$, let $(i = i_0, i_1, \ldots, i_{q-1}, i_q = j)$ be a path connecting i to j in the directed graph of the irreducible matrix H. Then

$$\frac{e_i^k}{e_j^k} = \frac{e_i^k}{e_{i_1}^k} \cdot \frac{e_{i_1}^k}{e_{i_2}^k} \cdot \cdots \cdot \frac{e_{i_{q-1}}^k}{e_j^k}.$$

and each of the right hand factors converges, showing that $\lim e_i^k / e_j^k$ again exists.

It remains to show (iv). Note first that by (26) we have

$$\lim_{k\to\infty} \arg\left(h_{ij}\frac{e_j^k}{e_i^k}\right) = 0 \bmod 2\pi \quad \text{whenever } h_{ij} \neq 0. \tag{27}$$

Let $\mathcal{T}$ denote the set of all the matrices $T^k = (1/c) \cdot \text{diag}(e^k),\ k = 0, 1, \ldots$. By (i), $\mathcal{T}$ is bounded, so that it has at least one accumulation point S. Each such matrix S is a signature matrix, again by (i). By (27), for such S we have $S^{-1}HS \geq 0$. This ends the proof of (iv) and of the whole theorem. □

It is interesting to compare the assumptions and assertions of Theorem 2 with those of our new theorem, Theorem 3. First, note that Theorem 2 proves convergence of the asynchronous iterates to non-trivial solutions of a singular system $(I - H)x = 0$, whereas in Theorem 3 we look at systems $(I - H)x = c$ with $(I - H)$ non-singular and we prove convergence of the iterates to the unique solution of that system. While Theorem 2 deals with irreducible nonnegative matrices H, the new Theorem 3 considers matrices H which are again irreducible, but will not be nonnegative (otherwise the assumption $\rho(H) < \rho(|H|)$ would not be fulfilled). Finally, Theorem 2 requires at least one diagonal component of H to be positive, whereas Theorem 3 requires all of them to be positive.

5. Conclusions

Our new result, Theorem 3, shows that asynchronous iterations converge for a larger class of matrices, if, in addition to (12) and (13), the natural condition (11) is also met. This last condition is crucial in this context, since the construction of [6], see Theorem 1 (ii), shows that without it we can always construct delays and active components still satisfying (12) and (13) such that there is divergence. We think that the major merit of our new result is therefore to having contributed to clarify the role of the various typical conditions present in practical and mathematical models for asynchronous iterations.

We wish to end this paper by a simple example which shows that at least some assumption on positive diagonal elements of H (as present in Theorem 3) is definitely necessary in order to get convergence. Let

$$H = \begin{pmatrix} -1/2 & -1/2 \\ 1/2 & -1/2 \end{pmatrix},$$

which has negative diagonal elements and is irreducible with $\rho(H) = 1/\sqrt{2} < 1 = \rho(|H|)$. If we take $e^0 = (1, 1)$ and $J_k = \{k \bmod 2\}$, $s_i(k) = k - 1$ for all i and k, then a simple calculation shows that the errors e^k go through a cycle of length 4, where

$$\begin{aligned} e^0 &= e^4 = e^8 = \cdots = (1, 1)^T, \\ e^1 &= e^5 = e^9 = \cdots = (-1, 1)^T, \\ e^2 &= e^6 = e^{10} = \cdots = (-1, -1)^T, \\ e^3 &= e^7 = e^{11} = \cdots = (1, -1)^T. \end{aligned}$$

Note that this example does not give a complete answer on the role of our assumption of positive diagonal elements in Theorem 3. The example shows that if all diagonal elements are negative we can have divergence. Theorem 3 proves convergence if all diagonal elements are positive. We were not able to prove convergence or find a counter example in any in-between situation where only some diagonal elements are positive. We therefore leave this topic as an open question for further investigation, together with the question (suggested by one of the referees) whether it is possible to generalize Theorem 3 to reducible matrices in a similar manner as [17, Theorem 5.4] generalized Theorem 2.

Acknowledgement

The work of A. Frommer was partially funded by the French ministry of research through a visiting grant at ENSEEIHT.

References

[1] Bahi, J.: Algorithmes parallèles asynchrones pours des systèmes singuliers. CR *326*, 1421–1425 (1998).
[2] Bahi, J.: Asynchronous iterative algorithms for nonexpansive linear systems. J. Parallel Distribut. Comput. *60*, 92–112 (2000).
[3] Baudet, G. M.: Asynchronous iterative methods for multiprocessors. J. ACM *25*, 226–244 (1978).
[4] Berman, A., Plemmons, R. J.: Nonnegative matrices in the mathematical sciences. Philadelphia: SIAM, 1994.
[5] Bertsekas, D. P., Tsitsiklis, J. N.: Parallel and distributed computation. Englewood Cliffs: Prentice Hall, 1989.
[6] Chazan, D., Miranker, W. L.: Chaotic relaxation. Linear Algebra Appl. *2*, 199–222 (1969).
[7] El Baz, D., Spiteri, P., Miellou, J.-C.: Asynchronous multisplitting methods with flexible communication for pseudolinear p.d.e. In: Proceedings of the Eighth International Colloquium on Differential Equations, pp. 145–152, Plovdiv, Romania, August 18–23, 1997. Utretcht: VSP International, 1997.
[8] El Baz, D., Spiteri, P., Miellou, J.-C., Gazen, D.: Asynchronous iterative algorithms with flexible communication for nonlinear network flow problems. J. Parallel Distribut. Comput. *38*, 1–15 (1996).
[9] Frommer, A.: Lösung linearer Gleichungssysteme auf Parallelrechnern. Wiesbaden: Vieweg, 1990.

[10] Frommer, A.: On asynchronous iterations in partially ordered spaces. Numer. Funct. Anal. Optimization *12*, 315–325 (1991).
[11] Frommer, A., Schwandt, H.: Asynchronous parallel methods for enclosing solutions of nonlinear equations. J. Comp. Appl. Math. *60*, 47–62 (1995).
[12] Frommer, A., Szyld, D. B.: On asynchronous iterations. J. Comp. Appl. Math. *123*, 201–216 (2000).
[13] Lubachevsky, B., Mitra, D.: A chaotic asynchronous algorithm for computing the fixed point of a nonnegative matrix of unit spectral radius. J. ACM *33*, 130–150 (1986).
[14] Miellou, J.-C.: Itérations chaotiqes à retards; études de la convergence dans le cas d'espaces partiellement ordonnés. C. R. Acad. Sci., Paris, Sér. I, Math. *280*, 233–236 (1975).
[15] Miellou, J.-C., El Baz, D., Spiteri, P.: A new class of asynchronous iterative algorithms with order intervals. Math. Comput. *67*, 237–255 (1998).
[16] Ortega, J. M., Rheinboldt, W. C.: Iterative solution of nonlinear equations in several variables. New York: Academic Press, 1970.
[17] Pott, M.: On the convergence of asynchronous iteration methods for nonlinear paracontractions and consistent linear systems. Linear Algebra Appl. *283*, 1–33 (1998).
[18] Szyld, D. B.: The mystery of asynchronous iterations convergence when the spectral radius is one. Research Report 98–102, Department of Mathematics, Temple University, http://www.math.temple.edu/~szyld/papers.html, 1998.
[19] Üresin, A., Dubois, M.: Sufficient conditions for the convergence of asynchronous iterations. Parallel Comput. *10*, 83–92 (1989).
[20] Varga, R. S.: Matrix iterative analysis. Englewood Cliffs: Prentice Hall, 1962. 2nd edition, revised and corrected: Berlin Heidelberg New York Tokyo: Springer, 2000.

A. Frommer
Fachbereich Mathematik
Universität Wuppertal
D-42097 Wuppertal
Germany
e-mail: frommer@math.uni-wuppertal.de

P. Spiteri
ENSEEIHT: Ecole Nationale Supérieure
d'Electrotechnique, d'Electronique,
d'Informatique et d'Hydraulique de Toulouse
2, rue Camichel, B.P. 7122
F-31071 Toulouse-Cédex 1
France
e-mail: pierre.spiteri@enseeiht.fr

Computing [Suppl] 15, 105–118 (2001)

Computing

Iterative Methods for Eigenvalue Problems with Nondifferentiable Normalized Condition of a General Complex Matrix

K. Ishihara, Sakai

Dedicated to Professor Tetsuro Yamamoto on the occasion of his 65th birthday

Abstract

In this paper, we consider iterative methods with line search for eigenvalue problems of a general complex matrix. The eigenvalue problem is written as a system of complex nonlinear equations with nondifferentiable normalized condition. Convergence theorems for iterations are established. Finally, some numerical examples are presented to demonstrate the effectiveness of the iterative methods.

AMS Subject Classifications: 65F15, 65H10.

Key Words: Eigenvalue problem of complex matrix, system of complex nonlinear equations, nondifferentiable normalized condition, iterative method with line search, convergence.

1. Introduction

Let **R** be the set of real numbers and **C** be the set of complex numbers. In the papers [2, 13], Dongarra, Moler and Wilkinson, and Yamamoto discussed accuracy and error bounds of numerical solutions for eigenvalue problems of a matrix by using the Newton method. In this paper, we consider iterative methods for the eigenvalue problem of a general complex $n \times n$ matrix A

$$Az = \lambda z, \quad \|z\|^2 = 1. \tag{1}$$

Here $\lambda \in \mathbf{C}$, $z = (z_1, z_2, \ldots, z_n)^T \in \mathbf{C}^n$, $\|z\| = \sqrt{z^H z} = \sqrt{\sum_{j=1}^n |z_j|^2}$, T and H denote transpose and conjugate transpose, respectively. Following [13] which considered a real matrix with a real eigenvalue and a corresponding real eigenvector, the eigenvalue problem (1) can be written as a system of complex nonlinear equations

$$F(Z) \equiv F(z, \lambda) = \begin{pmatrix} Az - \lambda z \\ -\frac{1}{2}(\|z\|^2 - 1) \end{pmatrix} = 0. \tag{2}$$

Here $Z = (z_1, z_2, \ldots, z_n, \lambda)^T \in \mathbf{C}^{n+1}$. When z is normalized such that $z_s = 1$ with some s $(1 \leq s \leq n)$, this excludes any exact eigenvector for which $z_s = 0$ [2, 11]. Therefore, as in [13], the normalized condition $\|z\|^2 = 1$ is employed. However, we note that $\|z\|^2 = \sum_{j=1}^{n} |z_j|^2$ is not a differentiable function of complex variables $z_1, z_2, \ldots, z_n$. Recently, several iterative methods are discussed for nonsmooth equations with nondifferentiable terms [1, 4, 5, 8, 9].

We start with some notations. For $w = x + iy \in \mathbf{C}$ with $x, y \in \mathbf{R}$, $i = \sqrt{-1}$ and $u, v \in \mathbf{C}^n$, we put

$$x = \text{Re}[w], \ y = \text{Im}[w], \ (u, v) = v^H u.$$

Further, $|||\cdot|||$ is the spectral norm induced by the vector norm $\|\cdot\|$ on $\mathbf{C}^n$, then it is well known [7] that

$$|||A||| \equiv \sup_{z \neq 0} \frac{\|Az\|}{\|z\|} = \sqrt{\rho(A^H A)},$$

where $\rho(A^H A)$ is the spectral radius of $A^H A$. Let

$$g(Z) = \frac{1}{2}\|F(Z)\|^2, \quad g'(Z; d) = \lim_{t \to +0} \frac{g(Z + td) - g(Z)}{t},$$

$$J(Z) \equiv J(z, \lambda) = \begin{pmatrix} A - \lambda I_n & -z \\ -z^H & 0 \end{pmatrix}.$$

Here $t > 0$, $d \in \mathbf{C}^{n+1}$ and I_n denotes the $n \times n$ identity matrix. We note that $g'(Z; d)$ is the directional derivative of g at $Z \in \mathbf{C}^{n+1}$ in the direction $d \in \mathbf{C}^{n+1}$. When A, λ, z are all real-valued, $J(Z)$ represents the Jacobian matrix of $F(Z)$. However, when A, λ, z are complex-valued, $J(Z)$ does not represent the Jacobian matrix of $F(Z)$ because of the nondifferentiable term $\|z\|^2$.

For numerically solving (2) we employ iterative methods with line search. We put $Z_k = (z_{1,k}, \ldots, z_{n,k}, \lambda_k)^T$, $k = 0, 1, 2, \ldots$ with initial vector Z_0. First, we note that in general the level set $\{Z = (z, \lambda) \in \mathbf{C}^{n+1}; g(Z) \leq g(Z_0)\}$ is not bounded, since $g(0, \lambda) = \frac{1}{8}$ for an arbitrary $\lambda \in \mathbf{C}$. In order to require the boundedness of the sequence $\{Z_k\}$, define open and closed balls with center at Z_0 and radius $r > 0$ as follows:

$$U_r = \{Z \in \mathbf{C}^{n+1}; \|Z - Z_0\| < r\}, \quad \bar{U}_r = \{Z \in \mathbf{C}^{n+1}; \|Z - Z_0\| \leq r\}.$$

We impose $Z_k \in U_r$, $k = 0, 1, 2, \ldots$ with U_r. Also, we may use $U_r = \{Z \in \mathbf{C}^{n+1}; \|Z\| < r\}$. Then we have the following iterative method GDNM (*generalized damped Newton method*) by choosing initial vector Z_0 and constants $0 < \beta < 1$, $0 < \sigma < 1$.

Step 1: By assuming that $F(Z_k) \neq 0$ and that $J(Z_k)$ is nonsingular, solve the system of linear equations $F(Z_k) + J(Z_k)d_k = 0$ to get d_k.

Step 2: Let m_k be the smallest nonnegative integer m such that $g(Z_k + \beta^m d_k) - g(Z_k) \leq \sigma\beta^m g'(Z_k; d_k)$, $Z_k + \beta^m d_k \in U_r$, $m = 0, 1, 2, \ldots$. Set $Z_{k+1} = Z_k + \alpha_k d_k$ with $\alpha_k = \beta^{m_k}$.

Step 3: Test Z_{k+1} for convergence.

When A, λ, z are all real-valued, GDNM with $\alpha_k = 1$ $(m_k = 0)$, $k = 0, 1, 2, \ldots$ reduces to the usual Newton method which locally converges quadratically [7].

Furthermore, in view of the case where $J(Z_k)$ is singular, we also have the following iterative method GDGNM (*generalized damped Gauss–Newton method*) by choosing initial vector Z_0 and constants $0 < \beta < 1$, $0 < \sigma < 1$, $\mu > 0$.

Step 1: By assuming that $J(Z_k)^H F(Z_k) \neq 0$, solve the system of linear equations $J(Z_k)^H F(Z_k) + (J(Z_k)^H J(Z_k) + \mu I_{n+1})d_k = 0$ to get d_k.

Step 2: Let m_k be the smallest nonnegative integer m such that $g(Z_k + \beta^m d_k) - g(Z_k) \leq \sigma \beta^m g'(Z_k; d_k)$, $Z_k + \beta^m d_k \in U_r$, $m = 0, 1, 2, \ldots$. Set $Z_{k+1} = Z_k + \alpha_k d_k$ with $\alpha_k = \beta^{m_k}$.

Step 3: Test Z_{k+1} for convergence.

It is clear that in the case where $J(Z_k)$, $k = 0, 1, 2, \ldots$ are nonsingular, GDGNM with $\mu = 0$ reduces to GDNM.

Remark 1. If $Z_* = (z_*, \lambda_*)$ is a solution of $F(Z) = 0$, then we have $\|Z_*\|^2 = 1 + |\lambda_*|^2$. Thus, in GDNM and GDGNM, we impose $Z_k \in U_r$, $k = 0, 1, 2, \ldots$ by choosing U_r. This means that $\{Z_k\}$ is bounded and there is an accumulation point $\tilde{Z}$ of $\{Z_k\}$. We expect that $\tilde{Z}$ is a solution of $F(Z) = 0$. See Theorems 1 and 2 in the next section.

2. Convergence Results

In this section, we consider the convergence of iterative methods GDNM and GDGNM. We prepare some lemmas which will be useful. By the same arguments as used in [13], we have the following Lemma 1, which is an easy extension of the real case to the complex case, so that we state without proof.

Lemma 1 [13]. *Let λ_* be an eigenvalue and z_* be a corresponding eigenvector of A. Then $J(z_*, \lambda_*)$ is nonsingular if and only if λ_* is a simple eigenvalue.*

Lemma 1 implies that when λ_* is a simple eigenvalue the iterative method GDNM is effective, because $J(Z)$ is nonsingular in some neighborhood of $Z_* = (z_*, \lambda_*)$.

Lemma 2. *For Z, $d \in \mathbf{C}^{n+1}$, it holds that*

$$g'(Z; d) = \mathrm{Re}[F(Z)^H J(Z) d].$$

Moreover, in GDNM *we have*

$$g'(Z_k; d_k) = -\|F(Z_k)\|^2 \leq 0.$$

In GDGNM, *we have*

$$g'(Z_k; d_k) = -\left(J(Z_k)^H F(Z_k)\right)^H \left(J(Z_k)^H J(Z_k) + \mu I_{n+1}\right)^{-1} J(Z_k)^H F(Z_k) \leq 0.$$

Proof: Let $t > 0$, $Z = (z, \lambda) \in \mathbf{C}^{n+1}$, $d = (b, \delta) \in \mathbf{C}^{n+1}$, λ, $\delta \in \mathbf{C}$. Direct computations yield

$$\begin{aligned} g(Z + td) - g(Z) = {} & t\mathrm{Re}[((A - \lambda I_n)b - \delta z - t\delta b, Az - \lambda z)] \\ & + \frac{t^2}{2}\|(A - \lambda I_n)b - \delta z - t\delta b\|^2 \\ & + \frac{1}{8}(\|z + tb\|^2 + \|z\|^2 - 2)(2t\mathrm{Re}[(b, z)] + t^2\|b\|^2). \end{aligned}$$

Therefore, it follows that

$$\begin{aligned} g'(Z; d) &= \lim_{t \to +0} \frac{g(Z + td) - g(Z)}{t} \\ &= \mathrm{Re}[((A - \lambda I_n)b - \delta z, Az - \lambda z)] + \frac{\|z\|^2 - 1}{2}\mathrm{Re}[(b, z)] \\ &= \mathrm{Re}\left[\begin{pmatrix} Az - \lambda z \\ -\frac{1}{2}(\|z\|^2 - 1) \end{pmatrix}^H \begin{pmatrix} A - \lambda I_n & -z \\ -z^H & 0 \end{pmatrix} \begin{pmatrix} b \\ \delta \end{pmatrix}\right] \\ &= \mathrm{Re}[F(Z)^H J(Z)d]. \end{aligned}$$

Hence, in GDNM we have

$$\begin{aligned} g'(Z_k; d_k) &= \mathrm{Re}[F(Z_k)^H J(Z_k)d_k] = \mathrm{Re}[F(Z_k)^H J(Z_k)(-J(Z_k)^{-1}F(Z_k))] \\ &= \mathrm{Re}[-F(Z_k)^H F(Z_k))] = -\|F(Z_k)\|^2 \leq 0. \end{aligned}$$

In GDGNM we have

$$\begin{aligned} g'(Z_k; d_k) &= \mathrm{Re}[F(Z_k)^H J(Z_k)d_k] \\ &= \mathrm{Re}\left[F(Z_k)^H J(Z_k)\left(-(J(Z_k)^H J(Z_k) + \mu I_{n+1})^{-1} J(Z_k)^H F(Z_k)\right)\right] \\ &= -(J(Z_k)^H F(Z_k))^H (J(Z_k)^H J(Z_k) + \mu I_{n+1})^{-1} J(Z_k)^H F(Z_k) \leq 0, \end{aligned}$$

since the Hermitian matrix $J(Z_k)^H J(Z_k) + \mu I_{n+1}$ is positive definite. This completes the proof.

Lemma 2 implies that the direction d_k of GDNM (or GDGNM) provides a decent direction for the function g.

Lemma 3 [4]. *Suppose that* $F(Z_k) \neq 0$ *in* GDNM (*or* $J(Z_k)^H F(Z_k) \neq 0$ *in* GDGNM). *Then there exists a scalar* $\tau_* > 0$ *such that for all* $\tau \in [0, \tau_*]$

$$g(Z_k + \tau d_k) - g(Z_k) \le \sigma\tau g'(Z_k; d_k).$$

Proof: From Lemma 2, we note that $g'(Z_k; d_k) < 0$. Suppose that no $\tau_* > 0$ as stated exists. Then there exists a sequence $\{\tau_p\}$ of positive scalars converging to zero such that for each p

$$g(Z_k + \tau_p d_k) - g(Z_k) > \sigma\tau_p g'(Z_k; d_k).$$

Dividing both sides by τ_p, passing to the limit $p \to \infty$, we have

$$g'(Z_k; d_k) \ge \sigma g'(Z_k; d_k).$$

Since $0 < \sigma < 1$, we get $g'(Z_k; d_k) \ge 0$. This is a contradiction. Thus, the proof is complete.

Lemma 3 implies that the integer m_k in *step 2* of GDNM (or GDGNM) can be determined after a finite number of trials starting with $m = 0, 1, 2, \ldots$. Thus, in actual computations, iterations GDNM and GDGNM are available.

We are now in a position to obtain the following results:

Theorem 1. *Let $Z_* = (z_*, \lambda_*)$ be a solution of $F(Z) = 0$, where λ_* is a simple eigenvalue of A, and V_* be the eigenspace corresponding to λ_* with $V_* = \text{span}[z_*] \subset \mathbf{C}^n$. Let U_r be an open ball with center at Z_0 and radius r. Suppose that $Z_* \in U_r$ and r is sufficiently small. Let $\{Z_k\}$ be a sequence given by* GDNM. *If $\tilde{Z}$ is any accumulation point of $\{Z_k\}$, then we have*

$$F(\tilde{Z}) = 0, \ \tilde{Z} = (cz_*, \lambda_*), \quad |c| = 1, \ c \in \mathbf{C}.$$

In addition, if A, Z_*, Z_k, $k = 0, 1, 2, \ldots$ are all real-valued with $V_* = \text{span}[z_*] \subset \mathbf{R}^n$, then the entire sequence $\{Z_k\}$ converges and

$$\lim_{k\to\infty} Z_k = Z_*.$$

Proof: Since λ_* is a simple eigenvalue, $J(Z_*)$ is nonsingular by Lemma 1. Hence there exists an open ball U_r with center at Z_0 so that $J(Z)$ is nonsingular for all $Z \in \bar{U}_r$,

$$|||J(Z)^{-1}||| \le M, \quad Z \in \bar{U}_r,$$

and $\bar{U}_r$ contains no solution $(z_\sharp, \lambda_\sharp)$ of $F(Z) = 0$ with $\lambda_\sharp \ne \lambda_*$, where M is a positive constant. We note that $\{Z_k\}$ is bounded since $Z_k \in U_r$, $k = 0, 1, 2, \ldots$. From Lemma 2, it follows that

$$g(Z_{k+1}) - g(Z_k) \le \sigma\alpha_k g'(Z_k; d_k) = -\sigma\alpha_k \|F(Z_k)\|^2 \le 0.$$

Therefore, the sequence $\{g(Z_k)\}$ is nonnegative and decreasing. Therefore, it converges and

$$\lim_{k\to\infty}(g(Z_{k+1}) - g(Z_k)) = 0,$$

$$\lim_{k\to\infty}\alpha_k\|F(Z_k)\| = 0.$$

If $\liminf \alpha_k > 0$, then we have $F(\tilde{Z}) = 0$. Assume that $\liminf \alpha_k = 0$ and that $\tilde{Z}$ is the limit of a subsequence $\{Z_k; k \in K\}$. Further we may assume that $\lim\{\alpha_k; k \in K\} = 0$. On the other hand, in GDNM we get

$$\|d_k\| = \| - J(Z_k)^{-1}F(Z_k)\| \leq |||J(Z_k)^{-1}||| \cdot \|F(Z_k)\| \leq M\|F(Z_k)\|.$$

Since $\{F(Z_k); k \in K\}$ is convergent, it follows that $\{d_k; k \in K\}$ is bounded. Thus $\{d_k; k \in K\}$ has an accumulation point $\tilde{d}$. Further we may assume that $\tilde{d} = \lim\{d_k; k \in K\}$. Consequently, for some $p \geq 0$ and all $k \in K$ with $k \geq p$, it follows that

$$Z_k + \alpha_k d_k \in U_r, \quad Z_k + \frac{\alpha_k}{\beta} d_k \in U_r,$$

since $\lim\{\alpha_k; k \in K\} = 0$ and $\lim\{d_k; k \in K\} = \tilde{d}$. Thus, by the definition of α_k, we obtain

$$g\left(Z_k + \frac{\alpha_k}{\beta} d_k\right) - g(Z_k) > \sigma\frac{\alpha_k}{\beta} g'(Z_k; d_k),$$

from which follows that

$$\frac{g(Z_k + (\alpha_k/\beta)d_k) - g(Z_k)}{\alpha_k/\beta} > \sigma g'(Z_k; d_k).$$

Passing to the limit as $k \to \infty$, we obtain

$$g'(\tilde{Z}; \tilde{d}) \geq \sigma g'(\tilde{Z}; \tilde{d}).$$

Hence, from Lemma 2, it follows that

$$-\|F(\tilde{Z})\|^2 \geq -\sigma\|F(\tilde{Z})\|^2.$$

Since $0 < \sigma < 1$, we have $F(\tilde{Z}) = 0$. Further, since $V_* = \text{span}[z_*] \subset \mathbf{C}^n$, we have $\tilde{Z} = (cz_*, \lambda_*) \in \bar{U}_r$, $|c| = 1$, $c \in \mathbf{C}$. In addition, if A, Z_*, Z_k, $k = 0, 1, 2, \ldots$ are all real-valued with $V_* = \text{span}[z_*] \subset \mathbf{R}^n$, $U_r \subset \mathbf{R}^{n+1}$, then Z_* is a unique solution of $F(Z) = 0$ in $\bar{U}_r$. Hence, $\tilde{Z} = Z_*$ is a unique accumulation point of $\{Z_k\}$ in $\bar{U}_r$. This means that the entire sequence $\{Z_k\}$ converges and $\lim_{k\to\infty} Z_k = Z_*$. Thus, the proof is complete.

Theorem 2. *Let $Z_* = (z_*, \lambda_*)$ be a solution of $F(Z) = 0$, where λ_* is an eigenvalue of multiplicity at least one, and V_* be the eigenspace of dimension at least one corresponding to λ_* with $V_* \subset \mathbf{C}^n$. Let U_r be an open ball with center at Z_0 and radius r, $Z_* \in U_r$ and $\{Z_k\}$ be a sequence given by* GDGNM. *If $\tilde{Z}$ is any accumulation point of $\{Z_k\}$, then, we have*

$$J(\tilde{Z})^H F(\tilde{Z}) = 0.$$

Moreover, for sufficiently small r, we have

$$F(\tilde{Z}) = 0, \quad \tilde{Z} = (\tilde{z}, \lambda_*), \quad \|\tilde{z}\| = 1, \quad \tilde{z} \in V_*.$$

Further, if V_* has dimension one, then we have $\tilde{z} = cz_*$, $|c| = 1$, $c \in \mathbf{C}$. In addition, if A, Z_*, Z_k, $k = 0, 1, 2, \ldots$ are all real-valued with $V_* =$ span$[z_*] \subset \mathbf{R}^n$, then the entire sequence $\{Z_k\}$ converges and

$$\lim_{k \to \infty} Z_k = Z_*.$$

Proof: We note that for $Z \in \mathbf{C}^{n+1}$ the Hermitian matrix $J(Z)^H J(Z) + \mu I_{n+1}$ is positive definte and

$$|||(J(Z)^H J(Z) + \mu I_{n+1})^{-1}||| \leq \frac{1}{\mu}.$$

By applying essentially the same proof of Theorem 1, we have

$$\lim_{k \to \infty} \alpha_k \|J(Z_k)^H F(Z_k)\| = 0,$$

$$J(\tilde{Z})^H F(\tilde{Z}) = 0.$$

Moreover, for sufficiently small r, there exists an open ball U_r with center at Z_0 so that U_r contains neither a solution $(z_\sharp, \lambda_\sharp)$ of $F(Z) = 0$ with $\lambda_\sharp \neq \lambda_*$ nor a solution of $J(Z)^H F(Z) = 0$ with $F(Z) \neq 0$. On the other hand, since

$$\begin{aligned} &\{Z \in \mathbf{C}^{n+1}; J(Z)^H F(Z) = 0\} \\ &\quad = \{Z \in \mathbf{C}^{n+1}; J(Z)^H F(Z) = 0,\ F(Z) \neq 0\} \cup \{Z \in \mathbf{C}^{n+1}; F(Z) = 0\}, \end{aligned}$$

we have that $\tilde{Z} \in \bar{U}_r$ is a solution of $F(Z) = 0$ with $\tilde{Z} = (\tilde{z}, \lambda_*)$, $\tilde{z} \in V_*$, $\|\tilde{z}\| = 1$. Moreover, if V_* has dimension one with $V_* =$ span$[z_*] \subset \mathbf{C}^n$, then we obtain the desired result $\tilde{z} = cz_*$, $|c| = 1$, $c \in \mathbf{C}$. In addition, if A, Z_*, Z_k, $k = 0, 1, 2, \ldots$ are all real-valued with $V_* =$ span$[z_*] \subset \mathbf{R}^n$, $U_r \subset \mathbf{R}^{n+1}$, then Z_* is a unique solution of $F(Z) = 0$ in $\bar{U}_r$. Hence, $\tilde{Z} = Z_*$ is a unique accumulation point of $\{Z_k\}$ in $\bar{U}_r$. This means that the entire sequence $\{Z_k\}$ converges and $\lim_{k \to \infty} Z_k = Z_*$. This completes the proof.

Remark 2. Because of nondifferentiable term $\|z\|^2$, we can not derive the order of convergence in Theorems 1 and 2. If A, Z_*, Z_k are all real-valued with $\alpha_k = 1$ $(m_k = 0)$, $k = 0, 1, 2, \ldots$, then GDNM is the Newton method with quadratic convergence. However, it seems that numerical results in the next section show linear or quadratic convergence. See Tables 5–10 below.

3. Numerical Results

In this section, some numerical results will be shown to indicate the effectiveness of the iterations GDNM and GDGNM. We deal with the following examples.

Example 1 [3, p.116]:

$$A = \begin{pmatrix} 5+9i & 5+5i & -6-6i & -7-7i \\ 3+3i & 6+10i & -5-5i & -6-6i \\ 2+2i & 3+3i & -1+3i & -5-5i \\ 1+i & 2+2i & -3-3i & 4i \end{pmatrix}$$

Eigenpair $\{\lambda_*, z_*\}$;

$$\left\{\lambda_{*,1} = 1+5i,\ \ z_{*,1} = \frac{1}{\sqrt{7}}\begin{pmatrix} 2 \\ 1 \\ 1 \\ 1 \end{pmatrix}\right\}, \quad \left\{\lambda_{*,2} = 2+6i,\ \ z_{*,2} = \frac{1}{\sqrt{7}}\begin{pmatrix} 1 \\ 2 \\ 1 \\ 1 \end{pmatrix}\right\},$$

$$\left\{\lambda_{*,3} = 3+7i,\ \ z_{*,3} = \frac{1}{\sqrt{3}}\begin{pmatrix} 1 \\ 1 \\ 0 \\ 1 \end{pmatrix}\right\}, \quad \left\{\lambda_{*,4} = 4+8i,\ \ z_{*,4} = \frac{1}{\sqrt{3}}\begin{pmatrix} 1 \\ 1 \\ 1 \\ 0 \end{pmatrix}\right\}.$$

Example 2 [3, p.117]:

$$A = \begin{pmatrix} 7 & 3 & 1+2i & -1+2i \\ 3 & 7 & 1-2i & -1-2i \\ 1-2i & 1+2i & 7 & -3 \\ -1-2i & -1+2i & -3 & 7 \end{pmatrix}$$

Eigenpair $\{\lambda_*, z_*\}$;

$$\left\{\lambda_{*,1} = 0,\ \ z_{*,1} = \frac{1}{2}\begin{pmatrix} -1 \\ 1 \\ -i \\ -i \end{pmatrix}\right\},$$

$$\left\{\lambda_{*,2} = \lambda_{*,3} = 8, \quad z_{*,2} = \frac{1}{2}\begin{pmatrix} -1+i \\ 0 \\ 1 \\ i \end{pmatrix}, \quad z_{*,3} = \frac{1}{2}\begin{pmatrix} i \\ 1 \\ 0 \\ 1+i \end{pmatrix}\right\},$$

$$\left\{\lambda_{*,4} = 12, \quad z_{*,4} = \frac{1}{2}\begin{pmatrix} 1 \\ 1 \\ 1 \\ -1 \end{pmatrix}\right\}.$$

We note that A of Example 2 is a Hermitian matrix, so that all the eigenvalues are real. The eigenspace corresponding to $\lambda_{*,2} = \lambda_{*,3} = 8$ of multiplicity two has dimension two, and $z_{*,2}$ and $z_{*,3}$ are two linearly independent eigenvectors in $\mathbf{C}^4$.

Example 3 [13]:

$$A = \begin{pmatrix} 14 & 9 & 6 & 4 & 2 \\ -9 & -4 & -3 & -2 & -1 \\ -2 & -2 & 0 & -1 & -1 \\ 3 & 3 & 3 & 5 & 3 \\ -9 & -9 & -9 & -9 & -4 \end{pmatrix}$$

Eigenpair $\{\lambda_*, z_*\}$;

$$\left\{\lambda_{*,1} = 5, \quad z_{*,1} = \frac{1}{\sqrt{2}}\begin{pmatrix} 1 \\ -1 \\ 0 \\ 0 \\ 0 \end{pmatrix}\right\}, \quad \left\{\lambda_{*,2} = \lambda_{*,3} = 2, \quad z_{*,2} = \frac{1}{\sqrt{6}}\begin{pmatrix} -1 \\ 2 \\ -1 \\ 0 \\ 0 \end{pmatrix}\right\},$$

$$\left\{\lambda_{*,4} = 1 + \sqrt{2}\, i, \quad z_{*,4} = \frac{1}{4}\begin{pmatrix} 0 \\ 0 \\ -1 \\ 2 - \sqrt{2}\, i \\ -1 + 2\sqrt{2}\, i \end{pmatrix}\right\},$$

$$\left\{\lambda_{*,5} = 1 - \sqrt{2}\, i, \quad z_{*,5} = \frac{1}{4}\begin{pmatrix} 0 \\ 0 \\ -1 \\ 2 + \sqrt{2}\, i \\ -1 - 2\sqrt{2}\, i \end{pmatrix}\right\}.$$

We note that the eigenspace V_* corresponding to $\lambda_{*,2} = \lambda_{*,3} = 2$ of multiplicity two has dimension one with $V_* = \mathrm{span}[z_{*,2}] \subset \mathbf{R}^5$.

Example 4 (10 × 10 complex matrix [3, p.121], [12]):

$$A=\begin{pmatrix} 2+3i & 3+i & 0 & 0 & 0 & 0 & 0 & 0 & 0 & 0 \\ 3+2i & -2-i & 1+2i & 0 & 0 & 0 & 0 & 0 & 0 & 0 \\ 5-3i & 1+2i & 2+i & -1+4i & 0 & 0 & 0 & 0 & 0 & 0 \\ 2+6i & -2+3i & 3-i & -4+2i & 5+5i & 0 & 0 & 0 & 0 & 0 \\ 1+4i & 2+2i & -3+7i & 1+5i & 2-3i & 1+6i & 0 & 0 & 0 & 0 \\ 5-i & 4i & 1+5i & -8-i & 4+7i & 7+i & 4-2i & 0 & 0 & 0 \\ 5+2i & 1+4i & 6-5i & 8+4i & 4-4i & -1+5i & 3 & -4+6i & 0 & 0 \\ -4-3i & 7+3i & 1+6i & 2-4i & 3+i & 1+2i & 1+4i & 6+3i & 7-i & 0 \\ 5 & 2+2i & 1+3i & 1+i & -4-2i & 1+6i & 1+2i & 2+5i & i & 3+2i \\ 5+2i & 2+6i & 1-3i & 7+4i & 4+i & -7 & 3-3i & 5-4i & 6+3i & 2+5i \end{pmatrix}$$

Eigenvalues;

$$\begin{aligned} &\lambda_{*,1} \approx 4.16174868 + 3.13751356i, && \lambda_{*,2} \approx 5.43644837 - 3.97142582i, \\ &\lambda_{*,3} \approx 2.38988759 + 7.26807071i, && \lambda_{*,4} \approx -1.93520144 - 3.97509382i, \\ &\lambda_{*,5} \approx -2.44755082 + 0.43726175i, && \lambda_{*,6} \approx -5.27950616 - 2.27596303i, \\ &\lambda_{*,7} \approx 1.03205812 + 9.29413278i, && \lambda_{*,8} \approx -4.96687009 - 8.08712475i, \\ &\lambda_{*,9} \approx 8.81130928 + 1.54938266i, && \lambda_{*,10} \approx 10.7976764 + 8.62338151i. \end{aligned}$$

As an initial vector, we use $Z_0 = (1+i, 1+i, \ldots, 1+i, 4+3i)^T \in \mathbf{C}^{11}$.

Example 5 ($n \times n$ real symmetric matrix [3, p.74]):

$$A=\begin{pmatrix} n & n-1 & n-2 & \cdots & 2 & 1 \\ n-1 & n-1 & n-2 & \cdots & 2 & 1 \\ n-2 & n-2 & n-2 & \cdots & 2 & 1 \\ \vdots & \vdots & \vdots & \ddots & \vdots & \vdots \\ 2 & 2 & 2 & \cdots & 2 & 1 \\ 1 & 1 & 1 & \cdots & 1 & 1 \end{pmatrix}$$

Eigenvalues;

$$\lambda_{*,j} = \frac{1}{2}\left[1 - \cos\frac{(2j-1)\pi}{2n+1}\right]^{-1}, \quad j = 1, 2, \ldots, n.$$

For a larger matrix, $n = 200$ is employed, so that the smallest eigenvalue is

$$\lambda_{*,200} = 0.25001534506667\cdots.$$

As an initial vector, we use $Z_0 = (1, -1, 1, -1, \ldots, 1, -1, \ldots 1, -1, 0.2)^T \in \mathbf{R}^{201}$.

In numerical experiments with GDNM and GDGNM, we employ $r = 300$ and the numerical convergence criterion is

$$\max\left\{\max_{1\le j\le n} |z_{j,k+1} - z_{j,k}|,\ |\lambda_{k+1} - \lambda_k|\right\} < 0.5 \times 10^{-8}.$$

Tables 1–3 give the comparative numbers of iterations necessary to achieve the numerical convergence criterion. For Example 3, Table 4 shows the numbers of iterations of GDGNM for various values of μ. We also present numerical solu-

Table 1. Number of iterations for Example 1 ($\beta = 0.8$, $\sigma = 0.4$, $\mu = 10^{-7}$)

$Z_0 = (z_0, \lambda_0)$	GDNM	GDGNM	$\lim_{k\to\infty} \lambda_k$
$(1+i, 1+i, 1+i, 1+i, 0)^T$	8	8	$\lambda_{*,1} = 1+5i$
$(1+i, 1+i, 1+i, 1+i, 2.5+2.5i)^T$	7	7	$\lambda_{*,2} = 2+6i$
$(1+i, 1+i, 1+i, 1+i, 3.5+6.5i)^T$	8	8	$\lambda_{*,3} = 3+7i$
$(1+i, 1+i, 1+i, 1+i, 4.5+7.5i)^T$	7	7	$\lambda_{*,4} = 4+8i$

Table 2. Number of iterations for Example 2 ($\beta = 0.8$, $\sigma = 0.4$, $\mu = 10^{-7}$)

$Z_0 = (z_0, \lambda_0)$	GDNM	GDGNM	$\lim_{k\to\infty} \lambda_k$
$(1+i, 1+i, 1+i, 1+i, 1)^T$	8	8	$\lambda_{*,1} = 0$
$(1+i, 1+i, 1+i, 1+i, 5)^T$	8	7	$\lambda_{*,2} = \lambda_{*,3} = 8$
$(1+i, 1+i, 1+i, 1+i, 15)^T$	7	7	$\lambda_{*,4} = 12$

Table 3. Number of iterations for Example 3 ($\beta = 0.8$, $\sigma = 0.4$, $\mu = 10^{-15}$)

$Z_0 = (z_0, \lambda_0)$	GDNM	GDGNM	$\lim_{k\to\infty} \lambda_k$
$(1, 1, 1, 1, 1, 6)^T$	8	8	$\lambda_{*,1} = 5$
$(1, 1, 1, 1, 1, 1)^T$	27	29	$\lambda_{*,2} = \lambda_{*,3} = 2$
$(1+i, 1+i, 1+i, 1+i, 1+i, 2+2i)^T$	9	9	$\lambda_{*,4} = 1+\sqrt{2}\,i$
$(1+i, 1+i, 1+i, 1+i, 1+i, 2-2i)^T$	9	9	$\lambda_{*,5} = 1-\sqrt{2}\,i$

Table 4. Number of iterations of GDGNM for Example 3
($Z_0 = (1+i, 1+i, 1+i, 1+i, 1+i, 2-2i)^T$, $\beta = 0.8$, $\sigma = 0.4$)

μ	10^{-1}	10^{-2}	10^{-3}	10^{-5}	10^{-7}	10^{-15}
Number of iterations	164	28	13	10	9	9

Table 5. Numerical solutions of Example 3 for GDNM
($Z_0 = (1, 1, 1, 1, 1, 6)^T$, $\beta = 0.8$, $\sigma = 0.4$)

k	m_k	λ_k	$\|F(Z_k)\|$
0	19	6.000000	$6.205643 \times 10^{+01}$
1	0	5.833238	$6.160122 \times 10^{+01}$
2	0	5.722243	$2.461971 \times 10^{+00}$
3	0	5.385764	6.158646×10^{-01}
4	0	5.113088	1.179964×10^{-01}
5	0	5.007389	6.746737×10^{-03}
6	0	5.000017	1.396670×10^{-05}
7	0	5.000000	2.985258×10^{-11}
8	*Convergent*	5.000000	1.079355×10^{-15}
	Exact $\lambda_{*,1}$	5.000000	

tions for Examples 3–5 in Tables 5–10. Thus, we can see that the iterative methods GDNM and GDGNM are effective and Theorems 1 and 2 are valid.

Table 6. Numerical solutions of Example 3 for GDNM
($Z_0 = (1,1,1,1,1,1)^T$, $\beta = 0.8$, $\sigma = 0.4$)

k	m_k	λ_k	$\|F(Z_k)\|$
0	3	1.000000	$5.954830 \times 10^{+01}$
1	0	1.170667	$4.047107 \times 10^{+01}$
2	0	1.284823	$8.054332 \times 10^{+00}$
3	0	1.555609	$2.817876 \times 10^{+00}$
4	0	1.696398	6.297022×10^{-01}
5	0	1.825814	1.011827×10^{-01}
6	0	1.919700	9.229459×10^{-03}
7	0	1.961583	1.156395×10^{-03}
8	0	1.980819	2.924320×10^{-04}
9	0	1.990409	7.316745×10^{-05}
10	0	1.995205	1.829157×10^{-05}
11	0	1.997602	4.572891×10^{-06}
12	0	1.998801	1.143223×10^{-06}
13	0	1.999401	2.858057×10^{-07}
14	0	1.999700	7.145142×10^{-08}
15	0	1.999850	1.786285×10^{-08}
16	0	1.999925	4.465716×10^{-09}
17	0	1.999963	1.116427×10^{-09}
18	0	1.999981	2.791062×10^{-10}
19	0	1.999991	6.977790×10^{-11}
20	0	1.999995	1.744560×10^{-11}
21	0	1.999998	4.363340×10^{-12}
22	0	1.999999	1.088382×10^{-12}
23	0	1.999999	2.750840×10^{-13}
24	0	2.000000	6.710154×10^{-14}
25	1	2.000000	1.612963×10^{-14}
26	0	2.000000	7.871975×10^{-15}
27	*Convergent*	2.000000	6.629921×10^{-16}
	Exact $\lambda_{*,2} = \lambda_{*,3}$	2.000000	

Table 7. Numerical solutions of Example 3 for GDNM
($Z_0 = (1+i, 1+i, 1+i, 1+i, 1+i, 2+2i)^T$, $\beta = 0.8$, $\sigma = 0.4$)

k	m_k	λ_k	$\|F(Z_k)\|$
0	2	$2.000000 + 2.000000i$	$8.500735 \times 10^{+01}$
1	0	$1.653234 + 2.274796i$	$4.993285 \times 10^{+01}$
2	0	$1.333469 + 1.998749i$	$1.351637 \times 10^{+01}$
3	0	$1.200091 + 1.736889i$	$3.371306 \times 10^{+00}$
4	0	$1.098347 + 1.556285i$	7.635614×10^{-01}
5	0	$1.030216 + 1.455280i$	1.210298×10^{-01}
6	0	$1.002658 + 1.417781i$	6.547362×10^{-03}
7	0	$1.000012 + 1.414230i$	2.362606×10^{-05}
8	0	$1.000000 + 1.414214i$	2.383196×10^{-10}
9	*Convergent*	$1.000000 + 1.414214i$	1.088752×10^{-15}
	Exact $\lambda_{*,4}$	$1.000000 + 1.414214i$	

Table 8. Numerical solutions of Example 3 for GDGNM
($Z_0 = (1+i, 1+i, 1+i, 1+i, 1+i, 2-2i)^T$, $\beta = 0.8$, $\sigma = 0.4$, $\mu = 10^{-15}$)

k	m_k	λ_k	$\|F(Z_k)\|$
0	2	$2.000000 - 2.000000i$	$8.500735 \times 10^{+01}$
1	0	$1.653234 - 2.274796i$	$4.993285 \times 10^{+01}$
2	0	$1.333469 - 1.998749i$	$1.351637 \times 10^{+01}$
3	0	$1.200091 - 1.736889i$	$3.371306 \times 10^{+00}$
4	0	$1.098347 - 1.556285i$	7.635614×10^{-01}
5	0	$1.030216 - 1.455280i$	1.210298×10^{-01}
6	0	$1.002658 - 1.417781i$	6.547362×10^{-03}
7	0	$1.000012 - 1.414230i$	2.362606×10^{-05}
8	0	$1.000000 - 1.414214i$	2.383190×10^{-10}
9	*Convergent*	$1.000000 - 1.414214i$	8.749051×10^{-16}
	Exact $\lambda_{*,5}$	$1.000000 - 1.414214i$	

Table 9. Numerical solutions of Example 4 for GDNM
($Z_0 = (1+i, 1+i, \ldots, 1+i, 4+3i)^T \in \mathbf{C}^{11}$, $\beta = 0.8$, $\sigma = 0.4$)

k	m_k	λ_k	$\|F(Z_k)\|$
0	0	$4.00000000 + 3.00000000i$	$7.172343 \times 10^{+01}$
1	0	$3.99645999 + 2.80256545i$	$2.720544 \times 10^{+01}$
2	0	$4.08027840 + 2.97451641i$	$6.717562 \times 10^{+00}$
3	0	$4.12388916 + 3.06189298i$	$1.563498 \times 10^{+00}$
4	0	$4.14743343 + 3.10891044i$	2.966783×10^{-01}
5	0	$4.15909886 + 3.13221883i$	2.787532×10^{-02}
6	0	$4.16168033 + 3.13737695i$	3.827680×10^{-04}
7	0	$4.16174867 + 3.13751351i$	8.158858×10^{-08}
8	0	$4.16174869 + 3.13751356i$	5.876846×10^{-15}
9	*Convergent*	$4.16174869 + 3.13751356i$	1.449947×10^{-15}
	$\lambda_{*,1}$ [3, 12]	$4.16174868 + 3.13751356i$	

Table 10. Numerical solutions of Example 5 with $n = 200$ for GDNM
($Z_0 = (1, -1, 1, -1, \ldots, 1, -1, \ldots, 1, -1, 0.2)^T \in \mathbf{R}^{201}$, $\beta = 0.8$, $\sigma = 0.4$)

k	m_k	λ_k	$\|F(Z_k)\|$
0	0	0.20000000000000	$8.225377 \times 10^{+02}$
1	0	0.22536477018903	$2.485524 \times 10^{+01}$
2	0	0.23801871721578	$6.101313 \times 10^{+00}$
3	0	0.24452352118938	$1.412691 \times 10^{+00}$
4	0	0.24802476809991	2.623877×10^{-01}
5	0	0.24970126460117	2.458541×10^{-02}
6	8	0.25002087810879	8.790451×10^{-03}
7	7	0.25001990725851	8.126051×10^{-03}
8	5	0.25001892352509	7.255321×10^{-03}
9	2	0.25001773383146	6.089624×10^{-03}
10	0	0.25001619986859	4.169796×10^{-03}
11	0	0.25001534764228	5.757059×10^{-04}
12	0	0.25001534506816	1.902029×10^{-07}
13	2	0.25001534506667	1.153154×10^{-13}
14	*Convergent*	0.25001534506667	5.149069×10^{-14}
	Exact $\lambda_{*,200}$	0.25001534506667	

All computations were performed on the SUN ULTRA 2 Model 1200 at Osaka Women's University by using double-precision complex arithmetic in FORTRAN which carries about 16 significant digits for real part and imaginary part.

Acknowledgement

The author would like to thank referees for their comments that improved this paper.

References

[1] Chen, X., Nashed, Z., Qi, L.: Convergence of Newton method for singular smooth and nonsmooth equations using adaptive outer inverses. SIAM J. Optim. *7*, 445–462 (1997).
[2] Dongarra, J. J., Moler, C. B., Wilkinson, J. H.: Improving the accuracy of computed eigenvalues and eigenvectors. SIAM Numer. Anal. *20*, 23–45 (1983).
[3] Gregory, R. T., Karney, D. L.: A collection of matrices for testing computational algorithms. New York: Wiley-Interscience, 1969.
[4] Han, S. P., Pang, J. S., Rangaraj, N.: Globally convergent Newton methods for nonsmooth equations. Math. Oper. Res. *17*, 586–607 (1992).
[5] Ishihara, K., Aizawa, N.: Newton-secant method for complex nonlinear equations with nondifferentiable terms. Math. Jpn. *49*, 123–137 (1999).
[6] Ortega, J. M.: Numerical analysis, a second course. New York: Academic Press, 1972.
[7] Ortega, J. M., Rheinboldt, W. C.: Iterative solution of nonlinear equations in several variables. New York: Academic Press, 1970.
[8] Pang, J. S.: Newton's method for B-differentiable equations. Math. Oper. Res. *15*, 311–341 (1990).
[9] Pang, J. S., Han, S. P., Rangaraj, N.: Minimization of locally Lipschitzian functions. SIAM J. Optim. *1*, 57–82 (1982).
[10] Petres, G., Wilkinson, J. H.: Inverse iteration, ill-conditioned equations and Newton's method. SIAM Rev. *21*, 339–360 (1979).
[11] Symm, H. J., Wilkinson, J. H.: Realistic error bounds for a simple eigenvalue and its associated eigenvector. Numer. Math. *35*, 113–126 (1980).
[12] Wilkinson, J. H.: Error analysis of floating-point computation. Numer. Math. *2*, 319–340 (1960).
[13] Yamamoto, T.: Error bounds for computed eigenvalues and eigenvectors. Numer. Math. *34*, 189–199 (1980).
[14] Yamamoto, T.: Error bounds for computed eigenvalues and eigenvectors. II. Numer. Math. *40*, 201–206 (1982).
[15] Yamamoto, T.: The Symm–Wilkinson method for improving an approximate eigenvalue and its associated eigenvector. Computing *33*, 179–184 (1984).

K. Ishihara
Department of Applied Mathematics
Osaka Women's University
Sakai, Osaka
590-0035 Japan
e-mail: ishi@appmath.osaka-wu.ac.jp

Computing [Suppl] 15, 119–132 (2001)

Computing

Global Optimization in Quadratic Semi-Infinite Programming

Y. Liu* and **K. L. Teo,** Hong Kong, and **S. Ito,** Tokyo

Dedicated to Professor Tetsuro Yamamoto on the occasion of his 65th birthday

Abstract

We consider a class of LQ semi-infinite programming (SIP) problems where the objective is positive quadratic and the linear infinite constraint functions continuously depend on their index variable on a compact set. It is known that, by using the dual parameterization technique, this SIP problem can be reduced to a finite nonlinear programming problem with special features. Any global solution of the nonlinear problem will give rise to the solution of the SIP problem. However, there exists no efficient method for finding the global solution of such nonlinear programming problems. In this paper, we present a method for global solution of the nonlinear programming problem. Combining this method with the dual parameterization, we obtain an efficient dual parameterization algorithm for the LQ SIP problem. Convergence results are established and several numerical examples are given to demonstrate the efficiency of the algorithm.

Key Words: LQ semi-infinite programming, global optimization, dual parameterization, algorithm.

1. Introduction

Consider the following semi-infinite programming (SIP) problem

Problem (P):

$$\min f(x) = (1/2)x^T Qx + b^T x, \tag{1}$$

$$\text{subject to } a(t)x - c(t) \leq 0 \quad \text{for } t \in T, \tag{2}$$

where $x = (x_1, x_2, \ldots, x_n)^T \in R^n$ is the decision vector, $b = (b_1, b_2, \ldots, b_n)^T \in R^n$ is a constant vector, $Q \in R^{n \times n}$ is a positive definite matrix, and $a(t) : t \to R^n$ and $c(t) : t \to R$ are continuously differentiable functions defined on a compact set $T \subset R^p$. The vector inequality (2) is to be understood as component-wise inequalities.

---

* Part of the research was done during this author's visit to the Institute of Statistical Mathematics, 4-6-7 Minami-Azabu, Minato-ku, Tokyo 106-8569, Japan, supported by the Japan Society for the Promotion of Science.

There are many methods available in the literature for semi-infinite programming problems. In general, these methods can be grouped into the following categories: discretization methods, cutting plane methods, local reduction methods, decent methods, and smoothing methods. For details, we refer to Fiacco and Ishizuka [2], Gustafson and Kortanek [4], Hettich [5], Hettich and Kortanek [6], Teo et al. [10] and X. Chen [1]. Various methods for linear SIPs are discussed in Goberna and Lopez [3]. Recently, this class of convex linear-quadratic SIP problems is studied in Liu et al. [7] and Ito et al. [8] where the so-called dual parameterization method is introduced. By using the dual parameterization method, the SIP problem is transformed into a finite nonlinear programming problem. It is shown that for the positive LQ case, a global solution of the nonlinear programming problem obtained from the dual parameterization method directly gives rise to the optimal solution of the SIP problem. However, finding a global solution of the nonlinear optimization problem is known to be difficult. In fact, there is still no efficient optimization method available for this purpose. This is so even for small sized problems. In this paper, by using some special features of the nonlinear programming problem obtained by dual parameterization, we propose an efficient algorithm for finding its global solution. The idea is to find an approximate global solution for the parameterized problem. Then a local search, starting from this approximate solution, is performed to locate the true global solution. Convergence results are established. Based on this global optimization method, an algorithm is developed for the LQ SIP problem. The efficiency of the algorithm is demonstrated by solving a number of numerical examples.

2. Dual Parameterization

In this section, we briefly introduce some relevant results in dual parameterization technique for positive LQ SIP problems. For the remaining of this paper, we assume that the following constraint qualification is satisfied.

Assumption 1 (Slater). There is an $x \in R^n$ such that

$$a(t)x - c(t) < 0 \quad \text{for all } t \in T. \tag{3}$$

We denote by $C(T)$ the Banach space of all continuous real functions on T equipped with the supremum norm, and by $M(T)$ the space of all signed finite regular Borel measures on T. It is known that $M(T)$ is the dual space of $C(T)$. Let V be the cone of $C(T)$ consisting of all the nonnegative functions in $C(T)$. The cone in $M(T)$ associated with V, denoted by V', consists of all the nonnegative elements (nonnegative as measure) of $M(T)$. Thus, $\Lambda \in V'$ if and only if $\Lambda(f) \geq 0$ for all $f \in V$. We will use the same symbol '$\geq$' to denote the partial orders in both $C(T)$ and $M(T)$ induced by V and V', respectively. To be more specific, if f and g are two elements in $C(T)$ (respectively, $M(T)$), we write $f \geq g$ (equivalently, $g \leq f$) if and only if $f - g \in V$ (respectively, V').

Let $A : R^n \rightarrow C(T)$ be an operator defined by

$$(Ax)(t) = a(t)x \quad \text{for } t \in T \tag{4}$$

and denote by A^* the dual of A. Using the above symbols, problem (P) can be stated as

$$\begin{aligned} &\min f(x) = (1/2)x^T Qx + b^T x, \\ &\text{subject to } Ax - c \leq 0. \end{aligned}$$

The Dorn's dual of problem (P) can be written as:

Problem (P′)

$$\begin{aligned} &\min_{x,\Lambda} (1/2)x^T Qx + \langle \Lambda, c \rangle \\ &\text{s.t. } Qx + b + A^* \Lambda = 0 \\ &\quad \Lambda \geq 0 \end{aligned} \tag{5}$$

where

$$\langle \Lambda, c \rangle = \int_T c(t) d\Lambda(t). \tag{6}$$

In the following, we summarize some of the results related to dual parameterization in Theorem 2.1 through Theorem 2.4. These results can be found in Liu et al. [7] and Ito et al. [8].

Theorem 2.1. *If (x^*, Λ^*) is a solution of Problem (P') then x^* is the solution of Problem (P). On the other hand, if x^* is the solution of Problem (P), then there exists a Λ^* such that (x^*, Λ^*) is a solution of Problem (P').*

The Karush–Kuhn–Tucker (KKT) optimality conditions for problem (P) are given in the following.

Theorem 2.2 (KKT conditions). *Let the Slater constraint qualification be satisfied. The minimum of problem (P) is achieved at $x^* \in R^n$ if and only if x^* is feasible and there exists a $\Lambda^* \in M(T)$ such that*

$$\begin{aligned} Qx^* + b + A^* \Lambda^* &= 0, \\ \langle \Lambda^*, Ax^* - c \rangle &= 0, \\ \Lambda^* &\geq 0 \end{aligned} \tag{7}$$

We note that the solution of problem (P) is unique but the solution pair (x^*, Λ^*) of problem (P′) may not be unique in Λ^*. In other words, the multiplier $\Lambda^* \in M(\Omega)$ satisfying the KKT conditions (7) is not unique in general. It is shown in Ito et al. [8] that among all the optimal measures in V' for problem (P′), there exists at least one measure that has a finite support of no more than n points.

Theorem 2.3. *Let Assumption* 1 *be satisfied, and assume that* $x^* \in R^n$ *is feasible. Then the set of multipliers satisfying the KKT conditions* (7) *necessarily includes a measure with finite support at no more than n points.*

The following important result is a consequence of Theorem 2.3.

Theorem 2.4. *Let Assumption* 1 *be satisfied, and assume that the minimum of problem* (*P*) *is achieved at* $x^* \in R^n$. *Then the solution set of the dual problem* (P') *contains a solution pair of which the measure has a finite support at no more than n points.*

Problem (P′) is an infinite dimensional problem. Thus it is, in general, not directly solvable in numerical sense. However, by Theorem 2.4, we see that it suffices to restrict our search for Λ^* to a subset V_k' of V', where V_k' consists of all elements of V' with a finite support of no more than k points. Here k is a suitably large integer of which the choice will be discussed later. We note that a measure $\Lambda \in V_k'$ is characterized by the location of its supporting points and the measure of each of the supporting points. In case when the number of supporting points of Λ is less than k, we can assume Λ has k supporting points by adding some arbitrary points from T while assuming the measure of each of these added points is 0. Thus, Λ is a measure in V_k' if and only if it has k supporting points $t_i \in T$, $i = 1, 2, \dots, k$, and $\Lambda(\{t_i\}) = \lambda_i \geq 0$, $i = 1, 2, \dots, k$. By restricting Λ to V_k', problem (P′) is reduced to:

Problem (P_k')

$$\begin{aligned} &\min_{x,\lambda,\tau} \frac{1}{2} x^T Q x + \sum_{i=1}^{k} c(t_i)\lambda_i \\ &\text{s.t. } Qx + b + \sum_{i=1}^{k} \lambda_i a(t_i)^T = 0 \\ &\qquad \lambda_i \geq 0, \quad i = 1, 2, \dots, k, \\ &\qquad t_i \in T, \quad i = 1, 2, \dots, k \end{aligned} \tag{8}$$

where $\lambda = (\lambda_1, \lambda_2, \dots, \lambda_k)$ and $\tau = (t_1, t_2, \dots, t_k)$. We call problem ($\mathrm{P}_k'$) the parameterized dual of problem (P). Let λ and τ be as above. We define two more problems which depend on τ.

Problem ($\mathrm{P}_k'(\tau)$)

$$\min_{x,\lambda} (1/2) x^T Q x + \sum_{i=1}^{k} c(t_i)\lambda_i \tag{9}$$

$$\text{s.t. } Qx + b + \sum_{i=1}^{k} \lambda_i a(t_i)^T = 0 \tag{10}$$

$$\lambda_i \geq 0, \quad i = 1, 2, \dots, k. \tag{11}$$

Problem ($P_k(\tau)$)

$$\min_x \frac{1}{2}x^T Qx + bx \tag{12}$$

$$\text{s.t.}\, a(t_i)x - c(t_i) \leq 0, \quad i = 1, 2, \ldots, k. \tag{13}$$

We note that problem ($P'_k(\tau)$) is the dual of problem ($P_k(\tau)$).

It is clear from Theorem 2.4 that when the integer k is large enough, a global solution of problem (P'_k) provides a solution to problem (P'), hence giving a solution to problem (P). Since

$$V'_k \subset V'_{k+1}, \quad k = 0, 1, 2, \ldots,$$

we see that the optimal value of problem (P'_k) is decreasing with respect to k. In fact, we have more about this value sequence.

Theorem 2.5. *For problem (P'_k), the following are true.*

(*a*) *The optimal value sequence $\{v(P'_k)\}$ is decreasing and there is a $k^* \geq 0$ such that*

$$v(P'_{k^*}) = v(P'_k), \quad \text{for all } k \geq k^*$$

and if $k^* \geq 1$

$$v(P'_{k^*-1}) > v(P'_{k^*}).$$

(*b*) *The number k^* in (a) is the minimum integer such that for $k \geq k^*$, a global solution of problem (P'_k) provides the solution of problem (P) in the sense that if (x^*, λ^*, τ^*) is a global solution of problem (P'_k), then x^* is the solution of problem (P).*

(*c*) *The number k^* satisfies $0 \leq k^* \leq n$.*

(*d*) *If $0 \leq k < k^*$, then*

$$v(P'_k) > v(P'_{k+1}). \tag{14}$$

Proof: (a) The sequence $\{v(P'_k)\}$ is decreasing since the sequence of the corresponding feasible sets is increasing. It is clear that for any $k \geq 0$, $v(P'_k) \geq v(P'_{k+1})$ and from Theorem 2.4 there exists a k' such that $v(P'_{k'}) = v(P'_k)$ and hence

$$v(P'_{k'}) = v(P'_k), \quad \text{for all } k \geq k'. \tag{15}$$

Let k^* be the minimum of k' such that the above holds. Then k^* satisfies the requirement.

(b) From (a), for $0 \leq k < k^*, v(P'_k) \leq v(P')$ and for $k \geq k^*, v(P'_k) = v(P')$. Thus k^* is the minimum integer such that for $k \geq k^*$, a solution of problem (P'_k) is a solution of problem (P'). Hence, by the Theorem 2.1, the statement in (b) is true.

(c) This is a direct consequence of Theorem 2.4 and (b).

(d) Let $(\bar{x}, \bar{\tau}, \bar{\lambda})$ be a global solution to problem (P'_k), where $\bar{\tau} = (\bar{t}_1, \bar{t}_2, \ldots, \bar{t}_k)$ and $\bar{\lambda} = (\bar{\lambda}_1, \bar{\lambda}_2, \ldots, \bar{\lambda}_k)$. It is clear that $(\bar{x}, \bar{\lambda})$ is a solution for problem $(\mathrm{P}'_k(\bar{\tau}))$. We note that problem $(\mathrm{P}'_k(\bar{\tau}))$ is the dual problem of the positive quadratic problem $(\mathrm{P}_k(\bar{\tau}))$. Therefore, $\bar{x}$ is the solution of problem $(\mathrm{P}_k(\bar{\tau}))$. Since $k < k^*$, from the definition of k^* we see that $\bar{x}$ is not feasible to problem (P). Thus, there exists a point $\bar{t}_{k+1} \in T$ such that

$$a(\bar{t}_{k+1})\bar{x} - c(\bar{t}_{k+1}) > 0. \tag{16}$$

Let $\hat{\tau} = (\bar{t}_1, \bar{t}_2, \ldots, \bar{t}_{k+1})$. Denote by $\hat{x}$ the solution of problem $(\mathrm{P}_{k+1}(\hat{\tau}))$. We see that

$$v(P_k(\bar{\tau})) < v(P_{k+1}(\hat{\tau})). \tag{17}$$

From the standard duality theory, we have

$$v(P_k(\bar{\tau})) = -v(P'_k(\bar{\tau})), \tag{18}$$

$$v(P_{k+1}(\hat{\tau})) = -v(P'_{k+1}(\hat{\tau})). \tag{19}$$

Thus,

$$v(P'_k) = v(P'_k(\bar{\tau})) = -v(P_k(\bar{\tau})) > -v(P_{k+1}(\hat{\tau})) = v(P'_{k+1}(\hat{\tau})) \geq v(P'_{k+1}). \tag{20}$$

The proof is complete. □

We call the number k^* the minimum parameterization number. We note that $k^* = 0$ is possible. This occurs when the optimal primal solution is an interior point of the feasible region. From the definition of problem (P'_k), a smaller k gives rise to a problem of lower dimension. Thus it is desirable to choose $k = k^*$ when solving problem (P'_k). However, the minimum parameterization number is not known a priori. In practice, we use an estimated value for k, hoping it is no smaller than k^* and yet not substantially larger than k^*. In this regard, we suggest that: (a) When n is reasonably small, choose $k = n$. (b) When n is very large, the minimum parameterization number k^* may be substantially smaller than n. In this case, choosing $k = n$ would make the dimension of problem (P'_k) unnecessarily large and hence unnecessary computational effort would be required. Therefore, we can start from a reasonably small k and increase it by 1 until the global optimal values satisfy $v(P'_k) = v(P'_{k-1})$. By Theorem 2.5, this will guarantee that the global solution for problem (P'_k) is a solution of problem (P').

3. Global Solution for Problem (P'_k)

From the previous section, in order to solve problem (P) we need to find a global solution of problem (P'_k). However, problem (P'_k) is a nonlinear programming problem due to the presence of the nonlinear functions $c(\cdot)$ in the cost and $a(\cdot)$ in the constraint. In general, global optimization problem is difficult to solve and there is no existing efficient algorithm available for computing a global solution of such problems. On the other hand, problem (P'_k) has one special feature, namely it is convex with respect to x and linear with respect to λ. In this section, we take advantage of this property of problem (P'_k) to devise a numerical scheme for computing a global solution. Convergence result will be given.

The basic idea behind the algorithm to be proposed is: Choose an integer k and a vector τ of k index points in T, $\tau = (t_1, t_2, \dots, t_k)$. Solve problem ($\mathrm{P}'_k(\tau)$). At the optimal solution, update k and $t_1, t_2, \dots, t_k$. Then solve problem ($\mathrm{P}'_k(\tau)$) again. Repeat this procedure until some stopping criteria is satisfied. At the termination, a set of approximate index points together with the corresponding optimal values for x and λ is obtained. This will provide an approximate solution to problem (P'_k). Then we perform a local search starting from this approximate solution, a global solution will be obtained provided that the technique used in the local search is stable in the sense that it always find the local solution of which shares the same basin with the starting point (initial solution). We note that problem ($\mathrm{P}'_k(\tau)$) is a convex quadratic problem with linear constraints and is easy to solve.

Suppose $\{k_i\}$ be a given sequence of integers. For each i, let $T_i = \{t^i_j : j = 1, 2, \dots, k_i\}$ be a given subset of T and let $\tau^i = (t^i_1, t^i_2, \dots, t^i_{k_i})$. Define the density distance between T_i and T as

$$d(T_i, T) \hat{=} \max_{t \in T} \min_{1 \le j \le k_i} |t - t^i_j|.$$

As stated above, we will consider the sequence of problems $\mathrm{P}'_{k_i}(\tau^i)$, $i = 1, 2, \dots$. The choice of $\{k_i\}$ and T_i depends on the way of updating the fixed index points in problem (P'_k). The most simple way of updating k and $t_1, t_2, \dots, t_k$ is to increase k and refine $t_1, t_2, \dots, t_k$ so that the density distance between the set $\{t_1, t_2, \dots, t_k\}$ and T tends to zero as the iteration goes to infinity. In this case, we have $k_i \le k_{i+1}$ and

$$d(T_i, T) \to 0 \quad \text{as } i \to \infty. \tag{21}$$

If (21) is satisfied, convergence is available.

Theorem 3.1. *Let the sequence* $\{T_i\}$ *of subsets of* T *be given as above. Let* $(\bar{x}^i, \bar{\lambda}^i)$ *be a solution of problem* $(P'_{k_i}(\tau^i))$. *If* (21) *is satisfied, then*

(a) $\{\bar{x}^i\}$ *converges to the solution of problem* (P).

(b) $v(P'_{k_i}(\tau^i)) \to v(P')$.

Proof: (a) Note that problem (P) has a positive quadratic cost and a convex feasible region. Thus it has a unique solution x^*. From Theorem 2.5, there exists an integer $k^* \le n$, and k^* points $t^*_1, t^*_2, \dots, t^*_{k^*}$ in T such that the solution of problem

$P_{k^*}(\tau^*)$ is x^*, where $\tau^* = (t_1^*, t_2^*, \ldots, t_{k^*}^*)$. It is easy to see from (21) that from each sequence $t_1^i, t_2^i, \ldots, t_{k_i}^i$ we can choose k^* points $s_1^i, s_2^i, \ldots, s_{k^*}^i$ such that

$$\lim_{i \to \infty} s_j^i = t_j^*, \qquad \text{for } j = 1, 2, \ldots, k^*. \tag{22}$$

Let $\tilde{\tau}^i = (s_1^i, s_2^i, \ldots, s_{k^*}^i)$, $i = 1, 2, \ldots$. The following inequalities is obvious.

$$v(P_{k^*}(\tilde{\tau}^i)) \leq v(P_{k_i}(\tau^i)) \leq v(P'). \tag{23}$$

Since problem $(P'_{k_i}(\tau^i))$ is the dual of problem $(P_{k_i}(\tau^i))$, by the KKT Theorem we see that $\bar{x}^i$ is the solution of problem $(P_{k_i}(\tau^i))$. We now prove that the sequence $\{\bar{x}^i\}$ converges to x^*. Suppose this is not true. Then there exist an $\varepsilon > 0$ and a subsequence of $\{\bar{x}^i\}$, denoted again by $\{\bar{x}^i\}$, such that

$$|\bar{x}^i - x^*| \geq \varepsilon, \quad i = 1, 2, \ldots \tag{24}$$

where $|\cdot|$ is the distance in R^n. The sequence $\{\bar{x}^i\}$ is bounded since the corresponding cost value sequence is bounded from above by $v(P')$. Thus it has a converging subsequence $\{\bar{x}^{i_j}\}$. Let $\bar{x}^{i_j} \to \bar{x}$ as $j \to \infty$. We have $|\bar{x} - x^*| \geq \varepsilon$ from (24), and hence $\bar{x}$ is not a feasible point for problem $(P_{k^*}(\tau^*))$ due to the fact that the cost value at $\bar{x}$ cannot be greater than at x^*. Therefore, there exists $1 \leq j_0 \leq k^*$ such that

$$a(t_{j_0}^*)\bar{x} - c(t_{j_0}^*) > 0. \tag{25}$$

On the other hand, we have

$$a(t_{j_0}^{i_j})\bar{x} - c(t_{j_0}^{i_j}) \leq 0, \quad j = 1, 2, \ldots. \tag{26}$$

Since both $a(t)$ and $c(t)$ are continuous on T, let $j \to \infty$ in (26) we obtain

$$a(t_{j_0}^*)\bar{x} - c(t_{j_0}^*) \leq 0 \tag{27}$$

which contradicts to (25). Thus, $\{\bar{x}^i\}$ converges to x^*.

(b) Since problem $(P'_{k_i}(\tau^i))$ is the dual of problem $(P_{k_i}(\tau^i))$, part (b) is a direct consequence of (a) and the KKT theorem. □

We now propose the following algorithm.

Algorithm

1. Choose a small number $\varepsilon > 0$, integer sequence k_i, $i = 1, 2, \ldots$, and discretization vector sequence $\{\tau^i\}$, satisfying (21). Set $i = 1$.
2. Solve problem $(P'_{k_i}(\tau^i))$ and denote the solution by $(\bar{x}^i, \bar{\lambda}^i)$.
3. If $i \geq 2$ and $|v(P'_{k_i}(\tau^i)) - v(P'_{k_{i-1}}(\tau^{i-1}))| \leq \varepsilon$, goto step 4, else $i = i + 1$, goto step 2.

4. Performing a local search for problem (P'_{k_i}) starting from $(\bar{x}^i, \bar{\lambda}^i, \tau^i)$. The local minimum is denoted by (x^*, λ^*, τ^*). Then x^* is taken as the solution for problem (P).

Remark. In solving problem ($P'_{k_i}(\tau^i)$), we can easily find a feasible initial guess (x_0, λ_0). First, we choose λ_0 such that its elements are nonnegative. Then an x_0 is obtained from λ_0 according to (10). Step 3 means we stop the iteration when improvement in the objective value is small. The choice of ε depends on experience.

4. Numerical Example

In this section, we present two numerical examples to show the efficiency of the proposed algorithm. The first one is a modification of an example problem from Liu et al. [7]. The modification is made by dropping the ineffective infinite constraint – one of the two infinite constraints that is not active for all index points at the optimal solution. The second example is the one sided approximation of the tangent function by polynomials of degree not exceeding a certain integer n. For these examples, the index sets are intervals of R^1 and the sequence $\{k_i\}$ is chosen this way: choose a k_1 and the k_1 index points are uniquely determined by requiring that they are equally spaced and include the end points of the interval. Then each subsequent k_i and the corresponding set of index points are determined by adding the midpoints of each pair of adjacent points.

Example 1

$$\begin{aligned}&\min x^T Qx\\ &\text{s.t. } a(t)x - c(t) \le 0,\end{aligned}$$

where $Q \in R^{16\times16}, a(t)$ and $c(t)$ are specified as follows:

$$Q = \begin{bmatrix} 4 & 1 & 0 & \dots & 0 \\ 1 & 4 & 1 & \dots & 0 \\ 0 & 1 & 4 & \dots & 0 \\ & & & \vdots & \\ 0 & 0 & 0 & \dots & 4 \end{bmatrix} \tag{28}$$

$$a(t) = (-e^{-(t-5/16)^2}, -e^{-(t-2.5/16)^2}, \dots, -e^{-(t-16.5/16)^2}) \tag{29}$$

$$c(t) = -3 - 4.5\sin(4.7\pi(t - 1.23)/8) \tag{30}$$

For this example, different k_1 values, such as $k_1 = 2,\ 4,\ 6,\ 8$ and 10, are tested. For each of these k_1 values, the algorithm can successfully compute a global solution for the dual problem and hence a solution for the primal problem. Computational details are shown in Table 1 for the case $k_1 = 3$ with $\varepsilon = 0.5$. The

Table 1. Results for example 1

k_1	3 (start searching for approximate solution)
k_2	5 (end searching for approximate solution)
Approximate x	[0.084044 0.142695 0.258183 0.436827 0.707666 1.046537 1.345125 1.453519 1.303190 0.987357 0.708656 0.664137 0.914220 1.451353 1.763188 2.606933]
Approximate λ	0.000000 1.975311 16.463732 0.000000 24.350055
Optimal x	[0.076461 0.204917 0.486116 0.927634 1.437805 1.808159 1.845673 1.530850 1.039785 0.610216 0.404887 0.495781 0.855648 1.458993 1.798618 2.668698]
Optimal objective	154.116154

infinite constraint at the computed optimal primal solution is shown in Fig. 1. We note that the active constraint index points is identified as $t_1^* = 2.06165$ and $t_2^* = 5.00000$ which coincide with the result in Liu et al. [7].

Example 2. The one-sided L^2 approximation of the tangent function on $[0, 1]$ by polynomials of degree not exceeding n is stated as (see Ito et al. [8])

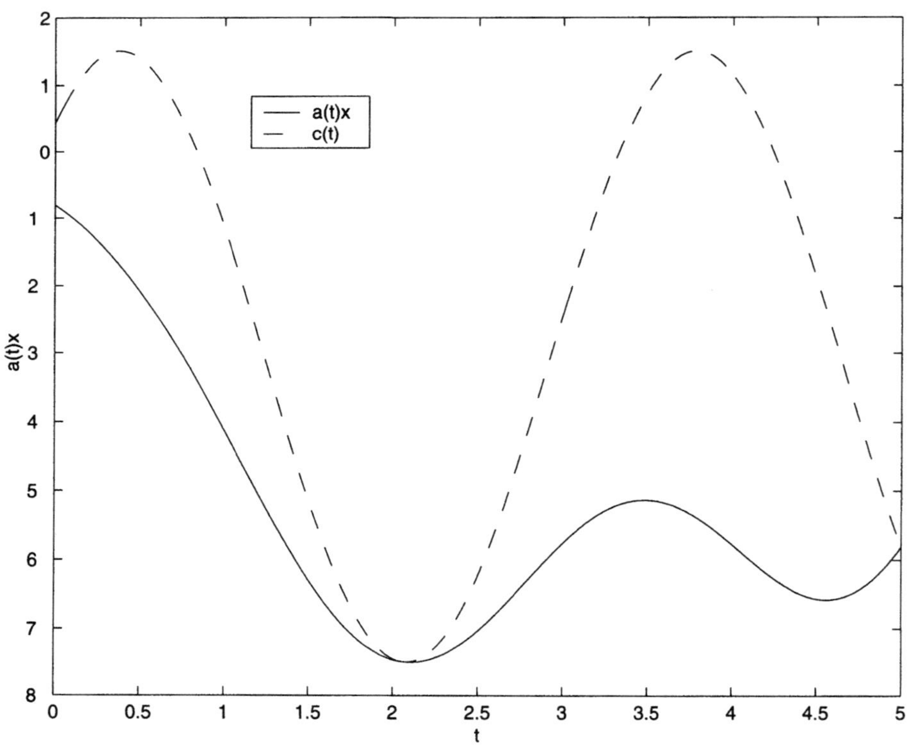

Figure 1. Example 1. Primal constraint at solution

$$\min_x \int_0^1 \left(\sum_1^n x_i t^{i-1} - \tan(t) \right)^2 dt$$

$$\text{s.t.} \sum_1^n -x_i t^{i-1} \le -\tan(t), \quad \text{for } t \in [0,1].$$

This problem can be written in the form of problem (P) (see (1)–(2)) with Q, b, a and c specified as follows:

$$Q = [2/(i+j-1)]_{n \times n}$$

$$b = \int_0^1 [1,\ t,\ \dots, t^{n-1}]^T \tan(t)\, dt$$

$$a(t) = [-1, -t, \dots, -t^{n-1}]$$

and

$$c(t) = -\tan(t).$$

We solved this problem for the cases from $n = 2$ to $n = 9$. For each case, we tested all even numbers between 2 and n as k_1. All are successful in finding a global solution for the parameterized dual problem. The numerical results obtained here coincide with those reported in Ito et al. [8] for cases from $n = 2$ to $n = 6$. Higher dimensional cases are very difficult to solve using the method of [8]. The proposed algorithm also successfully applied to cases of $n = 7, 8$ and 9. We present the computational results for the case of $n = 9$ in Table 2. The active constraint index points at the computed primal solution are [0.083358, 0.166655, 0.333329, 0.416638, 0.583346, 0.666662, 0.833346, 0.916640, 1.000000]. The primal constraints for $n = 5$ and $n = 6$ are shown in Figs. 2 and 3 respectively. For cases of dimension higher than 6, the computed polynomial $a(t)x$ is even more close to $\tan(t)$. We note that, by using the method of [8], it is very difficult to solve cases $n > 6$.

Table 2. Results for example 2

n	9
k_1	4 (start searching for approximate solution)
k_4	25 (end searching for approximate solution)
Approximate x	[0.002592 0.919776 0.606729 −1.335390 1.516574 0.484291 −0.285031 −1.410201 1.059898]
Approximate λ	[0.000000 0.000000 0.008609 0.000007 0.018036 0.000000 0.000048 0.000000 0.051664 0.000000 0.054516 0.000000 0.000069 0.000000 0.047213 0.000000 0.148703 0.000000 0.000063 0.0000000 0.103081 0.000000 0.142654 0.000011 0.041001]
Optimal x	[0.002560 0.919760 0.606718 −1.335399 1.516566 0.484284 −0.285038 −1.410207 1.059893]
Optimal objective	0.278718

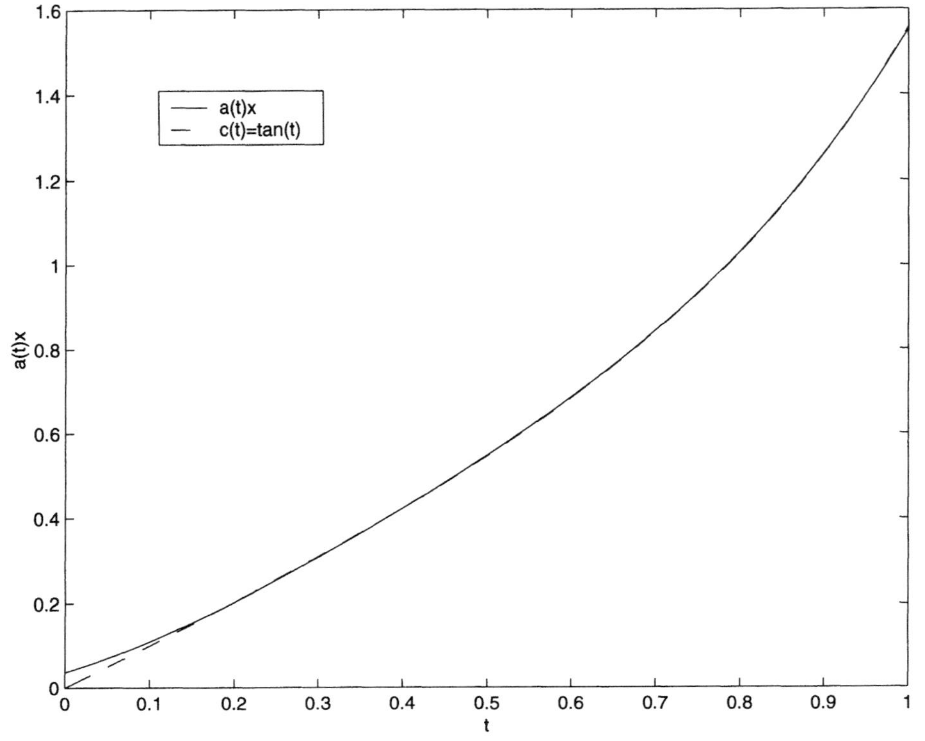

Figure 2. Example 2. $n = 5$. Primal constraint at solution

In general, the parameterized dual problem has local solutions. In earlier works (see [7] and [8]), local minimization technique was used to find global solutions of the parameterized dual problems. It was necessary in some cases to try different initial guesses in order to obtain a global solution. This can be successful in some cases. However, we do have situations (for example, Example 2 cases $n > 6$) where finding a good initial guess is very difficult. In fact, we were not able to solve Example 2, cases $n > 6$ using local minimization method. The efficiency of the present algorithm in finding a global solution has been demonstrated by results for the two numerical examples (including different cases for Example 2). It is also different from discretization methods. As we have seen from the two examples, only a small number of discretization points are needed. Traditional discretization method can only compute an approximate solution. The present method is to compute the exact solution (up to computer precision).

5. Comments

In this paper, we developed a global minimization algorithm for the class of nonlinear programming problems obtained from positive LQ SIP problems by the

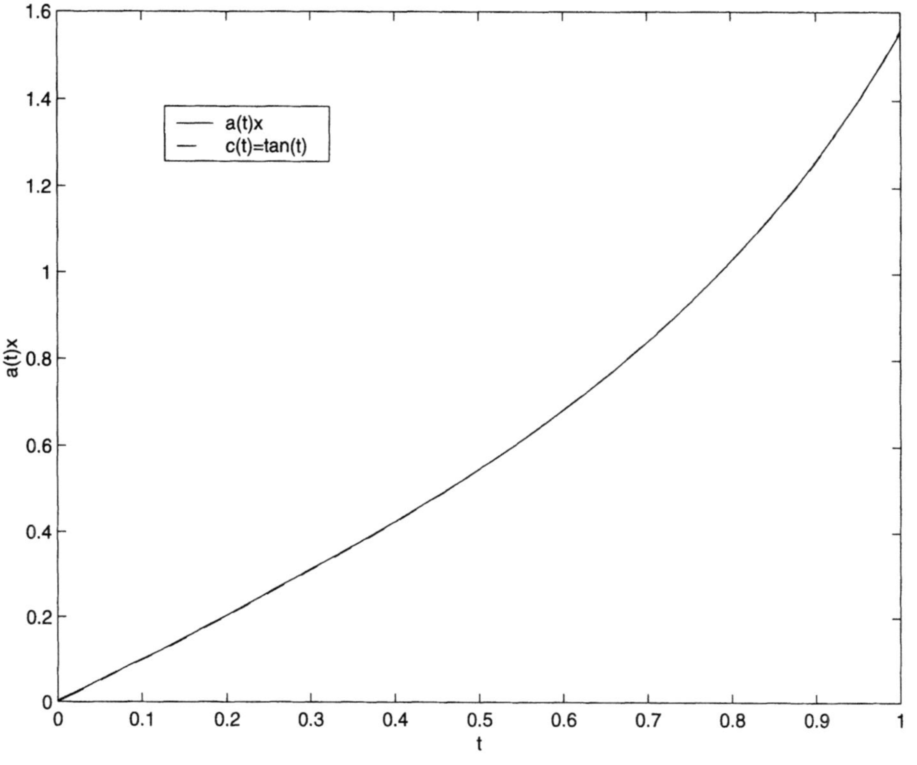

Figure 3. Example 2. $n = 6$. Primal constraint at solution

dual parameterization method. Convergence result is given and some useful properties regarding the relation between the primal SIP and transformed non-linear programming problem are discussed. Two existing examples are solved by the algorithm and the numerical results show that the algorithm is efficient in finding the global solution. This algorithm, on the other hand, suffers from a drawback, namely the number of parameterization points is not limited in theory. Therefore, for some complicated SIP problems, we may have to deal with high dimensional nonlinear global optimization problems. It is expected that some adaptive technique would be necessary to reduce the computing task. We will discuss this in a separate paper.

References

[1] Chen, X.: Smoothing methods for complementarity problems and their applications: a survey, J. Oper. Res. Soc. Jpn. *43*, 32–47 (2000).

[2] Fiacco, A. V., Ishizuka, Y.: Suggested research topics in sensitivity and stability analysis for semi-infinite programming. Annals Oper. Res. *27*, 65–76 (1990).

[3] Goberna, M. A., Lopez, M. A.: Linear semi-infinite optimization. Chichester: Wiley, 1998.

[4] Gustafson, S.-Å., Kortanek, K. O.: Semi-infinite programming and applications. In: Mathematical Programming. The State of the Art. (Bachen, A., et al., eds.), pp. 138–157. Berlin Heidelberg New York Tokyo: Springer, 1983.

[5] Hettich, R.: A review of numerical methods for semi-infinite optimization. In: Semi-infinite programming and applications (Fiacco, A. V., Kortanek, K. O., eds.), pp. 158–178. Berlin Heidelberg New York Tokyo: Springer, 1983.
[6] Hettich, R., Kortanek, K. O.: Semi-infinite programming: Theory, method, and applications. SIAM Rev. *35*, 380–429 (1993).
[7] Liu, Y., Teo, K. L., Ito, S.: A dual parametrization approach to linear-quadratic semi-infinite programming problems. Optim. Methods Software *10*, 471–491 (1999).
[8] Ito, S., Liu, Y., Teo, K. L.: A dual parameterization method for convex semi-infinite programming. Optim. Theory Appl. Ann. Oper. Res. (to appear).
[9] Luenberger, D. G.: Optimization by vector space method. New York: Wiley, 1969.
[10] Teo, K. L., Rehbock, V., Jennings, L. S.: A new computational algorithm for functional inequality constrained optimization problems. Automatica *29*, 789–792 (1999).

Y. Liu
K. L. Teo
Department of Applied Mathematics
The Hong Kong Polytechnic University
Hung Hom, Kowloon
Hong Kong
e-mail: mayliu@inet.polyu.edu.hk

S. Ito
The Institute of Statistical Mathematics
4-6-7 Minami-Azabu
Minato-ku, Tokyo 106-8569
Japan

Computing [Suppl] 15, 133–150 (2001)

Computing

Aggregation/Disaggregation Methods for *p*-cyclic Markov Chains*

I. Marek and **P. Mayer**, Praha

Dedicated to Professor Tetsuro Yamamoto on the occasion of his 65th birthday

Abstract

We present a new application of some aggregation/disaggregation iteration methods designed for finding stationary probability vectors of *p*-stochastic matrices. Our main goal consists of proving convergence of an aggregation/disaggregation iteration process based on the Gauss-Seidel splitting of the matrix considered algorithm. It should be noted that besides *p*-cyclicity the matrix under consideration may be quite general, in particular, hypotheses concerning spectral properties of the matrix examined and its powers such as reality of its spectrum etc. and consistent ordering of the matrix itself are not required.

AMS Subject Classifications: 65F10, 65F15, 15A51.

Key Words: Markov chain, *p*-cyclic stochastic matrix, stationary probability vector, aggregation/disaggregation algorithms.

1. Introduction

This contribution is tightly related to [7] and, more generally, directed towards mathematical modeling of reliable safety systems in railway transportation [7, 8]. The demands of reliability go hand in hand with complexity of the models and the speed of the adequate computational algorithms. In addition, the concept of reliability leads to very strict demands on the preciseness of the computed quantities modeled. In fact, the probabilities of some events are to be computed up to 10^{-12}. It is thus desirable to develop fast and stable algorithms that would avoid the difficulties connected with the above requirements. Obviously, one has to expect that some of the requirements mentioned would be possible to satisfy only on the account of generality. In this contribution we present some computational means that are efficient for models characterized by Markov chains whose transition matrices are p^*-cyclic.

It is quite natural that special classes of stochastic matrices deserve particular treatment. In this contribution we focus to the class of p^*-cyclic matrices. An attention paid in the past to the problems involving p^*-cyclic matrices is remarkable. A fast development started by D.M. Young in his study of SOR

*Research supported partially by the Grant Agency of the Czech Republic, grant Nr. 201/98/0528 grant CEZ J04/98:210000010 and grant CEZ J13/98:113200007.

methods of solving linear systems arising in discretizing diffusion systems [21, 22], where a central role has been played by matrices with *property A* (for details see [22]) and implicitely by two-cyclic matrices. Further generalization to p-cyclic matrices appeared in R.S. Varga's fundamental paper [19]. Crucial requirements in Varga's theory as well as in that of Young are on the one hand the *consistent ordering* of the matrix considered, let us denote it by B, and, on the other hand, positivity of the spectrum $\sigma(B^p)$ relaxed later to its reality [4]. Further generalizations to complex spectra of B^p are contained in [5] and [6].

Our main aim is to compute all stationary probability vectors of block p-cyclic matrices with no restriction on their structure. In particular, our theory admits any stochastic matrix B regardless whether the spectrum of its B^p is or is not located on the real axis as well as whether B is or is not consistently ordered.

We show that our iterative aggregation/disaggregation (IAD) method based on the Gauss-Seidel splitting of matrix $A = I - B$ is convergent to the required stationary probability vector (SPV). This goal is achieved on the account of our general error-vector formula (see Subsection 7.1 and also [12]) in conjunction with an important result of Courtois and Semal [3]. We also show that the aggregation/disaggregation step is theoretically to be performed just once. Thus, the whole process is in fact the Gauss-Seidel process terminated by one IAD sweep. This theoretical fact can be exploited in a computationally safe, and concerning complexity, cheap way.

The paper is organized as follows. *Section 2 contains definitions*, notation and some auxilliary results, Section 3 surveys some *properties of p-cyclic stochastic matrices* needed in the following text. In Section 4 a new class of matrices is introduced generalizing the *concept a stochastic matrix*. Definitions and some useful relations describing the aggregation/disaggregation procedures are presented in Section 5. IAD methods are introduced in Section 6. Main results represented by convergence theorems are contained in Section 7. The paper is concluded by Section 8 presenting some applications accompanied by several computational experiments confirming the conclusions of theoretical investigations.

2. Definitions, Notation and some Auxiliary Results

Objects of our investigation are matrices whose elements are real numbers. An $N \times N$ matrix $C = (c_{jk})$ with $c_{jk} \in \mathscr{R}^1$, is called nonnegative if $c_{jk} \geq 0,\ j, k = 1, \ldots, N$. Symbol I denotes the identity $N \times N$ matrix.

Let $N \geq 1$ be a positive integer. We let $\mathscr{R}^N$ to denote the standard arithmetic space of N-tuples of real numbers. Let $[.,.]$ denote the standard inner product on $\mathscr{R}^N$:

$$[x, y],\ x = (x_1, \ldots, x_N)^T, \quad y = (y_1, \ldots, y_N)^T \in \mathscr{R}^N.$$

A collection of all distinct eigenvalues of a square matrix A is called *spectrum* of A and it is denoted by $\sigma(A)$. We let

$$r(A) = \mathrm{Max}\{|\lambda|: \lambda \in \sigma(A)\}$$

and call it *spectral radius* of A.

Let I denote the identity map on $\{\mathscr{R}^N\}$. Let $\{M, W\}$ be a splitting of $A = I - B$, i.e. let M^{-1} exists and

$$A = M - W, \quad T = M^{-1}W. \tag{2.1}$$

A splitting (2.1) is called *convergent*, if $\{T^k\}$ is convergent, i.e.

$$\lim_{k\to\infty} T^k = Q.$$

If $Q = 0$ then T is called *zero-convergent*. Let Y be an $N \times N$ matrix. A splitting (2.1) is called *Y-convergent* and *Y-zero-convergent* if the sequence $\{YT^k\}$ is convergent and zero-convergent respectively.

If there exists an open set $\Omega \subset \mathscr{R}^N$ such that for every $x \in \Omega$ there is a positive integer $q = q(x)$ such that $T^k x \in \mathscr{R}^N_+$ for $k \geq q$, we call T *locally nonnegative*. If an iteration matrix T is nonnegative (locally nonnegative), the splitting $\{M, W\}$ is called *of nonnegative type* (of locally nonnegative type).

The next example shows that the SOR splittings with relaxation factors outside the range of the classical convergence offer locally nonnegative iteration operators.

Example 1. Let $B = D + L + U$ with D diagonal, L and U lower and upper triangular matrices respectively be *column stochastic*. Let $M = I - D - \omega L$ and $W = (1 - \omega)(I - D) + \omega U$. As usual, we denote the iteration matrix corresponding to this splitting as £(ω), i.e. £$(\omega) = (I - \omega(I - D)^{-1}L)^{-1}[(1 - \omega)(I - D) + \omega U]$.

Let spectrum $\sigma(B^p) \subset \mathscr{R}^N_+$. According to [9] we let

$$£(\omega) = \rho(\omega)Q(\omega) + Z(\omega), \quad [Q(\omega)]^2 = Q(\omega), \quad Q(\omega)Z(\omega) = Z(\omega)Q(\omega) = 0,$$
$$r(Z(\omega)) < \rho(\omega), \quad \omega \in (p/(p-1), +\infty). \tag{2.2}$$

and

$$[T(\omega)]^k = \left(\frac{1}{\rho(\omega)} £(\omega)\right)^k = Q(\omega) + \frac{1}{\rho(\omega)^k}[Z(\omega)]^k$$

and thus,

$$\lim_{k\to\infty}\left(\frac{1}{\rho(\omega)} Z(\omega)\right)^k = 0.$$

It follows that for $x \in \mathscr{R}^N$

$$Q(\omega)x = [x, x'(\omega)]x(\omega), \ [x(\omega), x'(\omega)] = 1,$$

where

$$x'(\omega) = [£(\omega)]^T x'(\omega), \ x'(\omega) \in \mathscr{R}^N, \ Q(\omega)x(\omega) = x(\omega) \in \mathscr{R}^N_+, \quad (x(\omega))_j > 0.$$

If $[x, x'(\omega)] > 0$, then $(Q(\omega)x)_j > 0,\ j = 1, \ldots, N$. Thus,

$$\left(\frac{1}{\rho(\omega)}\pounds(\omega)\right)^k x \in \mathscr{R}^N_+$$

for k sufficiently large. Therefore, $T(\omega)$ is locally nonnegative. Here,

$$\Omega = \left\{x \in \mathscr{R}^N_+ : [x, x'(\omega)] > 0\right\}. \tag{2.3}$$

Furthermore, it can be easily shown that $T(\omega)$ is *g-stochastic according to Definition* 1.

3. p*-cyclic Stochastic Matrices

Let N, p be positive integers, $n_j,\ j = 1, \ldots, p$, such that $n_1 + \cdots + n_p = N$. Let $A = I - B$, where

$$B = (b_{jk}), \quad b_{jk} \geq 0,\ \sum_{j=1}^{N} b_{jk} = 1,\ k = 1, \ldots, N, \tag{3.1}$$

is a (column) *stochastic matrix*. Furthermore, let

$$HBH^T = \begin{pmatrix} B_{11} & 0 & . & . & . & 0 & B_{1p} \\ B_{21} & B_{22} & . & . & . & 0 & 0 \\ & & & . & & & \\ & & & . & & & \\ & & & . & & & \\ 0 & 0 & . & . & . & B_{p-1p-1} & 0 \\ 0 & 0 & . & . & . & B_{pp-1} & B_{pp} \end{pmatrix}, \tag{3.2}$$

where H is a permutation matrix and the blocks B_{jk} are $n_j \times n_k$ matrices such that $I - B_{jj},\ j, k = 1, \ldots, p$, are invertible.

Assume that

$$\begin{cases} \left(I_{n_j} - B^T_{jj}\right)^{-1} B^T_{j+1j} e(n_{j+1}) = e(n_j), \quad j = 1, \ldots, p-1, \\ \left(I_{n_p} - B^T_{pp}\right)^{-1} B^T_{1p} e(n_1) = e(n_p), \end{cases} \tag{3.3}$$

where $e(n_j)^T = (1, \ldots, 1)^T \in \mathscr{R}^{n_j}$. Evidently,

$$e = e(N) = (1, \ldots, 1)^T = \left(e(n_1)^T, \ldots, e(n_p)^T\right) \in \mathscr{R}^N.$$

A block matrix $A = I - B$ satisfying relations (3.1)–(3.3) is called p^*-cyclic stochastic.

We further assume (without loss of generality) that B is irreducible. Then, there is a unique vector $\hat{x} \in \mathscr{R}^N$ – a *stationary probability vector of B* – such that

$$B\hat{x} = \hat{x}, \quad [\hat{x}, e] = 1.$$

Since, by hypothesis,

$$B^T e = e,$$

the following operator P defined by

$$Px = [x, e]\hat{x},$$

is the spectral projection called *Perron projection.*

4. Generalized Stochastic Matrices

The algorithms introduced in Section 6 are analogs of similar aggregation algorithms studied in [12] and [13]. The generalization consists of broadening the class of matrices admissible for utilizing: In contrast to [13] where the methods were applicable for the class of all stochastic matrices, our generalization admits the class of generalized stochastic matrices defined below.

Definition 1. *An $N \times N$ matrix T is said to be generalized stochastic matrix, shortly g-stochastic, or more precisely, $(\mathscr{K}, \hat{x}')$ – stochastic matrix, if there exists a pair $\mathscr{K}, \hat{x}'$, where $\mathscr{K}$ is a closed normal generating cone in $\mathscr{R}^N$ and $\hat{x}'$ a strictly positive linear form, i.e. $\hat{x}' \in \mathscr{K}'$, where $\mathscr{K}'$ is the dual cone with respect to the cone $\mathscr{K}$ (for definitions of $\mathscr{K}$ and $\hat{x}'$ see* [11]), *such that $T\mathscr{K} \subset \mathscr{K}$ and*

$$[Tx, \hat{x}'] = [x, \hat{x}'] \quad \forall x \in \mathscr{R}^N. \tag{4.1}$$

Remark 1. Note that the standard concept of a stochastic matrix is obtained by setting

$$\mathscr{K} = \mathscr{R}^N_+, \quad \hat{x}' = e(N).$$

Proposition 1. *Let T be a g-stochastic matrix.*

Then

$$r(T) = 1 \tag{4.2}$$

and

$$\text{ind}(I - T) = 1, \tag{4.3}$$

where ind(C) *denotes the maximal size of Jordan blocks corresponding to eigenvalue* 0 *of matrix C. We write* ind$(C) = 0$, *if* 0 *is not an eigenvalue of C.*

Proof: The validity of (4.2) and (4.3) follows from [15].

Example 2. Let $\pounds(\omega)$ be the SOR iteration matrix, $\omega \notin [0, 2)$. Then matrix $T(\omega)$ from Example 1 is g-stochastic.

Proof: We check easily that with $\mathscr{K} = \{x = \alpha x(\omega) + u \in \mathscr{R}^N : \|u\|_\varepsilon \leq \alpha\}$, where $\|.\|_\varepsilon$ is any norm on $\mathscr{R}^N$ for which $\|Z(\omega)\|_\varepsilon \leq r(Z) + \varepsilon < 1$ with ε sufficiently small and $\hat{x}' = T(\omega)^T\hat{x}'$, $[x(\omega), \hat{x}'] = 1$ relations (4.1) hold.

We can check also that $\mathscr{K}' = \{x' = \alpha\hat{x}' + u' \in \mathscr{R}^N : \|u'\|'_\varepsilon \leq \alpha\}$, where $\|.\|'_\varepsilon$ denotes the dual norm with respect to the norm $\|.\|_\varepsilon$.

Problem 1. To find vector $\hat{x}$ such that

$$Tx = x, \quad [x, e] = 1, \tag{4.4}$$

where T is a g-stochastic matrix with $\mathscr{K}$ and $\hat{x}'$ and $e \in \mathscr{R}^N$ being given.

Theorem 1. *Problem* 1 *possesses a solution* $\hat{x}$ *if and only if*

$$[\hat{x}, e] \neq 0, \tag{4.5}$$

where $\hat{x} = T\hat{x}$.

Moreover, if T *is* $\mathscr{K}$*-irreducible (for definition see* [11]), *then there is at most one solution to Problem* 1.

Proof: It is obvious that if there are no eigenvectors of T corresponding to value 1 satisfying (4.5), then there are no solutions to Problem 1.

On the other hand, since, by Proposition 1 spectrum of T always contains value 1, a solution to problem 1 exists as soon as (4.5) holds as stated.

Let T be $\mathscr{K}$-irreducible (for definition see [11]). It is known that then the generalized eigenspace of T corresponding to the spectral radius is one-dimensional (see e.g. [11]) and thus Problem 1 may possess at most one solution.

5. Aggregation/Disaggregation Iteration Methods

Let $\mathscr{G}$ be a map of $\{1, \ldots, N\}$ onto $\{\bar{1}, \ldots, \bar{p}\}$. We define communication operators R mapping $\mathscr{R}^N$ into $\mathscr{R}^p$ and $S(x)$ mapping $\mathscr{R}^p$ into $\mathscr{R}^N$ respectively by setting

$$(Ru)_{\bar{j}} = \sum_{\mathscr{G}(j)=\bar{j}} u_j, \quad u \in \mathscr{R}^N, \; u^T = (u_{\bar{1}}^T, \ldots, u_{\bar{p}}^T), \; u_{\bar{j}} \in \mathscr{R}^{n_j} \tag{5.1}$$

and

$$(S(x)z)_j = \frac{x_j}{(Rx)_{\bar{j}}} z_{\bar{j}}, \quad z \in \mathscr{R}^p, \; z_{\bar{j}} \in \mathscr{R}^1, \; \mathscr{G}(j) = \bar{j}, \; j = 1, \ldots, N, \tag{5.2}$$

for $x \in \mathscr{D}$, where

$$\mathscr{D} = \{x \in \mathscr{R}^N: x^T = (x_1, \ldots, x_N),\ x_j > 0,\ j = 1, \ldots, N\}.$$

We check immediately that

$$RS(x)z = z, \quad \forall x \in \mathscr{D}, \quad z \in \mathscr{R}^p.$$

Therefore,

$$P(x) = S(x)R$$

is a projection

$$[P(x)]^2 = P(x).$$

Moreover,

$$P(x)x = x, \quad \forall x \in \mathscr{D}$$

and

$$P(x)^T e = e, \quad \forall x \in \mathscr{D}.$$

We define matrix $\mathscr{B}(x) = RBS(x)$, $x \in \mathscr{D}$, and call it *aggregated matrix* (with respect to B).

In the following considerations we assume that g is such that each block in (3.2) is reduced just to a 1×1 matrix i.e.

$$(Ru)_{\bar{j}} = \sum_{g(j)=\bar{j}} u_j = \sum_{l=n_1+\cdots+n_{l-1}+1}^{n_1+\cdots+n_l} (Hu)_l, \quad \bar{j} = 1, \ldots, n.$$

To guarantee that the proposed two-level algorithms can be unlimitedly realized we need the following two statements proven in [12].

A natural choice of the communication operators is available if matrix B is given in its block form

$$B = \begin{pmatrix} B_{11} & B_{12} & \ldots & B_{1p-1} & B_{1p} \\ B_{21} & B_{22} & \ldots & B_{2p-1} & B_{2p} \\ & & \ddots & & \\ B_{p-11} & B_{p-12} & \ldots & B_{p-1p-1} & B_{p-1p} \\ B_{p1} & B_{p2} & \ldots & B_{pp-1} & B_{pp} \end{pmatrix}. \tag{5.3}$$

The following and analogous reduction maps with $p = \bar{p}$ are utilized in the whole of this paper with an obvious exploiting the p-cyclic structure given in (3.2).

Let $(x^T = (x_{\bar{1}}, \ldots, x_{\bar{p}})$ be the block representation of x compatible with (5.3). Then

$$(Rx)_{\bar{j}} = \sum_{\mathcal{G}(j)=\bar{j}} (x_{\bar{j}})_j.$$

In this way, each block B_{jk} in B is represented by a 1×1 matrix in the aggregated matrix $\mathscr{B}(x)$ for any admissible x.

Proposition 2. [12] *Let matrix B be stochastic. Then its aggregated matrix $\mathscr{B}(x), x \in \mathscr{D}$, is stochastic too.*

Proposition 3. [12] *Let stochastic matrix B be irreducible. Then its aggregated matrix $\mathscr{B}(x), x \in \mathscr{D}$, is irreducible too.*

6. SPV Algorithms

A central object of our study is the following

Algorithm 1. **SPV**$(B; M, W; t, s; x^{(0)})$. Let B be an $N \times N$ irreducible stochastic matrix, let M and W be a splitting of matrix A and $T = M^{-1}W$, $s, t \geq 1$ positive integers.

Let $\epsilon > 0$ be a given tolerance and let $y^{(0)}$ with $(y^{(0)})_j > 0$, $[y^{(0)}, e(N)] = 1$, $j = 1, \ldots, N$, be an otherwise arbitrary vector.

Step 1. Set $0 \to k$.

Step 2. Construct the matrix

$$\mathscr{B}(x^{(k)}) = RB^s S(x^{(k)}).$$

Step 3. Find the unique solution vector $\tilde{z}^{(k)}$ to the problem

$$B(x^{(k)})\tilde{z}^{(k)} = \tilde{z}^{(k)}, \tag{6.1}$$

$$[\tilde{z}^{(k)},\ e(n)]_n = 1. \tag{6.2}$$

Step 4. Disaggregate by setting

$$v^{(k+1)} = S(x^{(k)})\tilde{z}^{(k)}.$$

Step 5. Let

$$x^{(k+1,m)} = Tx^{(k+1,m-1)},\ x^{(k+1,0)} = v^{(k+1)}, \quad m = 1, \ldots, t,$$

$$x^{(k+1)} = x^{(k+1,t)},\ [x^{(k+1)}, e(N)]_{N}, = 1.$$

Step 6. Test whether

$$\|x^{(k+1)} - x^{(k)}\| < \epsilon.$$

Step 7. If NO in Step 6, then let

$$k + 1 \rightarrow k$$

and GO TO Step 2.

Step 8. If YES in Step 6, then set

$$\hat{x} := x^{(k+1)}$$

and STOP.

In fact, the following algorithm is realized during the computations if an SOR splitting is utilized for problems with p^*-cyclic Markov chains.

Algorithm 2. $\mathbf{SPV}(\pounds(\omega); SOR, \omega, \rho_k(\omega); t, s; y^{(0)})$.

Let B be an $N \times N$ irreducible p^*-cyclic column stochastic matrix, let $M = I - \omega L, W = (1-\omega)I + \omega U$ $\omega > p/(p-1)$ be an SOR splitting of $A = I - B$ and let $T(\omega) = \frac{1}{\rho(\omega)}\pounds(\omega), \rho(\omega) = r(\pounds(\rho(\omega))$ and $Y = [T(\omega)]^s$, with t and s being positive integers.

Furthermore, let

$$\rho_k = \frac{[\pounds(\omega)\rho_{k-1}y^{(k)}, e]_N}{[\rho_{k-1}y^{(k)}, e]_N}, \quad \rho_0 = 1.$$

Step 1. Set $0 \rightarrow k$.

Step 2. Construct the matrix

$$\mathscr{T}(\omega)(y^{(k)}) = R[T(\omega)]^s S(y^{(k)}).$$

Step 3. Find the stationary probability vector $\tilde{z}^{(k)}$ of $\mathscr{T}(\omega)(y^{(k)})$, i.e. the unique solution to the problem

$$\mathscr{T}(\omega)(y^{(k)})\tilde{z}^{(k)} = \tilde{z}^{(k)}, \tag{6.3}$$

$$[\tilde{z}^{(k)},\ e(n)]_n = 1. \tag{6.4}$$

Step 4. Disaggregate by setting

$$v^{(k+1)} = S(y^{(k)})\tilde{z}^{(k)}.$$

Step 5. Let

$$My^{(k+1,m)} = Wy^{(k+1,m-1)}, \quad y^{(k+1,0)} = v^{(k+1)}, \quad m = 1, \ldots, t,$$

$$y^{(k+1)} = (1/\rho_k)^t y^{(k+1,t)}, \quad [y^{(k+1)}, e(N)]_N = 1.$$

Step 6. Test whether

$$\|y^{(k+1)} - y^{(k)}\| < \epsilon.^{1}$$

Step 7. If NO in Step 6, then let

$$k + 1 \to k$$

and GO TO Step 2.

Step 8. If YES in Step 6, then set

$$x(\omega) := y^{(k+1)}$$

and STOP.

7. Convergence Analysis

7.1. Error-Vector Formula

Proposition 4. *The error-vector formula for the sequence of approximants $\{x^{(k)}\}$ returned by the* **SPV**$(B; M, W; t, s; x^{(0)})$*-algorithm reads*

$$x^{(k+1)} - \hat{x} = J_t(x^{(k)})(x^{(k)} - \hat{x}), \tag{7.1}$$

where

$$J_t(x^{(k)}) = T^t[I - P(x^{(k)})Z]^{-1}(I - P(x^{(k)})) \tag{7.2}$$

with Z coming from the spectral decomposition of matrix B:

$$B = P + Z, \quad P^2 = P, \quad PZ = ZP = 0, \ \ 1 \notin \sigma(Z),$$

where $\sigma(Z)$ denotes the spectrum of Z.

Consequently, $J_t(x) = T^{t-1}J_1(x),\ t \geq 1$, holds for any x with all components positive.

Proof: By definition of the **SPV**$(B; M, W; t, s; x^{(0)})$-algorithm,

$$Mx^{(k+1,1)} = WS(x^{(k)})z^{(k)}, \tag{7.3}$$

where

$$z^{(k)} = (I_{\mathscr{F}} - RZS(x^{(k)}))^{-1}R\hat{x}, \quad \hat{x} = (I - Z)\hat{x},$$

[1]Here the symbol $\|.\|$ denotes any norm on $\mathbb{R}^N$. We recommend the l_1-norm.

$I_{\mathscr{F}}$ being the identity map on $\mathscr{F} = \mathscr{R}^n$. It follows that

$$\begin{aligned} Mx^{(k+1,1)} &= W\left\{P(x^{(k)})\hat{x} + \sum_{t=0}^{\infty}(P(x^{(k)})Z)^t P(x^{(k)})\hat{x}\right\} \\ &= W(I - P(x^{(k)})Z)^{-1}P(x^{(k)})(I - Z)\hat{x} \\ &= W(I - P(x^{(k)})Z)^{-1}(P(x^{(k)}) - I + I - P(x^{(k)})Z)\hat{x} \\ &= W\hat{x} - W(I - P(x^{(k)})Z)^{-1}(I - P(x^{(k)}))\hat{x} \end{aligned}$$

and, since $M\hat{x} = W\hat{x}$,

$$\begin{aligned} Mx^{(k+1,1)} - W\hat{x} &= M(x^{(k+1,1)} - \hat{x}) \\ &= W(I - P(x^{(k)})Z)^{-1}(I - P(x^{(k)}))(x^{(k)} - \hat{x}). \end{aligned}$$

Finally,

$$x^{(k+1,1)} - \hat{x} = T([I - P(x^{(k)})]Z)^{-1}(I - P(x^{(k)}))(x^{(k)} - \hat{x}).$$

This is just formula (7.2) for $t = 1$. To obtain (7.2) for arbitrary $t \geq 1$ one needs to apply T^{t-1} to $J_1(x^{(k)})$. It is obvious that algorithm $\mathbf{SPV}(B; M, W\colon t, s; x^0)$ achieves this purpose by applying the iteration procedure determined by the splitting $\{M, W\}$.

Proposition 5. *The spectra of $J_t(\hat{x})$ and $(I - P(\hat{x}))J_t(\hat{x})$ are related as follows*

$$\sigma(J_t(\hat{x})) \subset \sigma((I - P(\hat{x}))J_t(\hat{x})) \cup \{0\}.$$

Consequently, $r(J_t(\hat{x})) = r((I - P(\hat{x}))J_t(\hat{x}))$.

Proof: Let $0 \neq \lambda \in \sigma(J_t(\hat{x}))$ and w be a corresponding eigenvector $J_t(\hat{x})w = \lambda w$, $w \neq 0$. According to the definition of $J_t(x)$ we see that $(I - P(\hat{x}))J_t(\hat{x})(I - P(\hat{x}))w = \lambda(I - P(\hat{x}))w$.

7.2. Convergence Results

Definition 2. *Let $M^{-1}W = T = \sum_{j=1}^{p} \lambda_j Q_j + Z$, $Q_j Q_k = Q_k Q_j = \delta_{jk} Q_j$, $Q_j Z = ZQ_j = 0, r(Z)\, r(T) = \|\lambda_j\|, j, k = 1, \ldots, p$, where δ_{jk} denotes the Kronecker "delta". A splitting $\{M, W\}$ of $A = I - B = M - W$ is called p-*shape convergent, *if the limits*

$$\lim_{k\to\infty}\left(\left[\frac{1}{\lambda_j}T\right]^k w\right)_{\bar{j}} = w_{\bar{j}}^{\infty}, \quad \bar{j} = 1, \ldots, p,$$

exist whenever $\sum_{j=1}^{p} Q_j w \neq 0$, where the blocks in $w^T = (w_{\bar{1}}^T, \ldots, w_{\bar{p}}^T)$ are consistent with the form of B in (3.2).

Remark 2. It is obvious that every convergent splitting is automatically p-shape convergent.

Definition 3. *A splitting* $\{M, W\}$ *of* $I - B = M - W$ *is called* aggregation-convergent *if* $T = M^{-1}W$ *is* Y *– convergent, where* $Y = I - P(\hat{x})$ *and* $P(\hat{x})$ *is the aggregation projection.*

Remark 3. Note that the splitting given by $M = I$ and $W = B$ is always aggregation-convergent. This is, according to a result of Courtois and Semal [3], because the subvectors $x_{\bar{j}}, \bar{j} = 1, \ldots, p$, where $x = (x_{\bar{1}}, \ldots, x_{\bar{p}})^T$ is an eigenvector of B corresponding to an eigenvalue $\lambda \in \sigma(B), |\lambda| = 1$, are parallel. This fact implies that $[I - P(\hat{x})]x = 0$ and followingly, $r([I - P(\hat{x})]B) < 1$.

Propositions 4 and 5 with respect to Definition 3 result in

Theorem 2. *Let* $\{M, W\}$ *be an aggregation-convergent splitting of* $A = I - B$*, where* B *is an irreducible stochastic matrix.*

Then there exists a positive integer $\hat{t}_1 \geq 1$*, a neighborhood* $\Omega(\hat{x})$*, positive real numbers* κ_1 *and* $q_1 < 1$ *independent of* k such that

$$\|x^{(k)} - \hat{x}\| \leq \kappa_1 q_1^k \|x^{(0)} - \hat{x}\|, \quad k = 0, 1, \ldots \tag{7.4}$$

where $x^{(0)} \in \Omega(\hat{x})$ *and* $\{x^{(k)}\}$ *is the sequence returned by Algorithm* 1 and $\|.\|$ *denotes any norm on* $\mathscr{R}^N$*. Thus,*

$$\lim_{k \to \infty} \|x^{(k)} - \hat{x}\| = 0. \tag{7.5}$$

Analogously, we have

Theorem 3. *Let* $\{M, W\}$ *be an SOR splitting of* $A = I - B$*, where* B *is an irreducible stochastic matrix. Let the corresponding iteration matrix* $T(\omega)$ *be aggregation-convergent.*

Then there exists a positive integer $\hat{t}_2 \geq 1$*, a neighborhood* $\Omega(\hat{x})$*, positive real numbers* κ_2 *and* $q_1 < 1$ *independent of* k *such that*

$$\|y^{(k)} - \hat{x}\| \leq \kappa_2 q_2^k \|y^{(0)} - \hat{x}\|, \quad k = 0, 1, \ldots \tag{7.6}$$

where $y^{(0)} \in \Omega(\hat{x})$ *and* $\{y^{(k)}\}$ *is the sequence returned by Algorithm* 2 *and* $\|.\|$ *denotes any norm on* $\mathscr{R}^N$*. Thus,*

$$\lim_{k \to \infty} \|y^{(k)} - \hat{x}\| = 0. \tag{7.7}$$

Remark 4. We are able to guarantee the aggregation convergence of $T(\omega)$ either for $0 < \omega < 2$ or under an additional hypothesis that the spectrum $\sigma(B^p)$ is real for $2 \leq \omega < +\infty$. A possibility that $T(\omega)$ is aggregation convergent without the additional hypothesis is not excluded, however, our experiments show that it is unlikely.

It is easy to see that Algorithm 2 with parameter $\rho(\omega)$ in place of ρ_k requires knowledge of the spectral radius of the relaxation operator and that this quantity is not available, in particular if the optimal relaxation parameter ω is to be utilized. On the other hand, the convergence proof is expected to be less demanding than that for Algorithm 2 with parameter $\rho(\omega)$ in place of ρ_k because the corresponding relaxation operator is independent of the index of iteration. However, we actually have

Proposition 6. *There is a relation between the sequences $\{y^{(k)}(\rho(\omega))\}$ and $\{y^{(k)}(\rho_k)\}$ returned by the Algorithm* 1 *with parameter $\rho(\omega)$ and Algorithm* 2 *with parameter ρ_k respectively and it reads*

$$y^{(k)}(\rho(\omega)) = \left(\frac{\rho(\omega)}{\rho_k}\right)^t y^{(k)}(\rho_k), \quad k = 1, 2, \dots \tag{7.8}$$

where

$$\rho_k = \frac{[\pounds(\omega)y^{(k)}(\rho(\omega)), e]_N^t}{[y^{(k)}(\rho_k), e]_N^t}.$$

Proof: Let $k = 1$. By definition we have

$$x^{(1)}(\rho(\omega)) = \left(\frac{1}{\rho(\omega)}\pounds(\omega)\right)^t S(y^{(0)})z^{(1)}$$

and

$$y^{(k)}(\rho_k) = \left(\frac{1}{\rho_k}\pounds(\omega)\right)^t S(y^{(0)})z^{(1)},$$

showing validity of (7.8) for $k = 1$.

Let (7.8) hold for $l \leq k$ and consider $y^{(k+1)}(\rho(\omega))$ and $y^{(k+1)}(\rho_k)$.

We see that

$$y^{(k+1)}(\rho(\omega)) = \left(\frac{1}{\rho(\omega)}\pounds(\omega)\right)^t S(y^{(k)}(\rho(\omega)))z^{(k+1)}$$

and

$$y^{(k+1)}(\rho_{k+1}) = \left(\frac{1}{\rho_k}\pounds(\omega)\right)^t S(y^{(k)}(\rho_k))\tilde{z}^{(k+1)}.$$

Since $\mathscr{B}(y^{(k)})(\rho(\omega)) = \mathscr{B}(y^{(k)})(\rho_k)$, it follows that

$$\tilde{z}^{(k+1)} = z^{(k+1)}$$

and, by homogeneity of $S(x)$ with respect to x,

$$S(y^{(k)}(\rho(\omega)))z^{(k+1)} = S(y^{(k)}(\rho_k))z^{(k+1)}$$

implying

$$y^{(k+1)}(\rho(\omega)) = \left(\frac{\rho(\omega)}{\rho_{k+1}}\right)^t y^{(k+1)}(\rho_{k+1}). \tag{7.9}$$

The proof is complete.

Corollary 1. *The sequences returned by the Algorithm 2 with $\rho(\omega)$ in place of ρ_k and Algorithm 2 with ρ_k are both simultaneously either convergent or divergent. In case of convergence,*

$$\lim_{k\to\infty} y^{(k)}(\rho(\omega)) = \lim_{k\to\infty} y^{(k)}(\rho_k) = x(\omega),$$

where

$$£(\omega)x(\omega) = \rho(\omega)x(\omega), \quad [x(\omega), e(N)]_N = 1.$$

Proof: The validity of the statement is a consequence of the fact that

$$\lim_{k\to\infty} \rho_k = \rho(\omega).$$

This completes the proof.

According to Proposition 6 it remains to prove convergence of the sequence returned by Algorithm 2 with $\rho(\omega)$ in place of ρ_k, where $T(\omega)$ is an aggregation-convergent matrix.

To this purpose, let us consider in place of B matrix $£(\omega)$ introduced in (2.2)

$$£(\omega) = \rho(\omega)Q(\omega) + Z(\omega).$$

According to [8]

$$\left(\frac{1}{\rho(\omega)}£(\omega)\right)^k = Q(\omega) + \left(\frac{1}{\rho(\omega)}Z(\omega)\right)^k \to Q(\omega)$$

and since

$$Q(\omega)x = [x, \tilde{e}]\tilde{x}$$

with $\tilde{x}$ and $\tilde{e}$ the unique appropriately normalized eigenvector of $£(\omega)$ and of the transpose to $£(\omega)$ corresponding to $r(£(\omega))$ respectively and using a result of [3] we can conclude that $\tilde{x} = (\tilde{x}_{\bar{j}}, \ldots, \tilde{x}_{\bar{p}})^T$ and $\hat{x} = (\hat{x}_{\bar{1}}, \ldots, \hat{x}_{\bar{p}})^T$ possess their "maxi-

components" parallel: There are nonzero real coefficients $\alpha_{\bar{1}}, \ldots, \alpha_{\bar{p}}$, such that $\tilde{x}_{\bar{j}} = \alpha_{\bar{j}} \hat{x}_{\bar{j}}, \bar{j} = 1, \ldots, p$. With these structural properties of the IAD process a convergence proof of Algorithm 2 can be given in the same manner as for Algorithm 1.

As a consequence of the previous results we can formulate our main result.

Theorem 4. *Let $\{M, W\}$ be a p-shape convergent splitting of $A = I - B$, where B is an irreducible p^*-cyclic stochastic matrix. Let $\lambda_1 = r(T), Q_1 v^{(0)} \neq 0$ and*

$$Mv^{(k+1)} = Wv^{(k)}, \quad k = 0, 1, \ldots$$

Then there exists an index $\hat{k} \geq 1$, positive real constants κ and $q < 1$, independent of k such that Algorithm 1 returns after one IAD sweep vector $w^{(1,k)} = w^{(1,k)}(Alg1)$ with the property that

$$\|w^{(1,k)} - \hat{x}\| \leq \kappa q^k \|w^{(0,k)} - \hat{x}\|, \quad k \geq \hat{k}, \tag{7.10}$$

where $w^{(0,k)} = v^{(k)}$.

An analogous result can be formulated for the case of Algorithm 2 as well.

Remark 5. Let in addition to hypotheses of Theorem 4 at least one of the off-diagonal blocks be formed by a tensor product of two vectors. Then the process described in this theorem terminates after at most two iterations (7.10) returning the exact solution $\hat{x}$ after one IAD sweep. This fact is a consequence of a result of [14].

8. An Application to IAD Methods Based on SOR Splittings. Computationl Experiments

The results of the previous sections in conjunction with the fact that any SOR iteration matrix is similar to a p^*-cyclic matrix ([20], [17, p. 380]) guarantee that both algorithms 1 and 2 are locally convergent for IAD methods based on *SOR* splittings with relaxation parameters ranging over the open real positive half-axis.

It might sound surprising that our results offer quite different optimality results in contrast to [19], [4] and [9] and also [5, 6]. Our computations show that the optimal relaxation parameter $\omega_{opt} = 1$. The explanation of this claim is based on the already mentioned result of Courtois and Semal [3]: The IAD based on the Gauss-Seidel splitting forms the subvectors of the computed stationary probability vector very well. The standard optimal methods converge faster with respect to a global norm but not with respect to the norms in the subspaces of the subvectors. In fact, a consistent application of the above properties to Algorithm 1 allows us, at least in principle, to realize the coarse grid correction just once. The claim concerning optimality of the IAD based on Gauss-Seidel splittings remains valid with respect to the standard as well as "extended" range of relaxation parameters.

To explain our position let us note that the optimality results known in the literature (e.g. cited above) are derived under the hypotheses that the spectrum $\sigma(B^p)$ is real and B is consistently ordered. Our experience is that these hypotheses are too restrictive. On the other hand, our convergence theory is free of such hypotheses.

From our huge collection of experiments we give a typical example showing some properties of various methods we were focusing to. Our main aim is the speed of each method. In order to compare the speed we present the number of iterations or better saying the number of sweeps of each of the method considered. This criterion is fair though it might not offer the computer time spent. Nevertheless, the winner – the IAD consisting of an appropriate number of Gauss-Seidel steps terminating by one coarse grid correction – is simultaneously the smallest computer time consumer.

We compare, on the one hand, the classical methods such as the power method applied to the original stochastic (convergent) matrix B, Jacobi method, Gauss-Seidel method and the SOR method with positive relaxation parameters ω (Table 1) and, on the other hand, the methods of Marek-Mayer [12], Vantilborgh [18], Koury-McAllister-Stewart [10], SOR method followed by just one coarse grid correction according to Algorithm 1 (Table 2) and Algorithm 2 (Table 3). The Algorithms 1 and 2 are applied with $t = s = 1$ and $x^{(0)} = y^{(0)} = (1, \dots, 1)^T$.

The matrices we are experimenting with are constructed blockwise. Each block is chosen randomly. The columns obtained are then normalized such that the whole matrix becomes column stochastic and p^*-cyclic. The matrices considered are thus general p^*-stochastic matrices possessing complex spectra as well as their powers, in particular the p-th power. Furthermore, we do not require the matrices to be consistently ordered.

The results confirm an already accepted fact that the IAD methods are efficient. One can even say that the gain of the IAD methods in comparison with other methods is the larger the worse conditioned are the matrices examined.

Table 1

Method	Number of iterations	Resulting error
power method	485 000	9.8e − 15
Jacobi	diverges	–
Gauss-Seidel	diverges	–

SOR method

ω	Number of iterations	Resulting error	ω	Number of iterations	Resulting error
0.1	691	9.9e − 15	1.4	380	8.e − 15
0.2	374	9.6e − 15	1.5	360	6.e − 15
0.3	271	9.5e − 15	1.6	340	4.9e − 15
0.4	225	9.8e − 15	1.7	320	7.2e − 15
0.5	206	9.9e − 15	1.8	320	5.6e − 15
0.55	204	8.1e − 15	1.9	305	5.6e − 15
0.6	206	8.0e − 15	2.0	310	3.1e − 15
0.7	228	8.9e − 15	3.0	290	4.6e − 15
0.8	292	9.1e − 15	4.0	280	4.8e − 15
0.9	517	8.8e − 15	5.0	280	7.0e − 15
1.1	1100	5.1e − 15	10.0	266	7.6e − 15
1.2	520	8.2e − 15	20.0	263	6.1e − 15
1.3	420	9.5e − 15	100.0	278	6.1e − 15

Table 2

Method	Number of iterations	Resulting error
Marek-Mayer	12	6.2e − 15
Vantilborgh	3	3.5e − 17
Koury-McAllister-Stewart	3	2.1e − 17

SOR method

ω	Number of iterations	Resulting error	ω	Number of iterations	Resulting error
0.1	281	9.6e − 15	1.7	27	3.4e − 15
0.2	133	8.9e − 15	1.8	28	3.3e − 15
0.3	83	9.6e − 15	1.9	28	9.5e − 15
0.4	58	9.4e − 15	2.0	29	6.6e − 15
0.5	43	7.8e − 15	2.1	30	4.3e − 15
0.6	33	5.1e − 15	2.2	30	8.1e − 15
0.7	25	5.8e − 15	2.3	31	4.5e − 15
0.8	19	3.6e − 15	2.4	31	7.4e − 15
0.9	13	6.9e − 15	2.5	32	3.7e − 15
1.0	3	2.1e − 17	3.0	33	5.8e − 15
1.1	13	6.8e − 15	4.0	34	8.9e − 15
1.2	19	1.4e − 15	5.0	35	5.9e − 15
1.3	21	2.6e − 15	10.0	35	7.2e − 15
1.4	23	2.8e − 15	20.0	34	6.8e − 15
1.5	24	7.2e − 15	100.0	31	3.3e − 15
1.6	26	2.8e − 15	200.0	29	5.6e − 15

Table 3

ω	Number of iterations	Resulting error	ω	Number of iterations	Resulting error
0.1	282	9.7e − 16	1.1	15	7.7e − 16
0.2	137	9.5e − 16	1.2	21	8.7e − 16
0.3	88	7.0e − 16	1.3	28	7.2e − 15
0.4	62	8.3e − 16	1.4	37	4.7e − 16
0.5	48	9.9e − 16	1.5	48	7.4e − 16
0.6	36	5.1e − 16	1.6	65	7.2e − 16
0.7	28	4.0e − 16	1.7	92	9.4e − 15
0.8	21	6.1e − 16	1.8	147	8.9e − 16
0.9	15	6.6e − 16	1.9	310	9.9e − 16
1.0	4	4.5e − 17	2.0	diverges	–

We present a numerical evidence of the claim that within the collection of methods considered the Gauss-Seidel method followed by just one IAD sweep is optimal with respect to the computer time spent for computing the stationary probability vectors of p-cyclic stochastic matrices.

The tables contain results of computations with a fixed p^*-cyclic matrix of order $N = 350$ with $p = 7$ block rows. Moreover, the relative size of the elements in the off-diagonal blocks is 0.0001. This explains the slow convergence rate of the power method.

Pay attention to the rows corresponding to value $\omega = 1$ in Tables 2 and 3, distinct minima!

References

[1] Ajmone Marsan, M., Balbo, G., Conte, G., Donatelli, S., Franceschinis, G.: Modelling with generalized stochastic Petri nets. New York: J. Wiley 1995.

[2] Ciardo, G., Blakmore, A., Chimento, P. F., Muppala, J. K., Trivedi, K. S.: Automated generation and analysis of Markov reward models using stochastic reward nets. In: Linear algebra, Markov chain and queueing models (Meyer, C. D., Plemmons, R. J., eds.), pp. 145–191. Berlin Heidelberg New York Tokyo: Springer 1993.

[3] Courtois, P. J., Semal, P.: Block iterative algorithms for stochastic matrices. Linear Algebra Appl. *76*, 59–80 (1986).

[4] Eiermann, M., Niethammer, W., Ruttan, A.: Optimal successive overrelaxation iterative methods for p-cyclic matrices. Numer. Math. *57*, 593–606 (1990).

[5] Galanis, S., Hadjidimos, A., Noutsos, D.: Optimal p-cyclic SOR for complex spectra. Linear Algebra Appl. *263*, 233–260 (1997).

[6] Galanis, S., Hadjidimos, A., Noutsos, D.: A Young-Eidson's type algorithm for complex p-cyclic SOR spectra. Linear Algebra Appl. *286*, 87–106 (1999).

[7] Klapka, Š., Mayer, P.: Reliability modeling of safety equipments. Proceedings of the Workshop Codes and Algorithms of Numerical Mathematics, Libverda June 2000 (Přikryl, P., ed.). (in Czech). Submitted 2000.

[8] Klapka, Š., Mayer, P.: Some aspects of modelling railway safety. Proc. SANM'99, University of West Bohemia, Nečtiny September 1999, 135–140 (1999).

[9] Kontovasilis, K., Plemmons, R. J., Stewart, W. J.: Block cyclic SOR for Markov chains with p-cyclic infinitesimal generator. Linear Algebra Appl. *154–156*, 145–223 (1991).

[10] Koury, R., McAllister, D. F., Stewart, W. J.: On the Raleigh-Ritz refinement technique for nearly uncoupled stochastic matrices. Linear Algebra Appl. *60*, 1–25 (1984).

[11] Marek, I.: Frobenius theory of positive operators. Comparison theorems and applications. SIAM J. Appl. Math. *19*, 607–628 (1970).

[12] Marek, I., Mayer, P.: Convergence analysis of an aggregation/disaggregation iterative method for computation stationary probability vectors of stochastic matrices. Numer. Linear Algebra Appl. *5*, 253–274 (1998).

[13] Marek, I., Mayer, P.: Convergence theory of a class of iterative aggregation/disaggregation methods for computing stationary probability vectors of stochastic matrices (submitted).

[14] Marek, I., Mayer, P.: Iterative aggregation/disaggregation methods for computing stationary probability vectors of stochastic matrices can be finitely terminating. J. Appl. Math. (submitted).

[15] Marek, I., Szyld, D.: Splittings of M-operators, irreducibility and the index of the iteration operators. Numer. Functional Anal. Optim. *11*, 529–553 (1990).

[16] Plateau, B., Atif, K.: Stochastic automata network for modelling parallel systems. IEEE Trans. Software Engrg *17*, 1093–1108 (1991).

[17] Stewart, W.: Introduction to numerical solution of Markov chains. Princeton: Princeton University Press 1994.

[18] Vantilborgh, H.: The error aggregation. A contribution to the theory of decomposable systems and applications. PhD Thesis, Faculty of Applied Sciences, Louvain Catholic University, Louvain-la Neuve, Belgium, 1981.

[19] Varga, R. S.: p-cyclic matrices: a generalization of the Young-Fraenkel successive over-relaxation scheme. Pacif. Math. J. *9*, 925–939 (1959).

[20] Wild, P., Niethammer, W.: Over- and underrelaxation for linear systems with weakly cyclic Jacobi matrices of index p. Linear Algebra Appl. *91*, 29–52 (1987).

[21] Young, D. M.: Iterative methods for solving difference equations of elliptic type. Trans. Amer. Math. Soc. *76*, 91–111 (1954).

[22] Young, D. M.: Iterative solution of large linear systems. Computer science and applied mathematics. New York, London: Academic Press 1971.

I. Marek
Chair of Mathematics
Faculty of Civil Engineering
Czech Technical University
Thákurova 7, 166 29 Praha 6
Czech Republic
e-mail: marek@ms.mff.cuni.cz

P. Mayer
Chair of Computational Mathematics
Charles University
Sokolovská 83
186 00 Praha 8
Czech Republic
e-mail: pmayer@ms.mff.cuni.cz

Computing [Suppl] 15, 151–163 (2001)

Computing

A New Way to Describe the Symmetric Solution Set S_{sym} of Linear Interval Systems

G. Mayer, Rostock

Dedicated to Professor Tetsuro Yamamoto on the occasion of his 65th birthday

Abstract

Using an elimination process based on intersections of intervals we indicate a new way for describing the set S_{sym} of all solutions of linear systems $Ax = b$ with a symmetric matrix A varying in a given $n \times n$ interval matrix $[A]$ and with b varying in a given interval vector $[b]$ with n components. The elimination process has the advantage that it can be applied independently of any orthant. It yields directly to Beeck's description for the general solution set S when it is applied to linear systems with any coefficient matrix $A \in [A]$ and any right-hand side $b \in [b]$.

AMS Subject Classifications: 65G10, 15A06, 15A18.

Key Words: Linear interval systems, solution set of linear interval systems, symmetric solution set, symmetric linear systems of equations, algebraic eigenvalue problem, eigenpair set, symmetric eigenpair set, elimination process, Fourier–Motzkin elimination.

1. Introduction

Given an $n \times n$ interval matrix $[A] = ([a]_{ij}) = ([\underline{a}_{ij}, \overline{a}_{ij}])$ and an interval vector $[b] = ([b]_i) = ([\underline{b}_i, \overline{b}_i])$ with n components a *linear interval system* $[A]x = [b]$ is defined as the variety of all linear systems $Ax = b$ with a real matrix $A \in [A]$ and $b \in [b]$. Restricting $A \in [A]$ to be symmetric it is an interesting problem to characterize the so-called *symmetric solution set*

$$S_{sym} := \{x \in \mathbb{R}^n \mid Ax = b,\ A = A^T \in [A],\ b \in [b]\} \tag{1}$$

and to enclose it by interval vectors. To our knowledge this latter problem was considered for the first time in [12], [13], where, in addition, inner enclosures were given. Cf. also [7], [14] and [16]. Numerical experiments indicated that in contrast to the ordinary *solution set*

$$S := \{x \in \mathbb{R}^n \mid Ax = b,\ A \in [A],\ b \in [b]\} \tag{2}$$

the boundary of S_{sym} can be curvilinear. This experimental behaviour was strictly proved for the first time in [8] for 2×2 interval matrices and in the general $n \times n$ case in [2]. (Cf. also [3], [4], [6] for generalizations.) The proof was essentially based on a Fourier–Motzkin elimination process [17] which was applied to the statement

$$\exists\, A=(a_{ij})\in\mathbb{R}^{n\times n},\ b=(b_i)\in\mathbb{R}^n\ :\quad \left\{\begin{array}{c} b_i\le (Ax)_i\le b_i \\ \underline{b}_i\le b_i\le \overline{b}_i \\ \underline{a}_{ij}\le a_{ij}=a_{ji}\le \overline{a}_{ij}\end{array}\right\}\quad i,j=1,\ldots,n.$$

Here, x had to be restricted to any fixed closed orthant O of $\mathbb{R}^n$. It was shown that the boundary of $S_{\mathrm{sym}}\cap O$ is piecewise quadratic, i.e., it is the union of pieces of hyperplanes and quadrics which depend on the bounds $\underline{a}_{ij}$, $\overline{a}_{ij}$, $\underline{b}_i$, $\overline{b}_i$. The use of these bounds and the restriction of S_{sym} to a fixed orthant grows out of the elimination process. Its result resembles part d) of the subsequent Theorem 1 which characterizes the unsymmetric solution set S from (2) in three equivalent manners. These characterizations are due to Oettli and Prager [15], Beeck [9] and Hartfiel [11], respectively.

Theorem 1. *For* $[A]$ *and* $[b]$ *as above the following assertions are equivalent:*

a) $x\in S$;

b) $|\check{b}-\check{A}x|\le \mathrm{rad}[A]\cdot|x|+\mathrm{rad}[b]$; (*Oettli and Prager 1964*)

c) $[b]\cap[A]x\ne\emptyset$; (*Beeck 1972*)

d) $\underline{b}_i-\sum_{j=1}^{n}a_{ij}^{+}x_j\le 0\le \overline{b}_i-\sum_{j=1}^{n}a_{ij}^{-}x_j,\quad i=1,\ldots,n,$

where a_{ij}^{-}, a_{ij}^{+} *are defined by*

$$[a]_{ij}=\begin{cases}[a_{ij}^{-},a_{ij}^{+}] & \text{if } x_j\ge 0\\ [a_{ij}^{+},a_{ij}^{-}] & \text{if } x_j<0\end{cases}.\qquad (\textit{Hartfiel } 1980)$$

Here $\check{A}=(\check{a}_{ij})$, $\mathrm{rad}[A]=(\mathrm{rad}[a]_{ij})\in\mathbb{R}^{n\times n}$, $\check{b}=(\check{b}_i)$, $\mathrm{rad}[b]=(\mathrm{rad}[b]_i)\in\mathbb{R}^n$ are defined by means of the midpoints $\check{a}_{ij}=0.5\cdot(\underline{a}_{ij}+\overline{a}_{ij})$, $\check{b}_i=0.5\cdot(\underline{b}_i+\overline{b}_i)$, and the radii $\mathrm{rad}[a]_{ij}=0.5\cdot(\overline{a}_{ij}-\underline{a}_{ij})$, $\mathrm{rad}[b]_i=0.5\cdot(\overline{b}_i-\underline{b}_i)$. The multiplication in c) is done according to the rules of interval arithmetic – cf. [1] or [14], e.g. In order to fix the coefficients in the inequalities of d) it is necessary to restrict x to an orthant. Then it is obvious that $a_{ij}^{\pm}$ and $b_i^{\pm}$ are fixed bounds of $[a]_{ij}$, $[b]_i$, respectively. Part c) of Theorem 1 seems to be the most elegant formulation. It uses neither orthants nor bounds of the interval entries. Particularly these two aspects led us to the central topic of the present paper, namely to express S_{sym} as a system of nonempty intersections of intervals. In order to end up with these intersections we need a lemma (Lemma 1) of purely auxiliary character which contains results on intersections in combination with interval arithmetic. Based on this lemma we develop a method of parameter elimination in an interval setting similar to that of Fourier and Motzkin. In fact this latter method is hidden behind our new way via the parts d) and e) of Lemma 1 . We show how the new elimination process can be used to characterize S and S_{sym}. Applied to S it leads directly to Beeck's criterion '*a)* ⇔ *c)*' of Theorem 1. Used for S_{sym} it allows a description which is more condensed than that in [2] and [3]. We illustrate this by three examples in which we consider the general case of 2×2 matrices, 3×3 matrices and 3×3 tridiagonal matrices. Here we also show how the representations of S_{sym} derived in [2] and [8] can be retrieved from ours. We conclude our paper by some remarks on the applicability of the method to further particular classes of linear systems and to the algebraic eigenvalue problem with general and with symmetric matrices.

2. Notations

In addition to the standard notation $\mathbb{R}^n$, $\mathbb{R}^{n\times n}$ of the set of real vectors with n components and the set of real $n \times n$ matrices, respectively, which we already used in Section 1, we denote by $\mathrm{I}(\mathbb{R})$, $\mathrm{I}(\mathbb{R}^n)$, $\mathrm{I}(\mathbb{R}^{n\times n})$ the set of intervals, the set of interval vectors with n components and the set of $n \times n$ interval matrices, respectively. By 'interval' we always mean a real compact interval. Interval vectors and interval matrices are vectors and matrices, respectively, with interval entries. We write intervals in brackets with the exception of degenerate intervals (so-called *point intervals*) which we identify with the element being contained, and we proceed similarly with interval vectors and interval matrices. As usual, we denote the lower and upper bound of an interval $[a]$ by $\underline{a}$ and $\overline{a}$, respectively. Similarly, we use $\underline{A}$, $\overline{A}$ for the bounds of interval matrices $[A] \in \mathrm{I}(\mathbb{R}^{n\times n})$. We write $[A] = [\underline{A}, \overline{A}] = ([a]_{ij}) = ([\underline{a}_{ij}, \overline{a}_{ij}]) \in \mathrm{I}(\mathbb{R}^{n\times n})$ simultaneously, without further reference, and we use a similar notation for interval vectors. By $[A]^T$ we mean the transposed matrix of $[A]$. We mention that $[A] = [A]^T$ is equivalent to $\underline{A} = \underline{A}^T$ and $\overline{A} = \overline{A}^T$.

For computations with interval quantities we refer to [1] and [14].

We call a nonempty intersection of two sets a *simple intersection* and a nonempty intersection of more than two sets a *multiple intersection*.

3. Results

We start with a lemma which is basic for the elimination process to follow.

Lemma 1. *Let* $[a]$, $[b]$, $[c]$ $\in \mathrm{I}(\mathbb{R})$, $[a]_i \in \mathrm{I}(\mathbb{R})$ *for* $i = 1, \ldots, n$, $c \in \mathbb{R}$, $\circ \in \{+, -, \cdot, /\}$. *Then the following assertions hold.*

a) $[a] \cap [b] \neq \emptyset \Leftrightarrow \max\{\underline{a}, \underline{b}\} \leq \min\{\overline{a}, \overline{b}\} \Leftrightarrow 0 \in [a] - [b] \Leftrightarrow ([a] \circ c) \cap ([b] \circ c) \neq \emptyset$, *where* $c \neq 0$ *in the cases* $\circ \in \{\cdot, /\}$.

b) $[a] \circ [b] \neq \emptyset \;\Rightarrow\; ([a] \circ [c]) \cap ([b] \circ [c]) \neq \emptyset$. *The converse is generally not true.*

c) $([a] + [b]) \cap [c] \neq \emptyset \;\Leftrightarrow\; [a] \cap ([c] - [b]) \neq \emptyset$.

d) Let $n \geq 3$. *Then*

$$\bigcap_{i=1}^{n} [a]_i \neq \emptyset \Leftrightarrow \max_{1\leq i\leq n} \underline{a}_i \leq \min_{1\leq i\leq n} \overline{a}_i$$
$$\Leftrightarrow [a]_i \cap [a]_j \neq \emptyset \quad \text{with } i < j,\ i, j = 1, \ldots, n.$$

This means, in particular, that a multiple intersection with n operands can be resolved into $\binom{n}{2}$ *simple intersections.*

e) $\exists\, c \in \mathbb{R}:\ \{c\} \cap [a]_i \neq \emptyset,\ i = 1, \ldots, n \ \Leftrightarrow\ \bigcap_{i=1}^{n} [a]_i \neq \emptyset.$

f) $[a] \cap [b] = [a] \cap [b] \cap \{([a] + [b])/2\}.$

g) $0 \in [c] \Rightarrow [a] \cap [b] = [a] \cap [b] \cap ([b] + [c]).$

Proof: a) The first two equivalences are trivial. In addition, we get

$$\begin{aligned}
[a] \cap [b] \neq \emptyset &\Leftrightarrow \exists\, a \in [a],\ b \in [b]:\ a = b \\
&\Leftrightarrow \exists\, a \in [a],\ b \in [b]: \\
&\qquad a \circ c \in [a] \circ c\ \wedge\ b \circ c \in [b] \circ c\ \wedge\ a \circ c = b \circ c \\
&\Leftrightarrow ([a] \circ c) \cap ([b] \circ c) \neq \emptyset.
\end{aligned}$$

It is for the second equivalence that we need the assumption $c \neq 0$ if $\circ \in \{\cdot, /\}$.

b) follows from a) by choosing any $c \in [c]$ and taking into account $[a] \circ c \subseteq [a] \circ [c]$ and $[b] \circ c \subseteq [b] \circ [c]$. The implication '$\Leftarrow$' fails as the example $[a] = 1$, $[b] = 2$, $[c] = [1, 2]$, $\circ = \cdot$ shows.

c) $([a] + [b]) \cap [c] \neq \emptyset \Leftrightarrow \exists\, a \in [a],\quad b \in [b],\quad c \in [c]:\ a + b = c \Leftrightarrow a = c - b \Leftrightarrow [a] \cap ([c] - [b]) \neq \emptyset.$

d) The first equivalence and the implication '$\Rightarrow$' of the second one are trivial. For the converse implication let, e.g.,

$$\underline{a}_1 = \max_{1 \leq i \leq n} \underline{a}_i \quad \text{and} \quad \overline{a}_2 = \min_{1 \leq i \leq n} \overline{a}_i. \tag{3}$$

Then the assumption $[a]_1 \cap [a]_2 \neq \emptyset$ together with a) implies $\underline{a}_1 \leq \overline{a}_2$, i.e., the assertion. Other choices in (3) terminate the proof similarly. The number $\binom{n}{2}$ of simple intersections is an elementary statement in combinatorics.

e) can be seen at once.

f) Let $x \in [a] \cap [b]$. Then $x = (x + x)/2 \in ([a] + [b])/2$, hence '$\subseteq$' is true. The converse '$\supseteq$' can be seen immediately.

g) Let $x \in [a] \cap [b]$. Then $x = x + 0 \in [b] \cap [c]$ whence '$\subseteq$' holds. The converse '$\supseteq$' is trivially true. □

Although e) is trivial it is the starting point of our elimination process for parameters. We will first demonstrate this process for unsymmetric linear interval systems $[A]x = [b]$. By this way we will deduce Beeck's characterization of the unsymmetric solution set S from (2):

$$x \in S \ \Leftrightarrow\ [A]x \cap [b] \neq \emptyset, \quad \text{(cf. Thm. 1a), c))}. \tag{4}$$

To this end we start with the trivial equivalences

$$x \in S \Leftrightarrow \exists\, A \in \mathbb{R}^{n\times n},\ b \in \mathbb{R}^n : \left\{ \begin{array}{c} \left\{ \sum_{j=1}^{n} a_{ij}x_j \right\} \cap \{b_i\} \neq \emptyset \\ \{a_{ij}\} \cap [a]_{ij} \neq \emptyset \\ \{b_i\} \cap [b]_i \neq \emptyset \end{array} \right\}, \quad i,j = 1,\ldots,n$$

$$\Leftrightarrow \exists\, a_{ij}, b_i \in \mathbb{R} : \left\{ \begin{array}{c} \left\{ \sum_{j=1}^{n} a_{ij}x_j \right\} \cap \{b_i\} \neq \emptyset \\ \{a_{ij}x_j\} \cap [a]_{ij}x_j \neq \emptyset \\ \{b_i\} \cap [b]_i \neq \emptyset \end{array} \right\}, \quad i,j = 1,\ldots,n \tag{5}$$

considering a_{ij}, b_i as parameters to be eliminated. Note that the second equivalence holds because of Lemma 1 a) with $c = x_j$ if $x_j \neq 0$. In the case $x_j = 0$ the implication '$\Rightarrow$' is trivial while the converse holds by choosing $a_{ij} \in [a]_{ij}$ arbitrarily. Its value does not matter since it is multiplied by $x_j = 0$ in the uppermost intersection of (5).

Step 1: Fix a parameter, say a_{11}, and consider those statements with intersections which contain a_{11}, i.e.,

$$\exists\, b_1 \in \mathbb{R}, a_{1j} \in \mathbb{R} \text{ for } j = 1,\ldots,n : \left\{ \begin{array}{c} \left\{ a_{11}x_1 + \sum_{j=2}^{n} a_{1j}x_j \right\} \cap \{b_1\} \neq \emptyset \\ \{a_{11}x_1\} \cap [a]_{11}x_1 \neq \emptyset \end{array} \right\}. \tag{6}$$

Step 2: Isolate the $a_{11}x_1$–terms according to Lemma 1 c) in order to get

$$\begin{array}{c} \exists\, b_1 \in \mathbb{R}, a_{1j} \in \mathbb{R} \quad \text{for } j = 1,\ldots,n : \\ \left\{ a_{11}x_1 + \sum_{j=2}^{n} a_{1j}x_j \right\} \cap \{b_1\} \neq \emptyset, \\ \Leftrightarrow \exists\, b_1 \in \mathbb{R}, a_{1j} \in \mathbb{R} \quad \text{for } j = 1,\ldots,n : \\ \{a_{11}x_1\} \cap \left\{ b_1 - \sum_{j=2}^{n} a_{1j}x_j \right\} \neq \emptyset. \end{array}$$

Step 3: Eliminate the $a_{11}x_1$–terms using Lemma 1 e) for the equivalence

$$\begin{array}{c} \exists\, b_1 \in \mathbb{R}, a_{1j} \in \mathbb{R} \quad \text{for } j = 1,\ldots,n : \\ \left\{ \begin{array}{c} \{a_{11}x_1\} \cap \left\{ b_1 - \sum_{j=2}^{n} a_{1j}x_j \right\} \neq \emptyset \\ \{a_{11}x_1\} \cap [a]_{11}x_1 \neq \emptyset \end{array} \right\}, \end{array}$$

$$\Leftrightarrow \exists\, b_1 \in \mathbb{R}, a_{1j} \in \mathbb{R} \quad \text{for } j = \mathbf{2}, \ldots, n:$$

$$\{b_1 - \sum_{j=2}^{n} a_{1j}x_j\} \cap [a]_{11}x_1 \neq \emptyset. \tag{7}$$

If necessary, bring (7) into an appropriate form using Lemma 1 c). For instance,

$$(7) \Leftrightarrow \exists\, b_1 \in \mathbb{R}, a_{1j} \in \mathbb{R} \quad \text{for } j = 2, \ldots, n:$$

$$\left\{ [a]_{11}x_1 + \sum_{j=2}^{n} a_{1j}x_j \right\} \cap \{b_1\} \neq \emptyset.$$

Repeat these steps for the remaining parameters. This finally yields to Beeck's criterion (4).

Note that up to now only simple intersections occur in the sense of Section 2.

The existence qualifier in the Steps 1–3 refers also to parameters which occur in some of the intersections (5) not longer listed during the elimination of a_{11}. In the equivalences of these steps these parameters can be thought to have fixed values. Otherwise one could get the impression that the value b_1, e.g., in (7) is different from that which guarantees $\{b_1\} \cap [b]_1 x_1 \neq \emptyset$.

In order to describe the symmetric solution set S_{sym} by means of nonempty intersections we assume $[A] = [A]^T$ without loss of generality. Otherwise we replace $[A]$ by the largest interval matrix $[B] \subseteq [A]$ such that $[B] = [B]^T$ holds. Since the matrices $A \in [A] \backslash [B]$ certainly are unsymmetric they are of no interest in view of S_{sym}. We start with the equivalence

$$x \in S_{\text{sym}} \Leftrightarrow x \in S \;\wedge\; \exists\, b_i \in \mathbb{R}, a_{ij} \in \mathbb{R} \quad \text{for } i, j = 1, \ldots, n:$$

$$\left(a_{ij} = a_{ji} \;\wedge\; \left\{ \begin{array}{c} \left\{ \sum_{j=1}^{n} a_{ij}x_i x_j \right\} \cap \{b_i x_i\} \neq \emptyset \\ \{a_{ij}x_i x_j\} \cap [a]_{ij} x_i x_j \neq \emptyset \\ \{b_i x_i\} \cap [b]_i x_i \neq \emptyset \end{array} \right\} \right). \tag{8}$$

The implication '$\Rightarrow$' is trivial. In order to prove the converse we assume that the right-hand statements in (8) hold. We will construct $\hat{A} = \hat{A}^T \in [A]$, $\hat{b} \in [b]$ such that $\hat{A}x = \hat{b}$ is fulfilled, hence $x \in S_{\text{sym}}$. With a_{ij}, b_i from the assumption we start with $\hat{a}_{ij} := a_{ij}$, $\hat{b}_i := b_i$. Then certainly $\hat{A} = \hat{A}^T$ but not necessarily $\hat{A} \in [A]$, $\hat{b} \in [b]$. Let $M := \{i | x_i = 0\}$. The nonempty intersections in (8) imply $a_{ij} \in [a]_{ij}$ if $i, j \notin M$ and

$$\sum_{j=1}^{n} a_{ij}x_j = b_i \in [b]_i \tag{9}$$

if $i \notin M$. In every equation of the form in (9) the entries a_{ij} can be replaced by any value if $j \in M$ since then a_{ij} is multiplied by $x_j = 0$. We will need this observation soon. Assume now $i_0 \in M$. Since $x \in S$ there are some $\tilde{A} \in [A]$, $\tilde{b} \in [b]$ such that $\tilde{A}x = \tilde{b}$, whence $\sum_{j=1}^{n} \tilde{a}_{i_0 j} x_j = \tilde{b}_{i_0} \in [b]_{i_0}$. Redefine $\hat{a}_{i_0 j}$ by $\tilde{a}_{i_0 j}$ and $\hat{b}_{i_0}$ by $\tilde{b}_{i_0}$. Based on our observation we also redefine $\hat{a}_{j i_0}$ by $\tilde{a}_{i_0 j}$ for $j = 1, \ldots, n$. Repeating this process for all i_0 in M yields to a symmetric matrix $\hat{A} \in [A]$ and a vector $\hat{b} \in [b]$ with the required property $\hat{A}x = \hat{b}$.

We now apply the Steps 1–3 above analogously to (8). The main difference in the elimination process occurs in (6) in which by the symmetry at least three different nonempty intersections occur when considering $a_{ij}x_i x_j$ for $i \neq j$. In order to eliminate a_{12}, e.g., as the first parameter we get

$$x \in S_{sym} \Leftrightarrow x \in S \;\wedge\; \exists\, a_{ij}, b_i \in \mathbb{R} \quad \text{for } i,j = 1, \ldots, n: \left(a_{ij} = a_{ji} \;\wedge\; \left\{ \begin{array}{c} \{a_{12}x_1x_2 + \sum\limits_{\substack{j=1 \\ j \neq 2}}^{n} a_{1j}x_1x_j\} \cap \{b_1 x_1\} \neq \emptyset \\ \{a_{12}x_1x_2 + \sum\limits_{j=2}^{n} a_{2j}x_2x_j\} \cap \{b_2 x_2\} \neq \emptyset \\ \{a_{12}x_1x_2\} \cap [a]_{12}x_1x_2 \neq \emptyset \\ \text{remaining } a_{12} - \text{free intersections} \end{array} \right\} \right),$$

where by 'a_{12} – free' we also mean 'a_{21} – free' here and in the sequel.

The isolation Step 2 contains the equivalence

$$x \in S_{sym} \Leftrightarrow x \in S \;\wedge\; \exists\, a_{ij}, b_i \in \mathbb{R} \quad \text{for } i,j = 1, \ldots, n: \left(a_{ij} = a_{ji} \;\wedge\; \left\{ \begin{array}{c} \{a_{12}x_1x_2\} \cap \{b_1 x_1 - \sum\limits_{\substack{j=1 \\ j \neq 2}}^{n} a_{1j}x_1x_j\} \neq \emptyset, \\ \{a_{12}x_1x_2\} \cap \{b_2 x_2 - \sum\limits_{j=2}^{n} a_{2j}x_2x_j\} \neq \emptyset, \\ \{a_{12}x_1x_2\} \cap [a]_{12}x_1x_2 \neq \emptyset \\ \text{remaining } a_{12} - \text{free intersections} \end{array} \right\} \right).$$

The elimination Step 3 yields to

$$x \in S_{sym} \Leftrightarrow x \in S \;\wedge\; \exists\, a_{ij}, b_i \in \mathbb{R} \quad \text{for } i,j = 1, \ldots, n,\; (i,j) \notin \{(1,2),(2,1)\}: \left(a_{ij} = a_{ji} \wedge \left\{ \begin{array}{c} \{[a]_{12}x_1x_2\} \cap \left\{ b_1x_1 - \sum\limits_{\substack{j=1 \\ j \neq 2}}^{n} a_{1j}x_1x_j \right\} \cap \left\{ b_2x_2 - \sum\limits_{j=2}^{n} a_{2j}x_2x_j \right\} \neq \emptyset \\ \text{remaining } a_{12} - \text{free intersections} \end{array} \right\} \right).$$

Here we used Lemma 1 e) with three intersections. In order to continue we resolve this multiple intersection into simple ones. The key for doing this is Lemma 1 d). We formulate this crucial part of the method as a particular step:

Step 4: If there is no parameter, say α, left to be eliminated then stop. Otherwise resolve those multiple intersections, in which α occurs, into simple ones according to Lemma 1 d). If $\bigcap_{i=1}^{m}$ is such a multiple intersection which is resolved, and if $S_{i_1}, \ldots, S_{i_k}$ are those of its operands in which α does not occur, then apply Lemma 1 d) again to replace the simple intersections $S_{i_j} \cap S_{i_l} \neq \emptyset$ by the multiple intersection $\bigcap_{l=1}^{k} S_{i_l} \neq \emptyset$ in which α does not occur either.

Goto Step 1.

Since the isolation Step 2 only uses $\pm$ –operations and multiplications/divisions by real numbers $c \neq 0$ the summands $a_{ij}x_ix_j$, b_ix_i are unchanged up to a constant factor so that after having eliminated all terms with parameters a_{ij}, b_i we end up with a number of intersections with operands of the form

$$[m] = \sum_{i,j} [c]_{ij} x_i x_j + \sum_i [d]_i x_i.$$

Since the bounds of $[c]_{ij}x_ix_j$ and of $[d]_ix_i$ depend only on the signs of x_i, x_j they remain the same when $x \in S_{\text{sym}}$ varies in a fixed orthant O. Therefore, $S_{\text{sym}} \cap O$ can finally be expressed by the intersection of appropriate sets whose boundary is a hyperplane or a quadric. This is the main result of [2] which we have derived in the present paper by a different way. Since this way does not need a fixed orthant during the elimination process it is not so costly as that in [2] based on the Fourier–Motzkin elimination process. Nevertheless, the total costs remain tremendous as is indicated by the subsequent Example 2.

Example 1. Let $[A] = [A]^T \in \mathrm{I}(\mathbb{R}^{2\times 2})$, $[b] \in \mathrm{I}(\mathbb{R}^2)$. In order to describe S_{sym} we first eliminate b_i, a_{ii} for $i = 1, 2$. This yields to the same result as for S. Eliminating $a_{12} = a_{21}$ as above ends with

$$[a]_{12}x_1x_2 \cap ([b]_1x_1 - [a]_{11}x_1^2) \cap ([b]_2x_2 - [a]_{22}x_2^2) \neq \emptyset. \tag{10}$$

Using Lemma 1 d) one can resolve (10) into three simple intersections. Since $x \in S$, Beeck's criterion implies directly two of these intersections, which therefore can be omitted. One finally obtains

$$x \in S_{\text{sym}} \Leftrightarrow x \in S \wedge \ ([b]_1x_1 - [a]_{11}x_1^2) \cap ([b]_2x_2 - [a]_{22}x_2^2) \neq \emptyset. \tag{11}$$

which, by Lemma 1 c), is equivalent to Theorem 4.2 in [8]. By the parts a) and c) of the same lemma the nonempty intersection in (11) can be transformed into

$$0 \in [b]_1x_1 - [b]_2x_2 - [a]_{11}x_1^2 + [a]_{22}x_2^2 = x^T \begin{pmatrix} -[a]_{11} & 0 \\ 0 & [a]_{22} \end{pmatrix} x + ([b]_1, -[b]_2)x.$$

If x is restricted to the first orthant O_1 this relation can equivalently be expressed by the two inequalities

$$\underline{b}_1x_1 - \overline{b}_2x_2 - \overline{a}_{11}x_1^2 + \underline{a}_{22}x_2^2 \leq 0$$
$$-\overline{b}_1x_1 + \underline{b}_2x_2 + \underline{a}_{11}x_1^2 - \overline{a}_{22}x_2^2 \leq 0$$

which, together with $x \in S$ characterized $S_{\text{sym}} \cap O_1$ in Section 3.5 of [2]. □

Example 2. Let $[A] = [A]^T \in I(\mathbb{R}^{3\times 3})$, $[b] \in I(\mathbb{R}^3)$ and consider again S_{sym}. Eliminating b_i, a_{ii} for $i = 1, 2, 3$ and a_{12} leads to

$$x \in S_{sym} \Leftrightarrow x \in S \ \wedge \ \exists\, a_{13},\ a_{23} \in \mathbb{R} : \Bigg(a_{13} = a_{31} \ \wedge \ a_{23} = a_{32}$$

$$\wedge \left\{ \begin{array}{lll} [a]_{12}x_1x_2 & \cap & ([b]_1x_1 - [a]_{11}x_1^2 - a_{13}x_1x_3) \\ & \cap & ([b]_2x_2 - [a]_{22}x_2^2 - a_{23}x_2x_3) \neq \emptyset \\ \{a_{13}x_1x_3\} & \cap & ([b]_3x_3 - a_{23}x_2x_3 - [a]_{33}x_3^2) \neq \emptyset \\ \{a_{13}x_1x_3\} & \cap & [a]_{13}x_1x_3 \neq \emptyset \\ \{a_{23}x_2x_3\} & \cap & [a]_{23}x_2x_3 \neq \emptyset \end{array} \right\} \Bigg). \tag{12}$$

After having resolved the triple intersection in (12) into simple intersections the isolation Step 2 for a_{13} and the corresponding elimination Step 3 yield to

$$x \in S_{sym} \Leftrightarrow x \in S \wedge \exists\, a_{23} \in \mathbb{R} : \Bigg(a_{23} = a_{32}$$

$$\wedge \left\{ \begin{array}{lll} [a]_{13}x_1x_3 & \cap & ([b]_1x_1 - [a]_{11}x_1^2 - [a]_{12}x_1x_2) \\ & \cap & ([b]_3x_3 - a_{23}x_2x_3 - [a]_{33}x_3^2) \\ & \cap & ([b]_1x_1 - [b]_2x_2 - [a]_{11}x_1^2 + [a]_{22}x_2^2 + a_{23}x_2x_3) \neq \emptyset \\ \{a_{23}x_2x_3\} & \cap & ([b]_2x_2 - [a]_{12}x_1x_2 - [a]_{22}x_2^2) \neq \emptyset \\ \{a_{23}x_2x_3\} & \cap & [a]_{23}x_2x_3 \neq \emptyset \end{array} \right\} \Bigg).$$

Here again we omitted an intersection which can be deduced from $x \in S$. Eliminating the last parameter a_{23} the fourfold intersection produces $\binom{4}{2}$ simple intersections among which only one is a_{23} – free. This one turns out to be a consequence of $x \in S$. Therefore, it can be omitted, and we are finally left with the equivalence

$$\begin{array}{lll} & & x \in S_{sym} \Leftrightarrow \ x \in S \\ \wedge\ [a]_{23}x_2x_3 & \cap & ([b]_2x_2 - [a]_{12}x_1x_2 - [a]_{22}x_2^2) \\ & \cap & ([b]_3x_3 - [a]_{13}x_1x_3 - [a]_{33}x_3^2) \\ & \cap & (-[b]_1x_1 + [b]_2x_2 + [a]_{11}x_1^2 - [a]_{22}x_2^2 + [a]_{13}x_1x_3) \\ & \cap & (-[b]_1x_1 + [b]_2x_2 + [b]_3x_3 + [a]_{11}x_1^2 - [a]_{22}x_2^2 - [a]_{33}x_3^2)/2 \\ & \cap & (-[b]_1x_1 + [b]_3x_3 + [a]_{11}x_1^2 + [a]_{12}x_1x_2 - [a]_{33}x_3^2) \\ & \cap & \{([b]_1 - [b]_1)x_1 + [b]_2x_2 + ([a]_{11} - [a]_{11})x_1^2 \\ & & -[a]_{12}x_1x_2 - [a]_{22}x_2^2\} \\ & \neq & \emptyset. \end{array} \tag{13}$$

Using Lemma 1 d) this multiple intersection can equivalently be written as $\binom{7}{2} = 21$ simple ones. Two of them can be deduced from the requiry $x \in S$. The remaining 19 ones together with the three occuring from Beeck's criterion (4) result in a total of 22 which all can be rewritten in the form '$0 \in \ldots$'. Introducing the interval bounds, fixing an orthant O and taking into account

$$0 \in [a] \Leftrightarrow \underline{a} \leq 0 \leq \overline{a} \tag{14}$$

finally yields to 44 inequalities for the characterization of $S_{\text{sym}} \cap O$ as was already mentioned without proof in [2]. □

Example 3. Let $[b] \in \mathrm{I}(\mathbb{R}^3)$ and let $[A] = [A]^T \in \mathrm{I}(\mathbb{R}^{3\times 3})$ be tridiagonal, i.e., $[a]_{13} = [a]_{31} = 0$. Then S_{sym} is certainly characterized by (13) with $[a]_{13} = 0$. We denote by I, II, III, ... the operands in (13) according to their appearence. Since $x \in S$ we get $0 \in x_i([b] - [A]x)_i$, $i = 1, 2, 3$ by Beeck's criterion (4). The operands VI and VII can be written as $VI = III - x_1([b] - [A]x)_1$ and $VII = IV + x_1([b]- [A]x)_1$, taking into account $[a]_{13} = 0$. Therefore, they can be omitted by Lemma 1 g). The operand V equals $(III + IV)/2$. This time Lemma 1 f) shows that V can be omitted, so that we end up with

$$\begin{aligned} & x \in S_{\text{sym}} \Leftrightarrow x \in S \\ \wedge\ [a]_{23}x_2x_3 \quad & \cap \quad ([b]_2x_2 - [a]_{12}x_1x_2 - [a]_{22}x_2^2) \\ & \cap \quad ([b]_3x_3 - [a]_{33}x_3^2) \\ & \cap \quad (-[b]_1x_1 + [b]_2x_2 + [a]_{11}x_1^2 - [a]_{22}x_2^2) \\ & \neq \quad \emptyset. \end{aligned} \tag{15}$$

As in Example 1 we can resolve (15) into $\binom{4}{2} = 6$ simple intersections. Obviously, two of them, namely $I \cap II$ and $I \cap III$ can also be derived from $x \in S$ and thus are redundant. By Lemma 1 d) the remaining four nonempty intersections $I \cap IV \neq \emptyset$, $III \cap IV \neq \emptyset$, $II \cap IV \neq \emptyset$, $II \cap III \neq \emptyset$ can equivalently be rewritten as

$$\left.\begin{aligned} 0 &\in [b]_1x_1 - [b]_2x_2 - [a]_{11}x_1^2 + [a]_{22}x_2^2 + [a]_{23}x_2x_3 \\ 0 &\in [b]_1x_1 - [b]_2x_2 + [b]_3x_3 - [a]_{11}x_1^2 + [a]_{22}x_2^2 - [a]_{33}x_3^2 \\ 0 &\in [b]_1x_1 + ([b]_2 - [b]_2)x_2 - [a]_{11}x_1^2 - [a]_{12}x_1x_2 + ([a]_{22} - [a]_{22})x_2^2 \\ 0 &\in [b]_2x_2 - [b]_3x_3 - [a]_{12}x_1x_2 - [a]_{22}x_2^2 + [a]_{33}x_3^2 \end{aligned}\right\}. \tag{16}$$

If x is restricted to the first orthant O_1 and if we express (16) by means of interval bounds we can directly retrieve the inequalities (3.21) in [2] by which we characterized $S_{\text{sym}} \cap O_1$. We only have to apply (14). □

For examples which are more specific we refer to [2] and [8].

It remains an open problem whether one can get a representation of S_{sym} which is more compact than ours and than that in [2] based on the Fourier–Motzkin elimination. A representation similar to (16) can always be achieved. The statements have the form

$$0 \in x^T[C]x + [d]^Tx + [e]$$

with $[C] \in \mathrm{I}(\mathbb{R}^{n\times n})$, $[d] \in \mathrm{I}(\mathbb{R}^n)$, $[e] \in \mathrm{I}(\mathbb{R})$. This corresponds to the hyperplanes and quadrics involved to describe the boundary of S and S_{sym}. For $x \in S$ we get, e.g., $[C] = O$, $[d]^T = (-[a]_{i1}, \dots, -[a]_{in})$, $[e] = [b]_i$. For $x \in S_{\text{sym}}$ we do not see *in advance* how $[C]$, $[d]$, $[e]$ are related to the entries of $[A]$ and $[b]$.

We conclude our paper by mentioning that our elimination process can also be applied to skew-symmetric matrices $A = -A^T \in [A] = -[A]^T$ and to per-symmetric matrices $PA = (PA)^T \in P[A] = (P[A])^T$ where $p_{ij} = 1$ if $j = n+1-i$ and $p_{ij} = 0$ otherwise. Furthermore, it can be used for the algebraic eigenvalue problem

$$Ax = \lambda x, \; A \in [A], \; x \neq 0$$

and its symmetric variant

$$Ax = \lambda x, \; A = A^T \in [A] = [A]^T, \; x \neq 0.$$

In order to describe the general *eigenpair set*

$$E := \{(x, \lambda) \in \mathbb{R}^{n+1} \,|\, Ax = \lambda x, \; x \neq 0, \; A \in [A]\}$$

one starts with

$$(x, \lambda) \in E \Leftrightarrow \exists\, a_{ij} \in \mathbb{R} \quad \text{for } i, j = 1, \dots, n: \quad \left\{ \begin{array}{c} \left\{ \sum_{j=1}^{n} a_{ij} x_j \right\} \cap \{\lambda x_i\} \neq \emptyset \\ \{a_{ij} x_j\} \cap [a]_{ij} x_j \neq \emptyset \end{array} \right\},$$

where we assume $x \neq 0$. For the *symmetric eigenpair set*

$$E_{\text{sym}} := \{(x, \lambda) \in \mathbb{R}^{n+1} \,|\, Ax = \lambda x, \; x \neq 0, \; A = A^T \in [A]\}$$

we begin with

$$\begin{array}{l} (x, \lambda) \in E_{\text{sym}} \Leftrightarrow (x, \lambda) \in E \;\wedge\; \exists\, a_{ij} \in \mathbb{R} \quad \text{for } i, j = 1, \dots, n: \\ \left(a_{ij} = a_{ji} \;\wedge\; \left\{ \begin{array}{c} \left\{ \sum_{j=1}^{n} a_{ij} x_i x_j \right\} \cap \{\lambda x_i^2\} \neq \emptyset \\ \{a_{ij} x_i x_j\} \cap [a]_{ij} x_i x_j \neq \emptyset \end{array} \right\} \right) \end{array}$$

prescribing again $x \neq 0$. The outputs are, of course, the same as in [5] and [10]: In the first case one obtains

$$(x, \lambda) \in E \Leftrightarrow [A]x \cap \{\lambda x\} \neq \emptyset \quad \text{and} \quad x \neq 0.$$

This can also be derived directly from Beeck's criterion (4) applied to the linear interval system $([A] - \lambda I)x = 0$ with λ assumed to be fixed for the moment. In the symmetric case the elimination process above yields to intersections with sets of which the boundary is an algebraic surface of order three at most. We leave the details to the reader.

4. Conclusion

We presented an alternative method to describe various solution sets, among them S, S_{sym}, E, E_{sym}. This method is based on the following steps:

Step 0: (Initialization)

Describe the solution set by means of obvious, nonempty intersections with two operands (= simple intersections).

Step 1: (Separation)

Fix a parameter to be eliminated, say α, and split the nonempty intersections into two groups: One of them, say the first, contains exactly those intersections in which α occurs. The second contains the remaining intersections.

Step 2: (Isolation)

Isolate the α-terms in the first group.

Step 3: (Elimination)

Eliminate the α-terms using (eventually multiple) intersections.

Step 4: (Stopping/Resolving)

If no further parameter is to be eliminated, then stop.

Otherwise resolve multiple intersections into simple ones and goto Step 1.

This alternative way is taylored for intervals and interval arithmetic. It allows a description of the above-mentioned solution sets which is more condensed than that based on inequalities and the Fourier–Motzkin process.

References

[1] Alefeld, G., Herzberger, J.: Introduction to interval computations. New York: Academic Press, 1983.

[2] Alefeld, G., Kreinovich, V., Mayer, G.: The shape of the symmetric solution set. In: Applications of interval computations (Kearfott, R. B., Kreinovich, V., eds.), pp. 61–79. Boston: Kluwer, 1996.

[3] Alefeld, G., Kreinovich, V., Mayer, G.: On the shape of the symmetric, persymmetric, and skew-symmetric solution set. SIAM J. Matrix Anal. Appl. *18*, 693–705 (1997).

[4] Alefeld, G., Kreinovich, V., Mayer, G.: The shape of the solution set of linear interval equations with dependent coefficients. Math. Nachr. *192*, 23–26 (1998).

[5] Alefeld, G., Kreinovich, V., Mayer, G.: Modifications of the Oettli–Prager theorem with application to the algebraic eigenvalue problem. In: Symbolic algebraic methods and verification methods – theory and applications (Alefeld, G., Rohn, J., Rump, S. M., Yamamoto, T., eds.), pp. 11–20. Wien: Springer, 2001.

[6] Alefeld, G., Kreinovich, V., Mayer, G.: On the solution set of particular classes of linear systems. Submitted for publication.

[7] Alefeld G., Mayer, G.: The Cholesky method for interval data. Linear Algebra Appl. *194*, 161–182 (1993).
[8] Alefeld, G., Mayer, G.: On the symmetric and unsymmetric solution set of interval systems. SIAM J. Matrix Anal. Appl. *16*, 1223–1240 (1995).
[9] Beeck, H.: Über Struktur und Abschätzungen der Lösungsmenge von linearen Gleichungssystemen mit Intervallkoeffizienten. Computing *10*, 231–244 (1972).
[10] Deif, A. S.: The interval eigenvalue problem. ZAMM *71*, 61–64 (1991).
[11] Hartfiel, D. J.: Concerning the solution set of $Ax = b$ where $P \leq A \leq Q$ and $p \leq b \leq q$. Numer. Math. *35*, 355–359 (1980).
[12] Jansson, C.: Rigorous sensitivity analysis for real symmetric matrices with uncertain data. In: Computer arithmetic, scientific computation and mathematical modelling (Kaucher, E., Markov, S. M., Mayer, G., eds.), pp. 293–316. Basel: Baltzer, 1991.
[13] Jansson, C.: Interval linear systems with symmetric matrices, skew-symmetric matrices and dependencies in the right hand side. Computing *46*, 265–274 (1991).
[14] Neumaier, A.: Interval methods for systems of equations. Cambridge: Cambridge University Press, 1990.
[15] Oettli, W., Prager, W.: Compatibility of approximate solution of linear equations with given error Bounds for coefficients and right-hand sides. Numer. Math. *6*, 405–409 (1964).
[16] Rump, S. M.: Verification methods for dense and sparse systems of equations. In: Topics in validated computations (Herzberger, J., ed.), pp. 63–135. Amsterdam: Elsevier, 1994.
[17] Schrijver, A.: Theory of linear and integer programming. New York: Wiley, 1986.

Günter Mayer
Fachbereich Mathematik
Universität Rostock
Universitätsplatz 1
D-18051 Rostock
Germany
e-mail: guenter.mayer@mathematik.uni-rostock.de

Computing [Suppl] 15, 165–173 (2001)

Computing

A Guaranteed Bound of the Optimal Constant in the Error Estimates for Linear Triangular Element

M. T. Nakao, Fukuoka, and **N. Yamamoto,** Tokyo

Dedicated to Professor Tetsuro Yamamoto on the occasion of his 65th birthday

Abstract

We consider a numerical method to get a guaranteed bound of the optimal constant in the error estimates of a finite element method with linear triangular elements in the plane. The problem is reduced to a kind of smallest eigenvalue problem for an elliptic operator in a certain function space on the reference triangle. In order to solve the problem, we formulate a numerical verification procedure based on finite element approximations and constuctive error estimates. Consequently, we obtain a sufficiently sharp bound of the desired constant by a computer assisted proof. In this paper, we provide the basic idea and outline the concept of verification procedures as well as show the final numerical result. The detailed description of procedures for actual computations will be presented in the forthcoming paper [11].

AMS Subject Classifications: 65N15, 65N30.

Key Words: Finite element method, error estimates, optimal constant, numerical verification method, computer assisted proof.

1. Introduction

In the error analysis of the finite element method on a polygonal domain in R^2, the convergence rates with order k are usually written of the form

$$||u - u_h|| \leq Ch^k|u|,$$

where C is a constant independent of the mesh size h. Here, $||\cdot||$ and $|\cdot|$ stand for some appropriate norm and seminorm in certain function spaces, respectively. We sometimes need to provide computable error bounds in the finite element approximation for various kinds of purposes. For example, in the numerical verification methods of solutions for nonlinear boundary value problems such as [6–8], [10] etc., it is very important, because the actual value for C much influences the possibility and the accuracy of verification. Also, the smallest value of C, which is denoted by C_0, is particularly desired because of the high efficiency of the verification process. For rectangular elements, several kinds of optimal constants have been found. Namely, in the error estimates for the H_0^1-projection on the rectangular domain, with H_0^1 norm and H^2 seminorm, it is known that $C_0 = 1/\pi$ ([13]) and $C_0 = 1/2\pi$

([9]) for the bilinear and biquadratic elements, respectively. But $C_0 > 1/3\pi$ for the bicubic case [9]. Here, the H_0^1 norm and H^2 seminorm are defined by

$$||\phi||_{H_0^1} \equiv ||\nabla\phi||_{L^2} \quad \text{and} \quad |\phi|_{H^2}^2 \equiv ||\phi_{xx}||_{L^2}^2 + 2||\phi_{xy}||_{L^2}^2 + ||\phi_{yy}||_{L^2}^2, \tag{1}$$

respectively.

In this paper, we consider upper and lower bounds of C_0 for the H_0^1-projection into the continuous linear triangular finite element space. As one can see (*cf.* [12]), this problem is reduced to the problem of finding the smallest positive constant C for the interpolation error satisfying, for any $\phi \in H^2(\Omega)$,

$$||\nabla(\phi - \mathscr{I}\phi)|| \leq C|\phi|_2. \tag{2}$$

Here, Ω is the reference triangle on the plane whose vertices consist of $a_1 = (0,0), a_2 = (1,0)$ and $a_3 = (0,1)$, and $\mathscr{I}\phi$ is the interpolant of ϕ at each vertex. Also, $||\cdot||$ and $|\cdot|_2$ stand for the L^2 norm, and H^2 seminorm on Ω defined by (1), respectively.

On this problem, Natterer [12] presented an upper bound: $C_0 \leq 0.81$ in 1975. After that, by some approximate computations, Arbenz [1] and Lehmann [5] obtained $C_0 \approx 0.4888$ and $C_0 \approx 0.5086$ in 1982 and 1986, respectively, but with no guarantee.

In order to get a lower bound, we now consider the following $H^2(\Omega)$ functions which vanish at each vertex of Ω:

$$u_1 = x - x^2 + y - y^2 \quad \text{and} \quad u_2 = xy.$$

Then, setting $u := \alpha u_1 + \beta u_2$ for any $\alpha, \beta \in R$, and noting that $\mathscr{I}u = 0$, by some simple calculations, we have

$$\max_{(\alpha,\beta)\neq(0,0)} \frac{||\nabla u||}{|u|_2} \geq 0.467,$$

which implies that $0.467 \leq C_0$.

In the following sections, we describe the formulation of a numerical verification method to get an improved upper bound of C_0. By the execution of an implementation of the procedure, we consequently obtain verified upper and lower bounds such that $0.467 \leq C_0 \leq 0.494$, which should be sufficiently narrow interval in the practical sense.

2. Formulation of the Problem

We define the set E by

$$E := \{v \in H^2(\Omega) | v(a_j) = 0, j = 1,2,3\}$$

and denote the boundary of Ω by $\Gamma_1 := a_1a_2, \Gamma_2 := a_2a_3$ and $\Gamma_3 := a_1a_3$.

First we have the following lemma for the smallest constant C_0 satisfying (2).

Lemma 1. *It holds that*

$$\frac{1}{C_0} = \min_{u \in E} \frac{|u|_2}{||\nabla u||} \geq \min_{u \in E} \frac{||\nabla u_y||}{||u_y||},$$

where u_y implies the y-directional derivative of u.

Proof: The first equality follows by considering that the interpolation $\mathscr{I}$ is linear. Next, let $u_1(x,y)$ be a function which attains $\min_{u \in E} \frac{|u|_2}{||\nabla u||}$ and set $u_2(x,y) := u_1(y,x)$. Then, by the symmetry of the domain, $u_2(x,y)$ also presents the same minimum value. Setting $u_0 := 1/2(u_1 + u_2)$, by using the fact that $|u_0|_2 = |u_1|_2$ and $||\nabla u_0|| = ||\nabla u_1||$, the function u_0 also attains the minimum.

Furthermore, we have

$$|u_0|_2^2 = 2||\nabla (u_0)_y||^2 \quad \text{and} \quad ||\nabla u_0||^2 = 2||(u_0)_y||^2.$$

Thus, we obtain

$$\frac{1}{C_0} = \frac{|u_0|_2}{||\nabla u_0||} = \frac{||\nabla (u_0)_y||}{||(u_0)_y||},$$

which means that the second inequality of the lemma holds. □

We now try to find the solution of the minimization problem:

$$\frac{1}{C_0'} \equiv \min_{u \in E} \frac{||\nabla u_y||}{||u_y||} (C_0 \leq C_0'). \tag{3}$$

Although this problem is different from the original one, it should be expected to present an upper bound closer to C_0 than the existing value, i.e., 0.81.

It is readily seen that (3) leads to the following eigenvalue problem finding λ and $u \in E$ satisfying:

$$(\nabla u_y, \nabla v_y) = \lambda (u_y, v_y) \quad \forall v \in E, \tag{4}$$

where $(\cdot, \cdot)$ stands for the usual inner product on $L^2(\Omega)$. Then C_0' is given by $1/\sqrt{\lambda}$ for the smallest eigenvalue λ of (4).

Now, for an arbitrary function $\psi \in H_0^2(\Omega)$, when we set $v := \int_0^y \psi \, dy$, it holds that $v \in E$. Therefore, for $v_y = \psi$ in (4), by considering that $H_0^2(\Omega)$ is dense in $L^2(\Omega)$, partial integration of (4) yields, in distributional sense, that

$$\begin{cases} -\Delta u_y = \lambda u_y, \\ \left\langle \frac{\partial u_y}{\partial n}, v_y \right\rangle = 0 \quad \forall v \in E, \end{cases} \tag{5}$$

where $\langle \cdot, \cdot \rangle$ means the inner product on $L^2(\partial\Omega)$ and $\partial u_y/\partial n$ is the outer normal derivative.

Now, for a function $\psi \in S = \{\psi \in H^2(\Omega) \mid \int_{\Gamma_3} \psi \, ds = 0\}$, setting $v := \int_0^y \psi \, dy$, we have $v \in E$. Then, by choosing ψ which vanishes on Γ_3, the latter part of (5) implies that

$$\frac{\partial u_y}{\partial n} = 0 \quad \text{on } \Gamma_1 \cup \Gamma_2. \tag{6}$$

Hence, for any $v \in E$ such that $v = 0$ on $\Gamma_1 \cup \Gamma_2$, integration by part in (5) yields that

$$\int_{\Gamma_3} \frac{\partial}{\partial y}\left(\frac{\partial u_y}{\partial n}\right) v \, dy = 0.$$

Therefore, we obtain

$$\frac{\partial u_y}{\partial n} = C \quad \text{on } \Gamma_3, \tag{7}$$

where C is a constant.

Furthermore, when we define $E' := \{v \in H^2(\Omega) \mid v(a_1) = v(a_3)\}$, it follows that

$$\min_{u \in E} \frac{||\nabla u_y||}{||u_y||} = \min_{v \in E'} \frac{||\nabla v_y||}{||v_y||}. \tag{8}$$

Indeed, for any $v \in E'$, setting $u(x, y) := v(x, y) - (v(1, 0) - v(0, 0))x - v(0, 0)$, we have $u \in E$ and $u_y = v_y$ which implies the above equality.

Since $u \in E'$ is equivalent to $\int_{\Gamma_3} u_y \, ds = 0$, from (5), (6) and (7), the eigenvalue problem (4) is rewritten as follows.

Find $\phi \in H^1(\Omega)$ and $\lambda \in R$ such that

$$\begin{cases} -\Delta\phi = \lambda\phi & \text{in } \Omega, \\ \frac{\partial\phi}{\partial n} = 0 & \text{on } \Gamma_1 \cup \Gamma_2, \\ \frac{\partial\phi}{\partial n} = C & \text{on } \Gamma_3, \\ \int_{\Gamma_3} \phi \, ds = 0. \end{cases} \tag{9}$$

Now, if $C = 0$ in (9), then it leads to the problem:

$$\begin{cases} -\Delta\phi = \lambda\phi & \text{in } \Omega, \\ \frac{\partial\phi}{\partial n} = 0 & \text{on } \partial\Omega, \\ \int_{\Gamma_3} \phi \, ds = 0. \end{cases} \tag{10}$$

Taking into account that the smallest positive eigenvalue of the operator $-\Delta$ with Neumann boundary condition on Ω is π^2, the smallest positive eigenvalue of (10) is not less than π^2. In the present situation, as we consider the case that $0 < \lambda < \pi^2$, it suffices to deal with the case $C \neq 0$. Furthermore, it is easily seen that we may assume $C = 1$. Defining the function $\psi(x, y) := \frac{1}{2}\{(1-x)^2 + y^2\} - \frac{2}{3}$, and setting $\hat{\phi} := \phi - \psi$, by some elementary calculations we have

$$\begin{cases} -\Delta\hat{\phi} = \lambda\hat{\phi} + \lambda\psi + \Delta\psi & \text{in } \Omega \\ \frac{\partial\hat{\phi}}{\partial n} = 0 & \text{on } \partial\Omega \\ \int_{\Gamma_3} \hat{\phi}\, ds = 0, \end{cases} \tag{11}$$

where $\Delta\psi = 2$.

We now introduce the operator $L_0 : L^2(\Omega) \longrightarrow L_0^2(\Omega)$, where $L_0^2(\Omega) = \{u \in L^2(\Omega) \mid \int_\Omega u\, d\Omega = 0\}$, by

$$L_0 u := u - 2\int_\Omega u\, d\Omega.$$

Setting $\xi := L_0\hat{\phi}$, we have

$$\hat{\phi} = \xi - \frac{2}{\lambda} + \frac{1}{3}. \tag{12}$$

Indeed, the fact that $\int_\Omega \Delta\hat{\phi}\, d\Omega = \int_{\partial\Omega} \frac{\partial\hat{\phi}}{\partial n}\, ds = 0$ implies $\Delta\hat{\phi} \in L_0^2(\Omega)$. Taking this into account, observe that

$$\lambda\hat{\phi} + \lambda\psi + \Delta\psi = -\Delta\hat{\phi} = -L_0(\Delta\hat{\phi}) = \lambda\xi + \lambda\psi_0,$$

where $\psi_0 \equiv L_0\psi = \frac{1}{2}\{(1-x)^2 + y^2\} - \frac{1}{3}$. Hence, we have the desired relation (12).

Consequently, we obtain the following problem.

(P) Find the smallest positive $\lambda \in R$ for which the following associated problem has a solution $\xi \in H^1(\Omega) \cap L_0^2(\Omega)$

$$\begin{cases} -\Delta\xi = \lambda\xi + \lambda\psi_0 & \text{in } \Omega, \\ \frac{\partial\xi}{\partial n} = 0 & \text{on } \partial\Omega, \end{cases} \tag{13}$$

with

$$\int_{\Gamma_3} \xi\, ds = \int_{\Gamma_3} \left(\frac{2}{\lambda} - \frac{1}{3}\right) ds. \tag{14}$$

Note that ψ_0 belongs to $L_0^2(\Omega)$.

3. Verification Procedures for Problem (P)

As mentioned before, it suffices to look for a solution λ of (P) in the interval $(0, \pi^2)$. We denote the smallest λ by λ^*. The outline of the verification strategy which we adopt is as follows.

1. Calculate an approximate solution $\hat{\lambda}$ to λ^*.
2. Try to prove that there exists no solution ξ which satisfies (13) and (14) from $\lambda = 0$ to near $\hat{\lambda}$. This procedure can be done by calculating ξ satisfying (13) for any λ in an interval Λ using the numerical verification method described below, as well as by numerically proving that (14) does not hold for any (λ, ξ). Then it implies that the given interval Λ contains no solution of (P). In the actual verification, we can start at $\lambda = \lambda_N := 1/0.81^2$ because of Natterer's result.
3. If there is an interval Λ, very close to $\hat{\lambda}$, for which we cannot prove 2, then we try to verify that there exists a solution of (P) near $\hat{\lambda}$. For the lower bound $\underline{\lambda}$ of such an interval, the value $1/\sqrt{\underline{\lambda}}$ gives an upper bound of the desired constant C_0.

We now describe the outline of the numerical verification procedure for enclosing the solution ξ of (13) for a given λ. The detailed computational procedures will be mentioned in the forthcoming paper [11]. For the present purpose, we use the numerical verification method for enclosing the solutions for elliptic problems which is based on the finite element method and the constructive error estimates ([7, 10] etc. for details).

Note that, by the regularity result for the solution of Poisson's equation (e.g., [2]), any solution ξ for (13) belongs to $H^2(\Omega)$. Also taking into account that the boundary $\partial\Omega$ consists of piecewise straight lines, one can obtain the following estimates by similar arguments as in Chapter 1 of [4]

$$|\xi|_2 \leq ||\Delta\xi||. \tag{15}$$

We define the functional subspace $H := H^1(\Omega) \cap L_0^2(\Omega)$ with norm $||\phi||_H \equiv ||\nabla\phi||$. Let $\mathscr{T}_1$ be a uniform triangulation of Ω with mesh size h, and let $S_h^* \subset H^1(\Omega)$ be a linear finite element subspace on $\mathscr{T}_1$. And we define the projection P_h from H to $S_h^* \bigcap L_0^2(\Omega)$ by

$$\begin{cases} (\nabla P_h u, \nabla v_h) = (\nabla u, \nabla v_h) & \forall\, v_h \in S_h^*, \\ (P_h u, 1) = 0, \end{cases} \tag{16}$$

for $u \in H^1(\Omega)$.

For any $f \in L_0^2(\Omega)$, let $u = Kf \in H$ denote a unique solution of the following problem:

$$\begin{cases} -\Delta u = f & \text{in } \Omega, \\ \frac{\partial u}{\partial n} = 0 & \text{on } \partial\Omega, \\ \int_\Omega u \, d\Omega = 0. \end{cases} \tag{17}$$

When we define $F\xi := K\lambda(\xi + \psi_0)$, the affine operator F is clearly a compact operator on H and (13) is rewritten as the following fixed point equation

$$\xi = F\xi. \tag{18}$$

Therefore, if we can get the constructive error estimates for the projection P_h of the form

$$\|\nabla(u - P_h u)\| \leq C_1 h |u|_2, \tag{19}$$

where C_1 is a numerically determined constant, then we can also apply the existing verification method, such as [6–8], [10] etc., to enclose solutions for (18).

We now have, by using the definition of P_h, for arbitrary $\chi \in S_h^*$

$$\begin{aligned} \|\nabla(u - P_h u)\|^2 &= (\nabla(u - P_h u), \nabla(u - P_h u)) = (\nabla(u - P_h u), \nabla(u - \chi)) \\ &\leq \|\nabla(u - P_h u)\| \, \|\nabla(u - \chi)\|. \end{aligned} \tag{20}$$

Hence, choosing $\chi = \mathscr{I}u$, by using Natterer's result, at least, we may use the constant $C_1 = 0.81$ in (19). Similarly, the well known Aubin–Nitsche's trick can also be applied to get the L^2 error by

$$\|u - P_h u\| \leq C_1^2 h^2 |u|_2. \tag{21}$$

As described below, the constant 0.81 can be iteratively improved as the verification proceeds.

We omit the detailed verification procedure based upon these estimates for the solution of (18) for a given λ, because it is quite similar to that in already published papers ([7, 8, 10] etc.).

In our enclosure method of solutions for (18), the solution ξ is enclosed in the set Ξ, which we call 'a candidate set', of the form

$$\Xi = \Xi_h + [\alpha],$$

where Ξ_h is represented as a linear combination of the base functions of S_h^* with interval coefficients and $[\alpha]$ stands for a ball in the orthogonal complement subspace of $S_h^* \cap L_0^2(\Omega)$ in H with radius α. Namely,

$$\Xi_h \equiv \sum_{i=1}^{M} [\underline{A}_i, \bar{A}_i] \phi_i$$

and

$$[\alpha] \equiv \{\phi \in S_h^* \cap L_0^2(\Omega)^{\perp} \mid \|\nabla \phi\| \leq \alpha\},$$

where $\{\phi\}$ is a basis of S_h^*. Therefore, after the verification of ξ, it is possible to check the following property for all $\xi \in \Xi$ and $\lambda \in \Lambda$:

$$\int_{\Gamma_3} \xi \, ds \neq \int_{\Gamma_3} \left(\frac{2}{\lambda} - \frac{1}{3} \right) ds. \tag{22}$$

The details of the efficient estimation of the left-hand side of (22) will be discussed in the forthcoming paper [11].

Thus, we can summarize the computational algorithm as follows.

1. Set $\lambda_0 := 1/0.81^2$, $C_1 = 0.81$ and $\Lambda := [\lambda_0, \lambda_0 + t_0]$ for some positive number t_0.
2. By using the error estimates (19) with C_1, enclose all solutions ξ of (13) for all $\lambda \in \Lambda$.
3. If the relation (22) holds, then set $C_1 := 1/\sqrt{\underline{\Lambda}}$ and $\Lambda := [\bar{\Lambda}, \bar{\Lambda} + t]$ for some positive number t which is dependent on the position of λ as well as on the mesh size h.

   Go to 2.

   Otherwise, enclose the solution pair (λ, ξ) for the simultaneous system (13) and (14). Here, λ has to be enclosed in the candidate set of the form : $[\underline{\Lambda}, \bar{\Lambda} + t_1]$ for some positive number t_1. Then, $C_1 := 1/\sqrt{\underline{\Lambda}}$ is a new upper bound for C_0. If it is impossible to enclose such a (λ, ξ), then refine the approximation space S_h^*, e.g., taking a finer mesh.

4. Computational Result and Conclusion

In the actual computation, we used a little bit of modified algorithm of the above to reduce the computing time, which will be described in [11].

In the excluding process, we started with mesh size $h = 1/40$ and $t_0 = 0.5$ but in later we needed $h = 1/120$ and very small width t. Finally, we enclosed the smallest value λ^* for the problem (P) in the interval $\Lambda^* \equiv [4.09940, 4.12611]$, which implies that $C_0 \leq 0.4939$. Thus we verified that the best constant $C \equiv C_0$ in (2) exists in the interval $[0.467, 0.494]$ which should be sufficiently accurate by the practical viewpoint.

In order to avoid the rounding error in the floating point computations, we used Fortran 90 with verified interval library INTLIB [3], coded by Kearfott, on the SUN Workstation Ultra Enterprise 450, Computing Center, Kyushu University.

References

[1] Arbenz, P.: Computable finite element error bounds for Poisson's equation. IMA J. Numer. Anal. *2*, 475–479 (1982).
[2] Grisvard, P.: Elliptic problems in nonsmooth domain. Boston: Pitman, 1985.
[3] Kearfott, R. B., Kreinovich, V. (eds.): Applications of interval computations. Dordrecht: Kluwer, 1996.
[4] Ladyzhenskaya, O. A.: The mathematical theory of viscous incompressible flow, 2nd edn. New York: Gordon and Breach, 1969.
[5] Lehmann, R.: Computable error bounds in the finite-element method. IMA J. Numer. Anal. *6*, 265–271 (1986).
[6] Nakao, M. T.: A numerical approach to the proof of existence of solutions for elliptic problems. Jpn. J. Appl. Math. *5*, 313–332 (1988).

[7] Nakao, M. T.: Solving nonlinear elliptic problems with result verification using an H^{-1} residual iteration. Computing [Suppl.] *9*, 161–173 (1993).

[8] Nakao, M. T., Yamamoto, N.: Numerical verification of solutions for nonlinear elliptic problems using L^∞ residual method. J. Math. Anal. Appl. *217*, 246–262 (1998).

[9] Nakao, M. T., Yamamoto, N., Kimura, S.: On best constant in the optimal error estimates for the H_0^1-projection into piecewise polynomial spaces. J. Approx. Theory *93*, 491–500 (1998).

[10] Nakao, M. T.: Numerical verification methods for solutions of ordinary and partial differential equations. Numer. Functional Anal. Optim. (to appear).

[11] Nakao, M. T., Yamamoto, N.: A guaranteed bound of the optimal constant in the error estimates for linear triangular element: Part II: Details, submitted.

[12] Natterer, F.: Berechenbare Fehlerschranken für die Methode der Finite Elemente. International Series of Numerical Mathematics, Vol. 28, 109–121 Basel: Birkhäuser, 1975.

[13] Schultz, M.H.: Spline anlysis. London: Prentice-Hall, 1973.

M. T. Nakao
Faculty of Mathematics
Kyushu University 33
Fukuoka 812-8581
Japan
e-mail: mtnakao@math.kyushu-u.ac.jp

N. Yamamoto
The University of Electro-Communications
Chofu
Tokyo 182-8585
Japan
e-mail: yamamoto@masuo.im.uec.ac.jp

Computing [Suppl] 15, 175–187 (2001)

Computing

Fast Verification of Solutions for Sparse Monotone Matrix Equations

T. Ogita and **S. Oishi,** Tokyo, and **Y. Ushiro,** Kanagawa

Dedicated to Professor Tetsuro Yamamoto on the occasion of his 65th birthday

Abstract

This paper concerns the problem of verifying the accuracy of a numerical solution for linear systems with a coefficient matrix being sparse and monotone. Fast algorithms calculate a verified error bound for numerical solutions of the linear systems and a guaranteed condition number of the coefficient matrix. The proposed verification method is based on iterative solution methods and rounding mode controlled computations. Numerical results have also been presented for illustrating that a cost for verification is less than that for calculating a numerical solution.

AMS Subject Classifications: 65G99, 65N22, 65F10, 65F35, 65F50.

Key Words: Numerical verification, monotone, sparse linear systems, rounding mode controlled computation, iterative solution methods, condition number.

1. Introduction

In this paper, we are concerned with an accuracy of a computed solution of the following matrix equation

$$A\mathbf{x} = \mathbf{b}, \tag{1}$$

where A is an $n \times n$ real matrix and $\mathbf{x}$ and $\mathbf{b}$ are n-vectors. Recently, as a result of the progress of interval analysis, various methods have been developed for calculating a rigorous error bound between an exact solution and a computed solution of Eq. (1). However, many of them are based on direct methods of solving Eq. (1).

In this paper, we restrict ourselves to the case where A is an $n \times n$ real sparse monotone matrix. It is well known that in this class of problems, there still remain various important problems including a numerical treatment of elliptic partial differential equation. Then, based on iterative methods, we propose a new method of calculating a rigorous error bound for a computed solution of Eq. (1).

A main point of this paper is to develop a fast validation method in which computations are executed keeping the sparsity of matrix A. Thus, we should not calculate A^{-1} nor an LU decomposition of A. For the purpose, we try to use iteration algorithms for verification. A main result of the article is to show that this is possible if we utilize a monotonicity of A and a new validation method based on *rounding mode controlled computation* invented by one of the authors (S. Oishi) and S. M. Rump [1]. Moreover, we will also show that fast verification of condition number of matrix A is possible easily if A is monotone.

We shall show results of numerical experiments to exhibit that computational cost of verifying an accuracy is less than calculating a solution of Eq. (1).

2. Rounding Mode Controlled Computation

In this section, we shall show that for given real column vectors $\mathbf{p}$ and $\mathbf{q}$ whose elements are floating point numbers, an inclusion of inner product $\mathbf{p}^T\mathbf{q}$ can be calculated rigorously with rounding mode controlled computation.

Let us start with discussions on floating point numbers. Let $\mathbb{R}$ be the set of real numbers. Let $\mathbb{F}$ be a set of floating point numbers. In the following, we assume that the floating point system considered in this paper satisfies the following assumptions. The floating point systems abiding by the IEEE 754 floating point standard satisfy these assumptions.

First, we assume that $\mathbb{F}$ is symmetric, i.e., $x \in \mathbb{F} \Rightarrow -x \in \mathbb{F}$, such that $|x|$ is exact for $x \in \mathbb{F}$. Let $c \in \mathbb{R}$. We next assume that the following two types of rounding modes are defined:

(1) rounding downward Round $c \in \mathbb{R}$ to the largest floating point number $f \in \mathbb{F}$ satisfying $f \leq c$. We represent it by $\bigtriangledown : \mathbb{R} \to \mathbb{F}$.

(2) rounding upward Round $c \in \mathbb{R}$ to the smallest floating point number f satisfying $f \geq c$. We represent it by $\triangle : \mathbb{R} \to \mathbb{F}$.

We denote the rounding operator as $\bigcirc : \mathbb{R} \to \mathbb{F}$. Then, $\bigcirc$ is one of $\triangle$ and $\bigtriangledown$.

Then, we assume that the floating point arithmetic is defined through the following formula: For $\cdot \in \{+, -, \times, /\}$ and $\bigcirc \in \{\triangle, \bigtriangledown\}$

$$x \odot y = \bigcirc(x \cdot y), \quad (\text{for any } x, y \in \mathbb{F}). \tag{2}$$

Here, in the left hand side of Eq. (2), $\odot$ means the floating point arithmetic and in the right hand side of this equation, $x \cdot y$ represents the ordinary real arithmetic.

We can now present an algorithm of calculating an inclusion of an inner product $\mathbf{p}^T\mathbf{q}$. Here, the instruction **setround** (*down*) means to adopt the rounding downward mode and the instruction **setround** (*up*) means to adopt the rounding upward mode. Then, we assume that once the rounding mode is changed, it remains unchanged until the next instruction **setround** appears. This assumption is ensured on a wide class of computer systems abiding by the IEEE 754. Under this assumption, the variable $\underline{c}$ becomes the lower bound of $\mathbf{p}^T\mathbf{q}$ and $\overline{c}$ becomes the upper bound of it. Thus, the following inequality holds:

```
function [c̲, c̄] = iprod(p, q);
setround (down);          % rounding downward mode
c̲ = p^T q;                % lower bound of p^T q
setround (up);            % rounding upward mode
c̄ = p^T q;                % upper bound of p^T q
```

Algorithm 1. *Rigorous inclusion of inner product*

$$\underline{c} \leq \mathbf{p}^T\mathbf{q} \leq \bar{c}.$$

Here, we present several remarks

1. In the first place, we note that we should not use uncertain software libraries in the verifying process, because such libraries may cause unintended changes of the rounding mode. In addition, we note that the implementation of **setround** is very dependent on the computer environment. Thus, we should choose carefully CPU, operating system and compiler. For example, we use Intel Pentium III machine with Linux and gcc (GNU C Compiler). This combination satisfies IEEE 754 requirement.

2. In our experiments of several ten thousands dimensional problems, it is usually possible to get a very sharp inclusion of inner products by our rounding mode controlled computation. However, it should be note that the relative error of an inner product computed in the ordinary floating point arithmetic known to be arbitrarily bad if ill-conditioned large-dimensional vectors are treated. In case of treating such ill-conditioned problems, we should use extra hardware or software for calculating inner product. Especially, the method invented by U. W. Kulisch and W. L. Miranker [3] using a long accumulator for calculating exact value of inner product is extremely useful for this problem. Moreover, if we can use such a long accumulator, inclusion of inner product can be obtained by calculating inner product once. Unfortunately, at the present time, the long accumulator of Kulisch and Miranker's approach is still not standardized. Since the computer systems satisfying IEEE 754 standard are easily available now and vast useful numerical tools have already been developed on such systems, we consider to use rounding mode controlled computation based on IEEE 754 standard for verification of numerical computation.

3. In our verification process, we assume that an inner product of two n-dimensional vectors $\mathbf{p} = (p_1, p_2, \ldots, p_n)^T$ and $\mathbf{q} = (q_1, q_2, \ldots, q_n)^T$ is calculated by the following usual formula

$$\mathbf{p}^T\mathbf{q} = \sum_{i=1}^{n} p_i \cdot q_i.$$

Namely, we assume that although the order of calculating the products $p_i \cdot q_i$'s and the summation is free, the inner product is calculated through multiplication and addition operations. Thus, we assume that an algorithm of calculating an inner product with minus or divide operations such as the fast algorithm using FFT (Fast Fourier Transform) is not used in the verification process.

3. Matrix Norm Inclusion

In this section, we shall show that for a given $n \times n$ real matrix A with $A^{-1} \geq O$, a rigorous inclusion of $||A^{-1}||_\infty$ can be calculated using rounding mode controlled computation.

3.1. Some Matrix Properties

In the first place, definitions of matrices used in this paper are in order. In the following, the notation $A \geq O$ means that all elements of A are nonnegative, and $\boldsymbol{v} \geq 0$ all elements of $\boldsymbol{v}$ are nonnegative.

Definition 1. *A real matrix A is called monotone if $A\mathbf{x} \geq 0$ implies $\mathbf{x} \geq 0$.*

Definition 2. *A real square matrix $A = (a_{ij})$ is called a (nonsingular) M-matrix if $a_{ij} \leq 0$ for $i \neq j$ and if it is monotone – i.e., (by Lemma 1 shown below) if $A^{-1} \geq O$.*

Of course, A is monotone if A is an M-matrix. The following well-known lemma plays an important role in this paper:

Lemma 1. *A is monotone if and only if A is nonsingular with $A^{-1} \geq O$.*

In addition, we shall also cite here, to unify the notations, some well-known properties of vector and matrix norms used in the following. Let $\boldsymbol{v} = (v_i)$ be a real n-vector, then

$$||\boldsymbol{v}||_\infty = \max_{1 \leq i \leq n} |v_i|, \tag{3}$$

and let $A = (a_{ij})$ be an $m \times n$ real matrix, then

$$||A||_1 = \max_{1 \leq j \leq n} \sum_{i=1}^{m} |a_{ij}|, \quad ||A||_\infty = \max_{1 \leq i \leq m} \sum_{j=1}^{n} |a_{ij}|. \tag{4}$$

3.2. Enclosure of Matrix Norm

In the following, we show a theorem, which becomes a key to enclose the infinity norm of the inverse of a monotone matrix.

Theorem 1. *Let A be an $n \times n$ real monotone matrix and $\mathbf{e}$ be n-vector where $\mathbf{e} = (1, 1, \ldots, 1)^T$. Let $\widetilde{\mathbf{y}}$ be a computed solution of the matrix equation $A\mathbf{y} = \mathbf{e}$. Let further $\mathbf{s} = A\widetilde{\mathbf{y}} - \mathbf{e}$ be the residual vector. Then,*

$$\frac{||\widetilde{\mathbf{y}}||_\infty}{1+||\mathbf{s}||_\infty} \le ||A^{-1}||_\infty. \tag{5}$$

Furthermore, if $||\mathbf{s}||_\infty < 1$, *then*

$$||A^{-1}||_\infty \le \frac{||\widetilde{\mathbf{y}}||_\infty}{1-||\mathbf{s}||_\infty}. \tag{6}$$

Proof: Since A is an $n \times n$ real monotone matrix, $A^{-1} = (\widetilde{a}_{ij}) \ge O$. From this, it follows that

$$||A^{-1}||_\infty = \max_{1\le i\le n} \sum_{j=1}^{n} |\widetilde{a}_{ij}| = \max_{1\le i\le n} \sum_{j=1}^{n} \widetilde{a}_{ij} = ||A^{-1}\mathbf{e}||_\infty, \tag{7}$$

where $\mathbf{e} = (1, 1, \ldots, 1)^T$. Let $\mathbf{y}^* = A^{-1}\mathbf{e}$, then we have from Eq. (7)

$$||A^{-1}||_\infty = ||A^{-1}\mathbf{e}||_\infty = ||\mathbf{y}^*||_\infty. \tag{8}$$

Let $\widetilde{\mathbf{y}}$ be a computed solution of the matrix equation $A\mathbf{y} = \mathbf{e}$, and $\mathbf{s} = A\widetilde{\mathbf{y}} - \mathbf{e}$ the residual vector. Then,

$$\begin{aligned} ||\widetilde{\mathbf{y}}||_\infty &= ||(\widetilde{\mathbf{y}} - \mathbf{y}^*) + \mathbf{y}^*||_\infty \\ &\le ||\widetilde{\mathbf{y}} - \mathbf{y}^*||_\infty + ||\mathbf{y}^*||_\infty = ||\widetilde{\mathbf{y}} - A^{-1}\mathbf{e}||_\infty + ||\mathbf{y}^*||_\infty \\ &= ||A^{-1}(A\widetilde{\mathbf{y}} - \mathbf{e})||_\infty + ||\mathbf{y}^*||_\infty \\ &\le ||A^{-1}||_\infty \cdot ||\mathbf{s}||_\infty + ||\mathbf{y}^*||_\infty, \end{aligned} \tag{9}$$

From Eq. (8) and Eq. (9), we have

$$||\widetilde{\mathbf{y}}||_\infty \le ||A^{-1}||_\infty \cdot ||\mathbf{s}||_\infty + ||A^{-1}||_\infty, \tag{10}$$

i.e.,

$$||\widetilde{\mathbf{y}}||_\infty \le (1 + ||\mathbf{s}||_\infty)||A^{-1}||_\infty, \tag{11}$$

which implies Eq. (5).

Similarly,

$$\begin{aligned} ||\mathbf{y}^*||_\infty &= ||(\mathbf{y}^* - \widetilde{\mathbf{y}}) + \widetilde{\mathbf{y}}||_\infty \\ &\le ||\widetilde{\mathbf{y}} - \mathbf{y}^*||_\infty + ||\widetilde{\mathbf{y}}||_\infty = ||\widetilde{\mathbf{y}} - A^{-1}\mathbf{e}||_\infty + ||\widetilde{\mathbf{y}}||_\infty \\ &= ||A^{-1}(A\widetilde{\mathbf{y}} - \mathbf{e})||_\infty + ||\widetilde{\mathbf{y}}||_\infty \\ &\le ||A^{-1}||_\infty \cdot ||\mathbf{s}||_\infty + ||\widetilde{\mathbf{y}}||_\infty. \end{aligned} \tag{12}$$

From Eq. (8) and Eq. (12), we have

$$||A^{-1}||_\infty \le ||A^{-1}||_\infty \cdot ||\mathbf{s}||_\infty + ||\widetilde{\mathbf{y}}||_\infty, \tag{13}$$

i.e.,

$$(1 - ||\mathbf{s}||_\infty)||A^{-1}||_\infty \le ||\widetilde{\mathbf{y}}||_\infty. \tag{14}$$

From this, it is seen that if $||\mathbf{s}||_\infty < 1$ is satisfied, then

$$||A^{-1}||_\infty \le \frac{||\widetilde{\mathbf{y}}||_\infty}{1 - ||\mathbf{s}||_\infty}. \tag{15}$$

Thus, we finally have Eq. (6). □

We now present Algorithm 2 which calculates the inclusion of $||A^{-1}||_\infty$ if A is monotone. Here, for given n-vector $\mathbf{v}$, the instruction **max(abs(**$\underline{\mathbf{v}}$**), abs(**$\overline{\mathbf{v}}$**))** produces the vector $\widehat{\mathbf{v}}$ with the i-th element $\widehat{v}_i = \max\{|\underline{v}_i|, |\overline{v}_i|\}$ and the instruction **norm**$(\mathbf{v}, \infty)$ is to calculate $||\mathbf{v}||_\infty$. Thus, if Algorithm 2 terminates with success ($\underline{d} > 0$), then the following inequality holds:

```
function [a_invinf_lower, a_invinf_upper] = ainvinf (A, ỹ);
setround (down);
s_lower = A ỹ − e;                    % lower bound for Aỹ − e
setround (up);
s_upper = A ỹ − e;                    % upper bound for Aỹ − e
ŝ = max (abs(s_lower), abs(s_upper)); % ŝ_i = max{|s_lower_i|, |s_upper_i|}
y_norm = norm (ỹ, ∞);                 % ||ỹ||_∞
s_norm_upper = norm (ŝ, ∞);           % upper bound of ||Aỹ − e||_∞
d_upper = 1 + s_norm_upper;
setround (down);
a_invinf_lower = y_norm / d_upper;    % lower bound of ||A^{-1}||_∞
d_lower = 1 − s_norm_upper;
if d_lower > 0;                       % ||s||_∞ < 1
   setround (up);
   a_invinf_upper = y_norm / d_lower; % upper bound of ||A^{-1}||_∞
else;
   print('inclusion failed');
end;
```

Algorithm 2. *Rigorous inclusion of* $||A^{-1}||_\infty$

$$\underline{a_{\text{invinf}}} \le ||A^{-1}||_\infty \le \overline{a_{\text{invinf}}}.$$

4. Verification of Solutions

In this section, we shall show that for a given $n \times n$ real matrix A with $A^{-1} \geq O$, a rigorous upper bound of $\|\tilde{\mathbf{x}} - \mathbf{x}^*\|_\infty$ can be calculated using iterative solution methods and rounding mode controlled computation. Here, $\|\tilde{\mathbf{x}} - \mathbf{x}^*\|_\infty$ is the maximum distance between the exact solution $\mathbf{x}^*$ and a computed solution $\tilde{\mathbf{x}}$ of Eq. (1).

4.1. Verification Theory

The following is a fundamental theorem for verifying the solution of sparse monotone matrix equations.

Theorem 2. *Let A be an $n \times n$ real monotone matrix and $\mathbf{x}$, $\mathbf{y}$, $\mathbf{b}$ and $\mathbf{e}$ n-vectors. Let $\mathbf{x}^*$ and $\tilde{\mathbf{x}}$ be the exact solution and a computed solution of a matrix equation $A\mathbf{x} = \mathbf{b}$, respectively. Let also $\tilde{\mathbf{y}}$ be a computed solution of a matrix equation $A\mathbf{y} = \mathbf{e}$. Here $\mathbf{e} = (1, 1, \ldots, 1)^T$. Let further $\mathbf{r} = A\tilde{\mathbf{x}} - \mathbf{b}$ and $\mathbf{s} = A\tilde{\mathbf{y}} - \mathbf{e}$ be residual vectors. If $\|\mathbf{s}\|_\infty < 1$, then*

$$\|\tilde{\mathbf{x}} - \mathbf{x}^*\|_\infty \leq \frac{\|\tilde{\mathbf{y}}\|_\infty \cdot \|\mathbf{r}\|_\infty}{1 - \|\mathbf{s}\|_\infty}. \tag{16}$$

Proof: Let $\tilde{\mathbf{x}}$ be a computed solution of $A\mathbf{x} = \mathbf{b}$, and $\mathbf{r} = A\tilde{\mathbf{x}} - \mathbf{b}$ the residual vector. Then, it is seen that

$$\begin{aligned}\|\tilde{\mathbf{x}} - \mathbf{x}^*\|_\infty &= \|\tilde{\mathbf{x}} - A^{-1}\mathbf{b}\|_\infty = \|A^{-1}(A\tilde{\mathbf{x}} - \mathbf{b})\|_\infty \\ &\leq \|A^{-1}\|_\infty \cdot \|\mathbf{r}\|_\infty.\end{aligned} \tag{17}$$

From Eq. (6) and Eq. (17), we have

$$\|\tilde{\mathbf{x}} - \mathbf{x}^*\|_\infty \leq \|A^{-1}\|_\infty \cdot \|\mathbf{r}\|_\infty \leq \frac{\|\tilde{\mathbf{y}}\|_\infty \cdot \|\mathbf{r}\|_\infty}{1 - \|\mathbf{s}\|_\infty}, \tag{18}$$

which implies the desired formula (16). □

Similarly, we can also derive a theorem for calculating a relative error bound of a computed solution as follows.

Theorem 3. *Let A be an $n \times n$ real matrix and $\mathbf{x}$ and $\mathbf{b}$ be n-vectors. Let $\mathbf{x}^*$ be the exact solution and $\tilde{\mathbf{x}}$ a computed solution of a matrix equation $A\mathbf{x} = \mathbf{b}$. If $\|\tilde{\mathbf{x}} - \mathbf{x}^*\|_\infty \leq \varepsilon < \|\tilde{\mathbf{x}}\|_\infty$, then*

$$\frac{\|\tilde{\mathbf{x}} - \mathbf{x}^*\|_\infty}{\|\mathbf{x}^*\|_\infty} \leq \frac{\varepsilon}{\|\tilde{\mathbf{x}}\|_\infty - \varepsilon}. \tag{19}$$

Proof: It is assumed that the inequality

$$\|\widetilde{\mathbf{x}} - \mathbf{x}^*\|_\infty \le \varepsilon \tag{20}$$

is satisfied. From a property of vector norm,

$$\|\widetilde{\mathbf{x}}\|_\infty = \|(\widetilde{\mathbf{x}} - \mathbf{x}^*) + \mathbf{x}^*\|_\infty \le \|\widetilde{\mathbf{x}} - \mathbf{x}^*\|_\infty + \|\mathbf{x}^*\|_\infty. \tag{21}$$

i.e.,

$$\|\mathbf{x}^*\|_\infty \ge \|\widetilde{\mathbf{x}}\|_\infty - \|\widetilde{\mathbf{x}} - \mathbf{x}^*\|_\infty. \tag{22}$$

From Eq. (20) and Eq. (22),

$$\|\mathbf{x}^*\|_\infty \ge \|\widetilde{\mathbf{x}}\|_\infty - \|\widetilde{\mathbf{x}} - \mathbf{x}^*\|_\infty \ge \|\widetilde{\mathbf{x}}\|_\infty - \varepsilon. \tag{23}$$

From this, it is seen that if $\|\widetilde{\mathbf{x}}\|_\infty - \varepsilon > 0$ is satisfied, then

$$\frac{1}{\|\mathbf{x}^*\|_\infty} \le \frac{1}{\|\widetilde{\mathbf{x}}\|_\infty - \varepsilon}. \tag{24}$$

From Eq. (20) and Eq. (24), we finally have

$$\frac{\|\widetilde{\mathbf{x}} - \mathbf{x}^*\|_\infty}{\|\mathbf{x}^*\|_\infty} \le \frac{\varepsilon}{\|\widetilde{\mathbf{x}}\|_\infty - \varepsilon}, \tag{25}$$

which implies the desired formula (19). □

We now present an algorithm of calculating an upper bound of $\|\widetilde{\mathbf{x}} - \mathbf{x}^*\|_\infty$ in Algorithm 3. If Algorithm 3 terminates with success, then the following estimate holds:

function $e_{\text{abs}} =$ **errabs** $(A, \mathbf{b}, \widetilde{\mathbf{x}}, \overline{a_{\text{invinf}}})$;
setround $(down)$;
$\underline{\mathbf{r}} = A\widetilde{\mathbf{x}} - \mathbf{b}$; % lower bound for $A\widetilde{\mathbf{x}} - \mathbf{b}$
setround (up);
$\overline{\mathbf{r}} = A\widetilde{\mathbf{x}} - \mathbf{b}$; % upper bound for $A\widetilde{\mathbf{x}} - \mathbf{b}$
$\widehat{\mathbf{r}} =$ **max** $($**abs**$(\underline{\mathbf{r}})$, **abs**$(\overline{\mathbf{r}}))$; % $\widehat{\mathbf{r}}_i = \max\{|\underline{\mathbf{r}}_i|, |\overline{\mathbf{r}}_i|\}$
$\overline{r_{\text{norm}}} =$ **norm** $(\widehat{\mathbf{r}}, \infty)$; % upper bound of $\|A\widetilde{\mathbf{x}} - \mathbf{b}\|_\infty$
$e_{\text{abs}} = \overline{a_{\text{invinf}}} * \overline{r_{\text{norm}}}$; % $\|\widetilde{\mathbf{x}} - \mathbf{x}^*\|_\infty \le e_{\text{abs}}$

Algorithm 3. *Upper bound of* $\|\widetilde{\mathbf{x}} - \mathbf{x}^*\|_\infty$

$$\|\widetilde{\mathbf{x}} - \mathbf{x}^*\|_\infty \le e_{\text{abs}}.$$

Furthermore, we present an algorithm of calculating an relative error bound of solution in Algorithm 4. If Algorithm 4 terminates with success ($\underline{e} > 0$), then the following estimate holds:

$$\frac{||\widetilde{\mathbf{x}} - \mathbf{x}^*||_\infty}{||\mathbf{x}^*||_\infty} \leq e_{\text{rel}}.$$

```
function e_rel = errrel (x̃, e_abs);
x_norm = norm (x̃, ∞);                 % ||x̃||_∞
setround (down);
e̲ = x_norm − e_abs;
if e̲ > 0;                             % ||x̃||_∞ − e_abs > 0
    setround (up);
    e_rel = e_abs/e̲;                  % ||x̃ − x*||_∞/||x*||_∞ ≤ e_rel
else;
    print ('verification failed');
end;
```

Algorithm 4. *Upper bound of* $||\widetilde{\mathbf{x}} - \mathbf{x}^*||_\infty/||\mathbf{x}^*||_\infty$

4.2. Fast Verification with Iterative Solution Method

In order to solve the sparse matrix equation $A\mathbf{x} = \mathbf{b}$ using the sparsity of A efficiently, iterative solution methods are frequently used. Algorithm 5 calculates a relative error bound of solution for the sparse monotone matrix equation with iteration methods. Here, the instruction **xsolve** $(A, \mathbf{b}, \varepsilon_r, p)$ produces a solution vector $\widetilde{\mathbf{x}}$ of the sparse matrix equation $A\mathbf{x} = \mathbf{b}$ using an iterative solution method. Furthermore, the constant ε_r is for a stopping criterion of the iteration such that $||A\widetilde{\mathbf{x}} - \mathbf{b}||_p/||\mathbf{b}||_p < \varepsilon_r$ where $p = 1, 2$ or ∞. Similarly, the instruction **ysolve** $(A, \mathbf{e}, \varepsilon_s, q)$ also produces a solution vector $\widetilde{\mathbf{y}}$ of $A\mathbf{y} = \mathbf{e}$, and ε_s is for the stopping criterion $||A\widetilde{\mathbf{y}} - \mathbf{e}||_q/||\mathbf{e}||_q < \varepsilon_s$.

```
x̃ = xsolve (A, b, ε_r, p);            % solving Ax = b with iterative solver
ỹ = ysolve (A, e, ε_s, q);            % solving Ay = e with iterative solver
ā_invinf = ainvinf (A, ỹ);
e_abs = errabs (A, b, x̃, ā_invinf);
e_rel = errrel (x̃, e_abs);
```

Algorithm 5. *Verification of solution with iterative solver*

From Eq. (6)

$$||A^{-1}||_\infty \leq \frac{||\widetilde{\mathbf{y}}||_\infty}{1 - ||\mathbf{s}||_\infty},$$

it is seen that $||\mathbf{s}||_\infty$ should be less than one but could be considerably bigger than $||\mathbf{r}||_\infty$, for example, $||\mathbf{s}||_\infty = 10^{-2}$ while $||\mathbf{r}||_\infty = 10^{-10}$, because an estimate for $||\mathbf{s}||_\infty$ is contributed to an estimation of the quantity $||A^{-1}||_\infty$ through the term $||\tilde{\mathbf{y}}||_\infty/(1 - ||\mathbf{s}||_\infty)$. Thus, it can be expected that to calculate a guaranteed error bound of a computed solution of matrix equation $A\mathbf{x} = \mathbf{b}$ is fairly faster than to calculate a numerical solution of this matrix equation provide that we choose ε_s considerably bigger than ε_r.

5. Condition Number Inclusion

In this section, we present an algorithm for verifying condition numbers of monotone matrices. For real square matrix A, the condition numbers in p-norms ($p = 1, 2, \ldots, \infty$) are defined by

$$\text{cond}_p(A) = ||A||_p \cdot ||A^{-1}||_p. \tag{26}$$

It is well known that from a relation between $||A||_1$ and $||A||_\infty$, it holds that

$$||A||_1 = ||A^T||_\infty \quad \text{and} \quad ||A^{-1}||_1 = ||A^{-T}||_\infty, \tag{27}$$

i.e.,

$$\text{cond}_1(A) = \text{cond}_\infty(A^T). \tag{28}$$

We now present an algorithm, Algorithm 6, for calculating a rigorous inclusion of $\text{cond}_\infty(A)$ in the following. Here, the instruction **norm** (A, ∞) is to calculate $||A||_\infty$. If Algorithm 6 terminates with success, then the following inequality holds:

> **function** $[\underline{c_{\text{inf}}}, \overline{c_{\text{inf}}}] =$ **condinf** $(A, \tilde{\mathbf{y}})$;
> $[\underline{a_{\text{invinf}}}, \overline{a_{\text{invinf}}}] =$ **ainvinf** $(A, \tilde{\mathbf{y}})$;
> **setround** $(down)$;
> $\underline{a_{\text{inf}}} =$ **norm** (A, ∞); % lower bound of $||A||_\infty$
> $\underline{c_{\text{inf}}} = \underline{a_{\text{inf}}} * \underline{a_{\text{invinf}}}$; % lower bound of $\text{cond}_\infty(A)$
> **setround** (up);
> $\overline{a_{\text{inf}}} =$ **norm** (A, ∞); % upper bound of $||A||_\infty$
> $\overline{c_{\text{inf}}} = \overline{a_{\text{inf}}} * \overline{a_{\text{invinf}}}$; % upper bound of $\text{cond}_\infty(A)$

Algorithm 6. *Rigorous inclusion of* $\text{cond}_\infty(A)$

$$\underline{c_{\text{inf}}} \le \text{cond}_\infty(A) \le \overline{c_{\text{inf}}}.$$

From Eq. (28), we can also obtain an inclusion of $\text{cond}_1(A)$ by solving the matrix equation $A^T\mathbf{y} = \mathbf{e}$ instead of solving $A\mathbf{y} = \mathbf{e}$ where $\mathbf{e} = (1, 1, \ldots, 1)^T$, and using the function **condinf** $(A^T, \tilde{\mathbf{y}})$.

6. Numerical Experiments

In this section, we shall show that the proposed verification method can be implemented using a usual iterative solver. We shall also report results of some numerical experiments.

In order to evaluate the proposed method, we consider the diffusion equation which often appear in heat transfer or electromagnetic problems. If we discretize a diffusion equation to a matrix equation by FDM (Finite Difference Method) or FEM (Finite Element Method), the derived matrix is known to become monotone. In particular, it is known that it becomes an M-matrix when FDM is used. Since the monotone matrix class includes the M-matrix class, we consider a monotone matrix equation which is a discretized form of a diffusion equation by FEM. Let us consider a two-dimensional heat transfer model as in Fig. 1. In this example, we have chosen that $k_1 = 1.0$, $f_1 = 0$, $k_2 = 0.1$, $f_2 = 0$, $k_3 = 1.0$, $f_3 = 20$ and $T_\infty = 0$. We discretize this model to a matrix equation $A\mathbf{x} = \mathbf{b}$ by FEM using linear triangular elements. We adopt MICCG (Modified Incomplete Cholesky Conjugate Gradient) method as an iterative solver to both of **xsolve** and **ysolve** in Algorithm 5, and we choose that $p = 2$ and $\varepsilon_r = 10^{-12}$ for **xsolve**, i.e., $||A\tilde{\mathbf{x}} - \mathbf{b}||_2/||\mathbf{b}||_2 < 10^{-12}$ as stopping criterion and that $q = 2$ and $\varepsilon_s = 10^{-3}$ for **ysolve**, i.e., $||A\tilde{\mathbf{y}} - \mathbf{e}||_2/||\mathbf{e}||_2 < 10^{-3}$ as stopping criterion. We use Intel Celeron 566 MHz CPU and Fortran and C compilers on our PC.

We have treated the following two cases:

	grid size (n)	**convection coefficient** (h)
Case 1 :	from 10,000 to 250,000	1.0 (fixed)
Case 2 :	10,000 (fixed)	from 1.0 to 1.0×10^{-8}

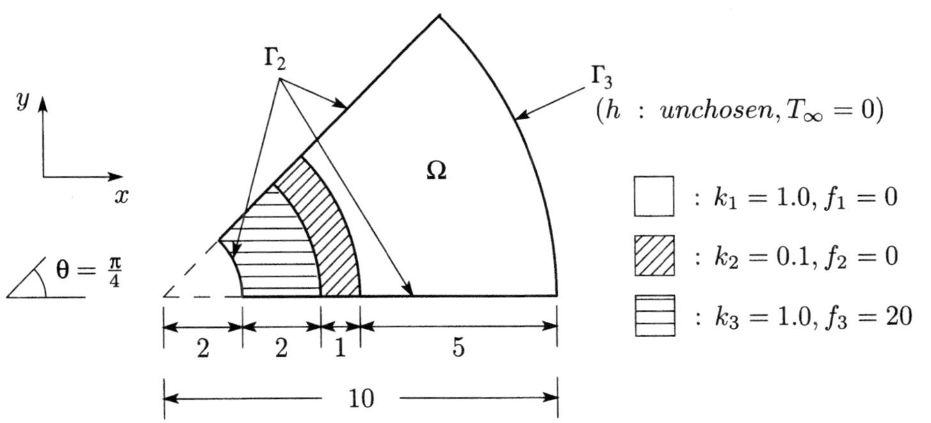

$$\begin{cases} \operatorname{div}\{-k \cdot \operatorname{grad}(u)\} = f & \text{in } \Omega \\ \{-k \cdot \operatorname{grad}(u)\} \times \mathbf{n} = 0 & \text{on } \Gamma_2 \\ \{-k \cdot \operatorname{grad}(u)\} \times \mathbf{n} = h(u - T_\infty) & \text{on } \Gamma_3 \end{cases}$$

Figure 1. Evaluation model

Table 1 and Table 2 present the relative error bound of solutions calculated by Algorithm 5. These tables also present the elapsed time of calculating a solution of $A\mathbf{x} = \mathbf{b}$ and that for calculating its error bound. These tables show that the elapsed time for verification of solution is less than that for calculation of solution in each case.

Furthermore, Table 3 and Table 4 present results of inclusion of condition number $\text{cond}_\infty(A)$ by Algorithm 6. These tables show that we can get a rigorous

Table 1. Results of verification of solution (**Case 1**)

Grid size (n)	Solving time (sec)	Verifying time (sec)	Relative error bound
10,000	3.3	1.7	4.1×10^{-10}
40,000	27.1	10.2	2.5×10^{-9}
90,000	90.7	32.3	7.1×10^{-9}
160,000	216.2	77.0	1.6×10^{-8}
250,000	458.5	146.8	3.3×10^{-8}

Table 2. Results of verification of solution (**Case 2**)

Convection coefficient (h)	Sol. time (sec)	Ver. time (sec)	Rel. error bound
1.0	3.3	1.7	4.1×10^{-10}
1.0×10^{-2}	3.6	2.0	3.4×10^{-9}
1.0×10^{-4}	5.1	2.3	2.3×10^{-7}
1.0×10^{-6}	6.0	2.7	3.1×10^{-5}
1.0×10^{-8}	7.0	2.9	3.3×10^{-3}

Table 3. Results of inclusion of condition number (**Case 1**)

Grid size (n)	Lower bound of $\text{cond}_\infty(A)$	Upper bound of $\text{cond}_\infty(A)$
10,000	5.92×10^5	6.03×10^5
40,000	2.49×10^6	2.56×10^6
90,000	5.71×10^6	5.87×10^6
160,000	1.02×10^7	1.06×10^7
250,000	1.61×10^7	1.66×10^7

Table 4. Results of inclusion of condition number (**Case 2**)

Convection coefficient (h)	Low. bound of $\text{cond}_\infty(A)$	Up. bound of $\text{cond}_\infty(A)$
1.0	5.92×10^5	6.03×10^5
1.0×10^{-2}	3.53×10^6	3.62×10^6
1.0×10^{-4}	2.98×10^8	3.04×10^8
1.0×10^{-6}	2.97×10^{10}	3.04×10^{10}
1.0×10^{-8}	2.99×10^{12}	3.03×10^{12}

inclusion of a condition number $\mathrm{cond}_\infty(A)$ sharply in each case. Similarly, we can also enclose $\mathrm{cond}_1(A)$.

7. Conclusion

We presented a method of verifying an accuracy of a numerical solution for linear systems with a coefficient matrix being sparse and monotone. This verification method does not need to calculate A^{-1} nor an LU decomposition, because it uses iterative solution methods. Using numerical experiments, it is illustrated that the method can be applied to fairly large size and ill-conditioned matrices and that computational cost of calculating an error bound between a computed solution and an exact solution is less than that for calculating a numerical solution. Moreover, a fast algorithm is also presented for enclosing condition numbers for monotone matrices.

References

[1] Oishi, S., Rump, S. M.: Fast verification of solutions of matrix equations. Numer. Math. (to appear).
[2] Oishi, S.: Fast enclosure of matrix eigenvalues and singular values via rounding mode controlled computation. Linear Alg. Appl. *324*, 133–146 (2001).
[3] Kulisch, U. W., Miranker, W. L.: Computer arithmetic in theory and practice. New York: Academic Press, 1981.
[4] Axelsson, O.: Iterative solution methods. Cambridge: Cambridge University Press, 1994.
[5] Golub, G. H., Van Loan, C. F.: Matrix computations, 3rd ed. Baltimore and London: The Johns Hopkins University Press, 1996.

Takeshi Ogita
School of Education
Waseda University
Tokyo 169-8050
Japan

Shin'ichi Oishi
Department of Computer and Information Science
School of Science and Engineering
Waseda University
Tokyo 169-8555
Japan

Yasunori Ushiro
Enterprise Server Division
Hitachi Ltd.
Kanagawa 243-0435
Japan

Computing [Suppl] 15, 189–209 (2001)

Computing

Laguerre-like Methods for the Simultaneous Approximation of Polynomial Zeros

M. S. Petković, L. Petković, and **D. Živković,** Niš

Dedicated to Professor Tetsuro Yamamoto on the occasion of his 65th birthday

Abstract

Using Newton's and Halley's corrections new simultaneous methods for solving polynomial equations, based on classical Laguerre's method, are obtained. The convergence order of the proposed methods is five and six, respectively. Further improvements of these methods are accomplished by applying the Gauss-Seidel approach. The lower bounds of the R-order of convergence of the accelerated (single-step) methods are derived. Faster convergence of all proposed methods is attained without additional operations, which provides a high computational efficiency of these methods. Convergence analysis and numerical results are given.

AMS Subject Classifications: 65H05.

Key Words: Determination of polynomial zeros, Laguerre's method, simultaneous iterative methods, accelerated convergence, R-order of convergence.

1. Introduction

Laguerre's method belongs to the most powerful methods for solving polynomial equations. Its convergence characteristics were investigated in details in the books [5] and [8] (see, also, [3, 4, 9]). Two modifications of Laguerre's method, which provide simultaneous determination of all simple zeros of a polynomial and have the order of convergence at least four, were presented in [4]. For these modified methods the implementation on parallel computers and the comparison with other methods were given.

In this paper we give another modifications of Laguerre's method which possess very fast convergence. The proposed methods have a high computational efficiency since the acceleration of convergence is attained with only a few additional computations.

The presentation of the paper is organized as follows. In Section 2 we derive a fixed point relation of Laguerre's type which is the basis for the construction of Laguerre-like simultaneous methods. Starting from this fixed point relation several modified methods of Laguerre's type in parallel and serial mode are derived in Section 3. Some estimates and inequalities, necessary for the convergence analysis, are given in Section 4. The main convergence theorems for the total-step as well as single-step Laguerre-like methods, are established in Section 5. The results of numerical experiments are presented in Section 6. Since the considered iterative formulas contain the sign $\pm$ in front of the square root involved in these formulas, a short analysis concerning the choice of the proper sign is given in Appendix.

2. Fixed-point Relation of Laguerre's Type

Let P be a monic polynomial with simple zeros $\zeta_1, \ldots, \zeta_n$ and let $z_1, \ldots, z_n$ be their mutually distinct approximations. For the point $z = z_i$ ($i \in I_n := \{1, \ldots, n\}$) let us introduce the notations:

$$\Sigma_{\lambda,i} = \sum_{\substack{j=1 \\ j \neq i}}^{n} \frac{1}{(z_i - \zeta_j)^{\lambda}}, \quad S_{\lambda,i} = \sum_{\substack{j=1 \\ j \neq i}}^{n} \frac{1}{(z_i - z_j)^{\lambda}} \quad (\lambda = 1, 2),$$

$$\delta_{1,i} = \frac{P'(z_i)}{P(z_i)}, \quad \delta_{2,i} = \frac{P'(z_i)^2 - P(z_i)P''(z_i)}{P(z_i)^2}, \tag{1}$$

$$f_i = nS_{2,i} - \frac{n}{n-1} S_{1,i}^2, \quad F_i = n\Sigma_{2,i} - \frac{n}{n-1} \Sigma_{1,i}^2,$$

$$\varepsilon_i = z_i - \zeta_i, \quad \varepsilon = \max_{1 \leq i \leq n} |\varepsilon_i|. \tag{2}$$

Lemma 1. *For $i \in I_n$ the following identity is valid*

$$n\delta_{2,i} - \delta_{1,i}^2 - F_i = \frac{1}{n-1} \left(\frac{n}{\varepsilon_i} - \delta_{1,i} \right)^2. \tag{3}$$

Proof: Starting from the identities

$$\frac{P'(z)}{P(z)} = \sum_{j=1}^{n} \frac{1}{z - \zeta_j} \tag{4}$$

and

$$\frac{P'(z)^2 - P(z)P''(z)}{P(z)^2} = -\left(\frac{P'(z)}{P(z)} \right)' = \sum_{j=1}^{n} \frac{1}{(z - \zeta_j)^2}, \tag{5}$$

we obtain

$$\begin{aligned} &n\delta_{2,i} - \delta_{1,i}^2 - F_i \\ &= n\left(\frac{1}{\varepsilon_i^2} + \Sigma_{2,i} \right) - \left(\frac{1}{\varepsilon_i} + \Sigma_{1,i} \right)^2 - n\Sigma_{2,i} + \frac{n}{n-1} \Sigma_{1,i}^2 = \frac{n-1}{\varepsilon_i^2} + \frac{1}{n-1} \Sigma_{1,i}^2 - \frac{2}{\varepsilon_i} \Sigma_{1,i} \\ &= \frac{1}{n-1} \left(\frac{1}{\varepsilon_i^2} + \Sigma_{1,i}^2 + \frac{2}{\varepsilon_i} \Sigma_{1,i} + \frac{n^2}{\varepsilon_i^2} - \frac{2n}{\varepsilon_i^2} - \frac{2n}{\varepsilon_i} \Sigma_{1,i} \right) = \frac{1}{n-1} \left(\frac{n}{\varepsilon_i} - \delta_{1,i} \right)^2. \quad \square \end{aligned}$$

From the identity (3) we obtain the following fixed point relation of Laguerre's type

$$\zeta_i = z_i - \frac{n}{\delta_{1,i} + [(n-1)(n\delta_{2,i} - \delta_{1,i}^2 - F_i)]^{1/2}} \quad (i \in I_n) \tag{6}$$

assuming that two values of the square root have to be taken in (6). The name comes from the fact that, neglecting the term F_i in (6), we obtain the classical Laguerre method of the third order for finding a single zero,

$$\hat{z}_i = z_i - \frac{n}{\delta_{1,i} + [(n-1)(n\delta_{2,i} - \delta_{1,i}^2)]^{1/2}}. \tag{7}$$

From the same reason, all methods derived from the fixed point relation (6) will be called Laguerre-like methods. We note that a different derivation of the fixed point relation (6) can be found in [4] and [9].

Remark 1. We assume that two values of the (complex) square root have to be taken in the last two formulas. Considering the fixed point relation (6), Laguerre's method (7) and all modified methods of Laguerre's type which will be presented below, we observe that a "proper" sign in front of the square root has to be chosen. We choose the sign so that a smaller step $|\hat{z}_i - z_i|$ is taken. A practical criterion for the choice of the proper sign is given in Section 6. In the sequel we will use the symbol $*$ to indicate the choice of the proper value of the square root involved in the presented iterative formulas and expressions appearing in the convergence analysis.

3. Some Modifications of Laguerre's Method

Let us introduce some notations:

1. The approximations $z_1^{(m)}, \ldots, z_n^{(m)}$ of the zeros at the mth iterative step will be briefly denoted by $z_1, \ldots, z_n$, and the new approximations $z_1^{(m+1)}, \ldots, z_n^{(m+1)}$, obtained by some simultaneous iterative method, by $\hat{z}_1, \ldots, \hat{z}_n$, respectively;

2.

$$N_i = N(z_i) = 1/\delta_{1,i} = \frac{P(z_i)}{P'(z_i)} \quad \text{(Newton's correction)}, \tag{8}$$

$$H_i = H(z_i) = \left[\frac{P'(z_i)}{P(z_i)} - \frac{P''(z_i)}{2P'(z_i)}\right]^{-1} = \frac{2\delta_{1,i}}{\delta_{1,i}^2 + \delta_{2,i}} \quad \text{(Halley's correction)}; \tag{9}$$

3.

$$S_{k,i}(\boldsymbol{a}, \boldsymbol{b}) = \sum_{j=1}^{i-1} \frac{1}{(z_i - a_j)^k} + \sum_{j=i+1}^{n} \frac{1}{(z_i - b_j)^k},$$

$$f_i(\boldsymbol{a}, \boldsymbol{b}) = nS_{2,i}(\boldsymbol{a}, \boldsymbol{b}) - \frac{n}{n-1} S_{1,i}^2(\boldsymbol{a}, \boldsymbol{b}),$$

where $\boldsymbol{a} = (a_1, \dots, a_n)$ and $\boldsymbol{b} = (b_1, \dots, b_n)$ are some vectors of distinct complex numbers. If $\boldsymbol{a} = \boldsymbol{b} = \boldsymbol{z} = (z_1, \dots, z_n)$, then we will write $S_{k,i}(\boldsymbol{z}, \boldsymbol{z}) = S_{k,i}$ and $f_i(\boldsymbol{z}, \boldsymbol{z}) = f_i$ as in (2).

4.

$$\boldsymbol{z} = (z_1, \dots, z_n) \quad \text{(the former approximations)},$$
$$\hat{\boldsymbol{z}} = (\hat{z}_1, \dots, \hat{z}_n) \quad \text{(the new approximations)},$$
$$\boldsymbol{z}_N = (z_{N,1}, \dots, z_{N,n}) \quad z_{N,i} = z_i - N(z_i) \quad \text{(the Newton approximations)},$$
$$\boldsymbol{z}_H = (z_{H,1}, \dots, z_{H,n}) \quad z_{H,i} = z_i - H(z_i) \quad \text{(the Halley approximations)}.$$

We recall that the correction terms (8) and (9) appear in the iterative formulas

$$\hat{z} = z - N(z) \quad \text{(Newton's method)}, \quad \hat{z} = z - H(z) \quad \text{(Halley's method)},$$

which have quadratic and cubic convergence, respectively.

For the total-step methods ("Jacobi" or parallel mode) and single-step methods (serial or "Gauss–Seidel" mode) the abbreviations TS and SS will be used.

Putting $\zeta_i := \hat{z}_i$ in (6), where $\hat{z}_i$ is a new approximation to the zero ζ_i, and taking certain approximations of ζ_j on the right-hand side of the fixed point relation (6), we obtain some modified iterative processes of Laguerre's type for the simultaneous determination of all simple zeros of a polynomial.

First, we will construct total-step methods of Laguerre's type:

Basic total-step method (TS):

$$\hat{z}_i = z_i - \frac{n}{\delta_{1,i} + [(n-1)(n\delta_{2,i} - \delta_{1,i}^2 - f_i(\boldsymbol{z}, \boldsymbol{z}))]_*^{1/2}} \quad (i \in I_n). \tag{TS}$$

Total-step method with Newton's correction (TSN):

$$\hat{z}_i = z_i - \frac{n}{\delta_{1,i} + [(n-1)(n\delta_{2,i} - \delta_{1,i}^2 - f_i(\boldsymbol{z}_N, \boldsymbol{z}_N))]_*^{1/2}} \quad (i \in I_n). \tag{TSN}$$

Total-step method with Halley's correction (TSH):

$$\hat{z}_i = z_i - \frac{n}{\delta_{1,i} + [(n-1)(n\delta_{2,i} - \delta_{1,i}^2 - f_i(\boldsymbol{z}_H, \boldsymbol{z}_H))]_*^{1/2}} \quad (i \in I_n). \tag{TSH}$$

The iterative method (TS) has been proposed by Hansen et al. [4]. It has been proved in [4] that the order of convergence of (TS)-method is four. The remaining two methods (TSN) and (TSH) are new.

The convergence rate of each of the three total-step methods presented above can be accelerated using any new approximation as soon as it is found. In this way we construct the following single-step methods:

Basic single-step method (SS):

$$\hat{z}_i = z_i - \frac{n}{\delta_{1,i} + [(n-1)(n\delta_{2,i} - \delta_{1,i}^2 - f_i(\hat{z}, z))]_*^{1/2}} \quad (i \in I_n). \tag{SS}$$

Single-step method with Newton's correction (SSN):

$$\hat{z}_i = z_i - \frac{n}{\delta_{1,i} + [(n-1)(n\delta_{2,i} - \delta_{1,i}^2 - f_i(\hat{z}, z_N))]_*^{1/2}} \quad (i \in I_n). \tag{SSN}$$

Single-step method with Halley's correction (SSH):

$$\hat{z}_i = z_i - \frac{n}{\delta_{1,i} + [(n-1)(n\delta_{2,i} - \delta_{1,i}^2 - f_i(\hat{z}, z_H))]_*^{1/2}} \quad (i \in I_n). \tag{SSH}$$

The basic single-step method (SS) has been proposed by Hansen et al. in [4] but without precise details about the order of convergence. The remaining two methods (SSN) and (SSH) are new. We recall that the symbol $*$ in the six listed methods denotes the proper value of the square root.

Comparing Laguerre's method (7) and its modifications we note that the correction term in the form of a sum enables (i) the increase of the convergence order and (ii) the determination of all zeros of a polynomial. Similar procedures have been considered in [6, 11] and [13] for some other type of iterative formulas.

4. Some Estimates

In this section we give some estimates necessary for the main convergence theorems presented in Section 5. For simplicity, we will often omit the iteration index m and denote quantities in the latter $(m+1)$-th iteration by $\hat{}$. Also, for the sums and products we will write

$$\sum_{j \neq i} \quad \text{and} \quad \prod_{j \neq i} \quad \text{instead of} \sum_{\substack{j=1 \\ j \neq i}}^{n} \quad \text{and} \quad \prod_{\substack{j=1 \\ j \neq i}}^{n}.$$

In our convergence analysis we will estimate some complex quantities using an approach by circular complex arithmetic which deal with disks. A disk Z with center $\operatorname{mid} Z = c$ and radius $\operatorname{rad} Z = r$, that is $Z := \{z : |z - c| \leq r\}$, will be denoted briefly by the parametric notation $Z = \{c; r\}$. For more details about properties of circular complex interval arithmetic see the books [2, Ch. 5] and [12, Ch. 1]. Our main attention will be paid to the enclosure (in the form of a disk) of the square root of a complex number and the inversion of a disk not containing 0.

Let z be a complex number such that $|z| \leq r < 1$. Considering only the principal branch of the complex square root, which will be indicated by $*$, and applying the binomial expansion, we obtain

$$(1+z)_*^{1/2} = 1 + \binom{1/2}{1}z + \binom{1/2}{2}z^2 + \binom{1/2}{3}z^3 + \cdots.$$

According to this we have

$$\begin{aligned}|(1+z)_*^{1/2} - 1| &= |\binom{1/2}{1}z + \binom{1/2}{2}z^2 + \binom{1/2}{3}z^3 + \cdots| \le \binom{1/2}{1}|z| - \binom{1/2}{2}|z|^2 + \binom{1/2}{3}|z|^3 - \cdots \\ &\le \binom{1/2}{1}r - \binom{1/2}{2}r^2 + \binom{1/2}{3}r^3 - \cdots = 1 - \sqrt{1-r}.\end{aligned}$$

Hence we conclude that

$$(1+z)_*^{1/2} \in \left\{1; 1 - \sqrt{1-r}\right\}, \tag{10}$$

in other words, the complex valued-set $\{(1+z)_*^{1/2} : |z| \le r\}$ is contained in the disk centered at 1 with the radius $1 - \sqrt{1-r}$.

Consider now the inversion of a disk $Z = \{c; r\}$ which does not contain the origin, that is, $|c| > r$ holds. Under the transformation $w(z) = 1/z$ this disk maps into the disk

$$Z^{-1} = \left\{\frac{1}{z} : z \in \{c; r\}\right\} = \left\{\frac{\bar{c}}{|c|^2 - r^2}; \frac{r}{|c|^2 - r^2}\right\}.$$

Together with the inverse disk Z^{-1} let us consider a disk of the form

$$Z^I = \left\{\frac{1}{c}; \frac{r}{|c|(|c| - r)}\right\}.$$

It is easy to check that

$$\operatorname{rad} Z^I > \operatorname{rad} Z^{-1} \quad \text{and} \quad |\operatorname{mid} Z^I - \operatorname{mid} Z^{-1}| = \operatorname{rad} Z^I - \operatorname{rad} Z^{-1}.$$

According to this and a geometric construction we infer that

$$\{c; r\}^{-1} = Z^{-1} \subset Z^I. \tag{11}$$

Although the inverse disk Z^{-1} is smaller in size, we will use the centered inverse disk Z^I for its simpler form.

Let us note that for disks the following simple properties are valid:

$$z \in \{c; r\} \;\Rightarrow\; |c| - r \le |z| \le |c| + r, \tag{12a}$$

$$\{c_1; r_1\} \pm \{c_2; r_2\} = \{c_1 \pm c_2; r_1 + r_2\}, \tag{12b}$$

$$\alpha\{c; r\} = \{\alpha c; |\alpha| r\} \quad (\alpha \in \mathbb{C}), \tag{12c}$$

$$\sqrt{\{a; \rho\}} := \{\pm\sqrt{a}; \sqrt{a} - \sqrt{a-\rho}\} \quad (0 < a \in \mathbb{R},\ a > \rho). \tag{12d}$$

Using the identities (4) and (5) and the introduced notations, we obtain

$$N_i = \frac{1}{\delta_{1,i}} = \left(\sum_{j=1}^{n} \frac{1}{z_i - \zeta_j}\right)^{-1} = \left(\frac{1}{z_i - \zeta_i} + \sum_{j \neq i} \frac{1}{z_i - \zeta_j}\right)^{-1} = \frac{\varepsilon_i}{1 + \varepsilon_i \Sigma_{1,i}} \tag{13}$$

and

$$\begin{aligned} H_i &= \frac{2\delta_{1,i}}{\delta_{1,i}^2 + \delta_{2,i}} = \frac{2\sum_{j=1}^{n}(z_i - \zeta_j)^{-1}}{\left(\sum_{j=1}^{n}(z_i - \zeta_j)^{-1}\right)^2 + \sum_{j=1}^{n}(z_i - \zeta_j)^{-2}} \\ &= \frac{2(1/\varepsilon_i + \Sigma_{1,i})}{(1/\varepsilon_i + \Sigma_{1,i})^2 + 1/\varepsilon_i^2 + \Sigma_{2,i}} = \frac{2\varepsilon_i(1 + \varepsilon_i \Sigma_{1,i})}{2 + 2\varepsilon_i \Sigma_{1,i} + \varepsilon_i^2(\Sigma_{1,i}^2 + \Sigma_{2,i})}. \end{aligned} \tag{14}$$

By (13) we find

$$z_i - N_i - \zeta_i = \varepsilon_i - N_i = \varepsilon_i - \frac{\varepsilon_i}{1 + \varepsilon_i \Sigma_{1,i}} = \frac{\varepsilon_i^2 \Sigma_{1,i}}{1 + \varepsilon_i \Sigma_{1,i}}. \tag{15}$$

Using (14) we obtain

$$\begin{aligned} z_i - H_i - \zeta_i = \varepsilon_i - H_i &= \varepsilon_i - \frac{2\varepsilon_i(1 + \varepsilon_i \Sigma_{1,i})}{2 + 2\varepsilon_i \Sigma_{1,i} + \varepsilon_i^2(\Sigma_{1,i}^2 + \Sigma_{2,i})} \\ &= \frac{\varepsilon_i^3(\Sigma_{1,i}^2 + \Sigma_{2,i})}{2 + 2\varepsilon_i \Sigma_{1,i} + \varepsilon_i^2(\Sigma_{1,i}^2 + \Sigma_{2,i})}. \end{aligned} \tag{16}$$

Studying the convergence analysis of the total-step methods (TS), (TSN) and (TSH), we will consider all three methods simultaneously. For this purpose we indicate these methods with the additional indices 1 (for (TS)), 2 (for (TSN)) and 3 (for (TSH)), and, in the same manner, we denote the corresponding vectors of approximations as follows:

$$\mathbf{z}^{(1)} = \mathbf{z} = (z_1, \ldots, z_n), \quad \mathbf{z}^{(2)} = \mathbf{z}_N = (z_{N,1}, \ldots, z_{N,n}), \quad \mathbf{z}^{(3)} = \mathbf{z}_H = (z_{H,1}, \ldots, z_{H,n}).$$

Now we are able to write all the mentioned total-step methods, denoted with (TS(k)) ($k = 1, 2, 3$), in the unique form as

$$\hat{z}_i = z_i - \frac{n}{\delta_{1,i} + [(n-1)(n\delta_{2,i} - \delta_{1,i}^2 - f_i(\mathbf{z}^{(k)}, \mathbf{z}^{(k)}))]_*^{1/2}} \quad (i \in I_n;\ k = 1, 2, 3). \tag{TS(k)}$$

We emphasize that, in the situation when the iteration index is omitted, the superscript index k will denote the corresponding method. Following this notation we introduce the corrections $\Delta_{k,i}$ ($k = 1, 2, 3$) by

$$\Delta_{1,i} = 0, \quad \Delta_{2,i} = N_i, \quad \Delta_{3,i} = H_i.$$

Introduce the notations

$$d = \min_{\substack{i,j \\ i \neq j}} |\zeta_i - \zeta_j|, \quad q = \frac{2n+1}{d}$$

and suppose that the conditions

$$|\varepsilon_i| < \frac{d}{2n+1} = \frac{1}{q} \quad (i = 1, \ldots, n) \tag{17}$$

are satisfied. In the sequel, we will always assume in our analysis and estimate procedures that $n \geq 3$.

Lemma 2. *Let $z_1, \ldots, z_n$ be distinct approximations to the zeros $\zeta_1, \ldots, \zeta_n$, and let $\varepsilon_i = z_i - \zeta_i$, $\hat{\varepsilon}_i = \hat{z}_i - \zeta_i$, where $\hat{z}_1, \ldots, \hat{z}_n$ are approximations produced by the iterative methods* TS(k). *If* (17) *holds, then*

(i) $|\hat{\varepsilon}_i| \leq \frac{q^{k+2}}{n-1} |\varepsilon_i|^3 \sum_{j \neq i} |\varepsilon_j|^k \quad (i \in I_n;\ k = 1, 2, 3)$;

(ii) $|\hat{\varepsilon}_i| < \frac{d}{2n+1} = \frac{1}{q} \quad (i = 1, \ldots, n)$.

Proof: Taking into account (17) we find

$$|z_i - \zeta_j| \geq |\zeta_i - \zeta_j| - |z_i - \zeta_i| > d - \frac{d}{2n+1} = \frac{2nd}{2n+1} = \frac{2n}{q}, \tag{18}$$

and, by (17) and (18),

$$|z_i - z_j| \geq |z_i - \zeta_j| - |z_j - \zeta_j| > \frac{2n}{q} - \frac{1}{q} = \frac{2n-1}{q}. \tag{19}$$

According to (18) we estimate

$$|\Sigma_{1,i}| \leq \sum_{j \neq i} \frac{1}{|z_i - \zeta_j|} < \frac{(n-1)q}{2n}, \quad |\Sigma_{2,i}| \leq \sum_{j \neq i} \frac{1}{|z_i - \zeta_j|^2} < \frac{(n-1)q^2}{4n^2}. \tag{20}$$

Having in mind the inequalities (17) and (20) let us find the upper bounds of the absolute values of the corrections $\Delta_{2,i} = N_i$ and $\Delta_{3,i} = H_i$. First, starting from (13), by (17) and (20) we estimate

$$|\Delta_{2,i}| = |N_i| \leq \frac{|\varepsilon_i|}{1 - |\varepsilon_i| \sum_{j \neq i} \frac{1}{|z_i - \zeta_j|}} < \frac{1/q}{1 - \frac{1}{q} \frac{(n-1)q}{2n}} = \frac{2n}{n+1} \cdot \frac{1}{q} < \frac{2}{q}. \tag{21}$$

By (17) and (20) we obtain

$$|2\varepsilon_i(1 + \varepsilon_i \Sigma_{1,i})| \leq 2|\varepsilon_i|(1 + |\varepsilon_i||\Sigma_{1,i}|) < \frac{2}{q}\left(1 + \frac{1}{q}\frac{(n-1)q}{2n}\right) = \frac{2}{q}\left(1 + \frac{n-1}{2n}\right)$$

and

$$\begin{aligned}|2+2\varepsilon_i\Sigma_{1,i}+\varepsilon_i^2(\Sigma_{1,i}^2+\Sigma_{2,i})| &\geq 2-2|\varepsilon_i|\,|\Sigma_{1,i}|-|\varepsilon_i|^2(|\Sigma_{1,i}|^2+|\Sigma_{2,i}|)\\ &> 2-\frac{2}{q}\frac{(n-1)q}{2n}-\frac{1}{q^2}\left(\frac{(n-1)^2q^2}{4n^2}+\frac{(n-1)q^2}{4n^2}\right)\\ &=\frac{3n+5}{4n}. \end{aligned} \tag{22}$$

Using the last two inequalities we find

$$|\Delta_{3,i}|=|H_i|\leq\frac{|2\varepsilon_i(1+\varepsilon_i\Sigma_{1,i})|}{|2+2\varepsilon_i\Sigma_{1,i}+\varepsilon_i^2(\Sigma_{1,i}^2+\Sigma_{2,i})|}<\frac{\frac{2}{q}(1+\frac{n-1}{2n})}{\frac{3n+5}{4n}}=\frac{4(3n-1)}{(3n+5)q}. \tag{23}$$

According to (19), (21) and (23) we get

$$|z_i-z_j+\Delta_{2,j}|\geq|z_i-z_j|-|\Delta_{2,j}|>\frac{2n-1}{q}-\frac{2}{q}=\frac{2n-3}{q} \tag{24}$$

and

$$|z_i-z_j+\Delta_{3,j}|\geq|z_i-z_j|-|\Delta_{3,j}|>\frac{2n-1}{q}-\frac{4(3n-1)}{(3n+5)q}=\frac{(n-1)(6n+1)}{(3n+5)q}. \tag{25}$$

By virtue of (15), (17) and (20) we find

$$|\varepsilon_i-\Delta_{2,i}|\leq\frac{|\varepsilon_i|^2\sum_{j\neq i}\frac{1}{|z_i-\zeta_j|}}{1-|\varepsilon_i|\sum_{j\neq i}\frac{1}{|z_i-\zeta_j|}}\leq\frac{|\varepsilon_i|^2\frac{(n-1)q}{2n}}{1-\frac{1}{q}\frac{(n-1)q}{2n}}=|\varepsilon_i|^2\frac{(n-1)q}{n+1}<q|\varepsilon_i|^2. \tag{26}$$

Starting from (16), and using (17), (20) and (22), we get

$$\begin{aligned}|\varepsilon_i-\Delta_{3,i}| &\leq\frac{|\varepsilon_i|^3(|\Sigma_{1,i}|^2+|\Sigma_{2,i}|)}{|2+2\varepsilon_i\Sigma_{1,i}+\varepsilon_i^2(\Sigma_{1,i}^2+\Sigma_{2,i})|}<\frac{|\varepsilon_i|^3\frac{n(n-1)q^2}{4n^2}}{\frac{3n+5}{4n}}\\ &=|\varepsilon_i|^3\frac{(n-1)q^2}{3n+5}<\frac{q}{3}|\varepsilon_i|^3. \end{aligned} \tag{27}$$

Following the introduced notations, we find

$$\begin{aligned}F_i-f_i(\mathbf{z}^{(k)},\mathbf{z}^{(k)}) &= -n\sum_{j\neq i}\frac{\varepsilon_j-\Delta_{k,j}}{(z_i-z_j+\Delta_{k,j})(z_i-\zeta_j)}\left(\frac{1}{z_i-z_j+\Delta_{k,j}}+\frac{1}{z_i-\zeta_j}\right)\\ &\quad+\frac{n}{n-1}\sum_{j\neq i}\frac{\varepsilon_j-\Delta_{k,j}}{(z_i-z_j+\Delta_{k,j})(z_i-\zeta_j)}\\ &\quad\times\left(\sum_{j\neq i}\frac{1}{z_i-z_j+\Delta_{k,j}}+\sum_{j\neq i}\frac{1}{z_i-\zeta_j}\right)\quad(k=1,2,3). \end{aligned}$$

Let

$$\frac{1}{|z_i - z_j + \Delta_{k,j}|\,|z_i - \zeta_j|} \le a_n^{(k)}, \quad \frac{1}{|z_i - z_j + \Delta_{k,j}|} + \frac{1}{|z_i - \zeta_j|} \le b_n^{(k)},$$

where $a_n^{(k)}$ and $b_n^{(k)}$ are the upper bounds which can be easily found using the inequalities (18), (19), (24) and (25). Then, from the previous expression for $F_i - f_i(z^{(k)}, z^{(k)})$, we obtain

$$\begin{aligned} |F_i - f_i(z^{(k)}, z^{(k)})| &\le n b_n^{(k)} a_n^{(k)} \sum_{j \ne i} |\varepsilon_j - \Delta_{k,j}| + \frac{n}{n-1} b_n^{(k)} (n-1) a_n^{(k)} \sum_{j \ne i} |\varepsilon_j - \Delta_{k,j}| \\ &\le 2n b_n^{(k)} a_n^{(k)} \sum_{j \ne i} |\varepsilon_j - \Delta_{k,j}|. \end{aligned} \tag{28}$$

Using the bounds (18), (19) and (24)–(27), from (28) we find

$$|F_i - f_i(z^{(k)}, z^{(k)})| < \alpha_n^{(k)} q^{k+2} \sum_{j \ne i} |\varepsilon_j|^k \quad (i \in I_n;\ k = 1, 2, 3), \tag{29}$$

where

$$\alpha_n^{(k)} = \begin{cases} \frac{4n-1}{2n(2n-1)^2} & k = 1 \text{ (TS)}, \\ \frac{4n-3}{2n(2n-3)^2} & k = 2 \text{ (TSN)}, \\ \frac{(3n+5)(12n^2+5n-1)}{2n(6n+1)^2(n-1)^2} & k = 3 \text{ (TSH)}. \end{cases}$$

Since

$$n/\varepsilon_i - \delta_{1,i} = \frac{n}{z_i - \zeta_i} - \sum_{j=1}^{n} \frac{1}{z_i - \zeta_j} = \frac{n-1}{\varepsilon_i} - \sum_{j \ne i} \frac{1}{z_i - \zeta_j},$$

by (17) and (20) we have

$$\begin{aligned} |n/\varepsilon_i - \delta_{1,i}| &\ge \frac{n-1}{|\varepsilon_i|} - \sum_{j \ne i} \frac{1}{|z_i - \zeta_j|} > \frac{n-1}{|\varepsilon_i|} - \frac{(n-1)q}{2n} \\ &= \frac{n-1}{|\varepsilon_i|}\left(1 - \frac{q|\varepsilon_i|}{2n}\right) > \frac{n-1}{|\varepsilon_i|}\left(1 - \frac{1}{2n}\right) = \frac{(n-1)(2n-1)}{2n|\varepsilon_i|} \end{aligned} \tag{30}$$

and

$$\begin{aligned} |n/\varepsilon_i - \delta_{1,i}|\,|\varepsilon_i| &= \left| n - 1 - \varepsilon_i \sum_{j \ne i} \frac{1}{z_i - \zeta_j} \right| \le n - 1 + |\varepsilon_i| \sum_{j \ne i} \frac{1}{|z_i - \zeta_j|} \\ &< n - 1 + \frac{1}{q} \frac{(n-1)q}{2n} = \frac{(n-1)(2n+1)}{2n} < n. \end{aligned} \tag{31}$$

Let us introduce the quantities $w_i^{(k)}$ and $v_i^{(k)}$ by

$$w_i^{(k)} = n\delta_{2,i} - \delta_{1,i}^2 - f_i(z^{(k)}, z^{(k)}), \quad v_i^{(k)} = \frac{(n-1)(F_i - f_i(z^{(k)}, z^{(k)}))}{(n/\varepsilon_i - \delta_{1,i})^2}.$$

Using the identity (3) we obtain

$$w_i^{(k)} = n\delta_{2,i} - \delta_{1,i}^2 - F_i + F_i - f_i(z^{(k)}, z^{(k)}) = \frac{1}{n-1}(n/\varepsilon_i - \delta_{1,i})^2(1 + v_i^{(k)}). \tag{32}$$

According to the bounds (29) and (30) of $|F_i - f_i(z^{(k)}, z^{(k)})|$ and $|n/\varepsilon_i - \delta_{1,i}|$, we obtain

$$|v_i^{(k)}| \leq \frac{(n-1)|F_i - f_i(z^{(k)}, z^{(k)})|}{|\delta_{1,i} - n/\varepsilon_i|^2} < \beta_n^{(k)} q^{k+2} |\varepsilon_i|^2 \sum_{j \neq i} |\varepsilon_j|^k \quad (i \in I_n;\ k = 1, 2, 3),$$

where

$$\beta_n^{(k)} = \begin{cases} \frac{2n(4n-1)}{(n-1)(2n-1)^4} & k = 1 \text{ (TS)}, \\ \frac{2n(4n-3)}{(n-1)(2n-1)^2(2n-3)^2} & k = 2 \text{ (TSN)}, \\ \frac{2n(3n+5)(12n^2+5n-1)}{3(6n+1)^2(n-1)^3(2n-1)^2} & k = 3 \text{ (TSH)}. \end{cases}$$

It is easy to prove that

$$\beta_n^{(k)} < \frac{1}{(n-1)^3} \quad \text{for all } n \geq 3 \quad \text{and all } k = 1, 2, 3,$$

so that the upper bound of $|v_i^{(k)}|$ becomes

$$|v_i^{(k)}| < \frac{1}{(n-1)^3} q^{k+2} |\varepsilon_i|^2 \sum_{j \neq i} |\varepsilon_j|^k =: h_i^{(k)} \quad (i \in I_n;\ k = 1, 2, 3). \tag{33}$$

Since $q|\varepsilon_i| < 1$ (according to (17)), we have

$$q^{k+2} |\varepsilon_i|^2 \sum_{j \neq i} |\varepsilon_j|^k = (q|\varepsilon_i|)^2 \sum_{j \neq i} (q|\varepsilon_j|)^k < n - 1, \tag{34}$$

so that $h_i^{(k)} < \frac{1}{(n-1)^2} \leq \frac{1}{4}$. Therefore

$$\frac{1}{1 + \sqrt{1 - h_i^{(k)}}} \leq \frac{1}{1 + \sqrt{1 - 1/4}} \approx 0.53 < \frac{3}{5}. \tag{35}$$

Since $|v_i^{(k)}| < h_i^{(k)}$, there follows $v_i^{(k)} \in V_k := \{0; h_i^{(k)}\}$, where V_k is the disk centered at the origin with the radius $h_i^{(k)}$. Now, using (10), (12b), (12d) (taking the principal branch) and the inequality (35), we find

$$[1 + v_i^{(k)}]_*^{1/2} \in [1 + \{0; h_i^{(k)}\}]_*^{1/2} = [\{1; h_i^{(k)}\}]_*^{1/2} = \left\{1; 1 - \sqrt{1 - h_i^{(k)}}\right\}$$
$$= \left\{1; \frac{h_i^{(k)}}{1 + \sqrt{1 - h_i^{(k)}}}\right\} \subset \left\{1; \frac{3}{5} h_i^{(k)}\right\}. \tag{36}$$

Let

$$u_i^{(k)} = \delta_{1,i} + [(n-1)w_i^{(k)}]_*^{1/2}.$$

Then, using (31), (32), (33) and (36), as well as the properties (12b) and (12c), we have

$$u_i^{(k)} = \delta_{1,i} + (n/\varepsilon_i - \delta_{1,i})[1 + v_i^{(k)}]_*^{1/2} \in \delta_{1,i} + (n/\varepsilon_i - \delta_{1,i})\left\{1; \frac{3}{5} h_i^{(k)}\right\}$$
$$= \left\{\frac{n}{\varepsilon_i}; \frac{3}{5}\left|\frac{n}{\varepsilon_i} - \delta_{1,i}\right| |\varepsilon_i| \frac{h_i^{(k)}}{|\varepsilon_i|}\right\} \subset \left\{\frac{n}{\varepsilon_i}; \frac{3n}{5(n-1)^3} q^{k+2} |\varepsilon_i| \sum_{j \neq i} |\varepsilon_j|^k\right\}$$
$$\in \left\{\frac{n}{\varepsilon_i}; R_i^{(k)}\right\} \quad (i \in I_n;\ k = 1, 2, 3), \tag{37}$$

where we put

$$R_i^{(k)} = \frac{3n}{5(n-1)^3} q^{k+2} |\varepsilon_i| \sum_{j \neq i} |\varepsilon_j|^k \quad (i \in I_n;\ k = 1, 2, 3).$$

The choice of the sign "+" in front of the square root in (37) is made according to the criterion discussed in Appendix, which enables the higher-order convergence.

Using (37), the centered inversion of a disk (11) and (12c), we obtain

$$\frac{n}{u_i^{(k)}} \in \frac{n}{\{n/\varepsilon_i; R_i^{(k)}\}} = n\left\{\frac{\varepsilon_i}{n}; \frac{R_i^{(k)}}{\frac{n}{|\varepsilon_i|}\left(\frac{n}{|\varepsilon_i|} - R_i^{(k)}\right)}\right\} = \left\{\varepsilon_i; \frac{R_i^{(k)} |\varepsilon_i|^2}{n - R_i^{(k)} |\varepsilon_i|}\right\}. \tag{38}$$

By (17) we find that $R_i^{(k)} |\varepsilon_i| < 1$, so that from (38) we get

$$\frac{n}{u_i^{(k)}} \in \left\{\varepsilon_i; \frac{R_i^{(k)} |\varepsilon_i|^2}{n-1}\right\}. \tag{39}$$

The iterative formula (TS(k)) can be written in the form

$$\hat{z}_i = z_i - \frac{n}{u_i^{(k)}},$$

wherefrom

$$\hat{z}_i - \zeta_i = z_i - \zeta_i - \frac{n}{u_i^{(k)}}, \quad \text{that is,} \quad \hat{\varepsilon}_i = \varepsilon_i - \frac{n}{u_i^{(k)}}.$$

Hence, by the inclusion (39), we find

$$\hat{\varepsilon}_i = \varepsilon_i - \frac{n}{u_i^{(k)}} \in \left\{0; \frac{R_i^{(k)}|\varepsilon_i|^2}{n-1}\right\},$$

wherefrom, by (12a),

$$|\hat{\varepsilon}_i| \le \frac{R_i^{(k)}|\varepsilon_i|^2}{n-1} = \frac{3n}{5(n-1)^4} q^{k+2}|\varepsilon_i|^3 \sum_{j\ne i} |\varepsilon_j|^k \quad (i \in I_n;\ k = 1, 2, 3). \tag{40}$$

Since

$$\frac{3n}{5(n-1)^3} < 1 \quad \text{for all } n \ge 3,$$

from (40) we obtain

$$|\hat{\varepsilon}_i| \le \frac{q^{k+2}}{n-1} |\varepsilon_i|^3 \sum_{j\ne i} |\varepsilon_j|^k \quad (i \in I_n;\ k = 1, 2, 3), \tag{41}$$

which proves the assertion (i) of Lemma 2.

Using again (34) from (41) it follows $|\hat{\varepsilon}_i| < 1/q$. Therefore, we have proved the implication

$$|\varepsilon_i| < \frac{d}{2n+1} = \frac{1}{q} \Rightarrow |\hat{\varepsilon}_i| < \frac{d}{2n+1} = \frac{1}{q},$$

which means that (ii) of Lemma 2 is valid. □

5. Convergence Theorems

Let $z_1^{(0)}, \ldots, z_n^{(0)}$ be reasonably close approximations to the zeros $\zeta_1, \ldots, \zeta_n$ of the polynomial P, and let

$$\varepsilon_i^{(m)} = z_i^{(m)} - \zeta_i, \quad \varepsilon^{(m)} = \max_{1\le i\le n} |\varepsilon_i^{(m)}|,$$

where $z_1^{(m)}, \ldots, z_n^{(m)}$ are approximations obtained in the mth iterative step. Suppose that the initial conditions

$$|\varepsilon_i^{(0)}| < \frac{d}{2n+1} = \frac{1}{q} \quad (i = 1, \dots, n) \tag{42}$$

hold.

Theorem 1. *Under the conditions* (42) *the total-step methods* (TS(k)) *are convergent with the convergence order equal to* $k+3$ $(k = 1, 2, 3)$.

Proof: In the previous section we have derived the relation (41) under the condition (17). Using the same argumentation, under the condition (42) we derive

$$|\varepsilon_i^{(1)}| \le \frac{q^{k+2}}{n-1} |\varepsilon_i^{(0)}|^3 \sum_{j \ne i} |\varepsilon_j^{(0)}|^k < \frac{1}{q} \quad (i \in I_n;\ k = 1, 2, 3),$$

which means that the implication

$$|\varepsilon_i^{(0)}| < \frac{d}{2n+1} = \frac{1}{q} \ \Rightarrow \ |\varepsilon_i^{(1)}| < \frac{d}{2n+1} = \frac{1}{q}$$

holds. Using the mathematical induction we can prove that the condition (42) implies

$$|\varepsilon_i^{(m+1)}| < \frac{q^{k+2}}{n-1} |\varepsilon_i^{(m)}|^3 \sum_{j \ne i} |\varepsilon_j^{(m)}|^k < \frac{1}{q} \quad (i \in I_n;\ k = 1, 2, 3) \tag{43}$$

for each $m = 0, 1, \dots$ and $i = 1, \dots, n$. Putting $|\varepsilon_i^{(m)}| = t_i^{(m)}/q$, the inequality (43) becomes

$$t_i^{(m+1)} < \frac{(t_i^{(m)})^3}{n-1} \sum_{j \ne i} (t_j^{(m)})^k \quad (i \in I_n;\ k = 1, 2, 3). \tag{44}$$

Let $t^{(m)} = \max_{1 \le i \le n} t_i^{(m)}$. By virtue of (42) it follows

$$q|\varepsilon_i^{(0)}| = t_i^{(0)} \le t^{(0)} < 1 \quad (i \in I_n)$$

and from (44) one obtains $t_i^{(m)} < 1$ for all $i = 1, \dots, n$ and $m = 1, 2, \dots$ According to this we obtain from (44)

$$t_i^{(m+1)} < (t_i^{(m)})^3 (t^{(m)})^k \le (t^{(m)})^{k+3} \quad (k = 1, 2, 3), \tag{45}$$

and conclude that the sequences $\{t_i^{(m)}\}$ $(i \in I_n)$ converge to 0. Consequently, the sequences $\{|\varepsilon_i^{(m)}|\}$ are also convergent which means that $z_i^{(m)} \to \zeta_i$ $(i \in I_n)$. Finally, from (45) we may infer that the total-step methods (TS(k)) have the convergence order $k+3$, that is, the total-step methods (TS) $(k = 1)$, (TSN) $(k = 2)$ and (TSH) $(k = 3)$ have the order of convergence *four*, *five* and *six*, respectively. □

Using the above notation for the arguments of f_i, the single-step methods (SS), (SSN) and (SSH), denoted commonly with (SS(k)), can be written in the unique form

$$\hat{z}_i = z_i - \frac{n}{\delta_{1,i} + [(n-1)(n\delta_{2,i} - \delta_{1,i}^2 - f_i(\hat{z}, z^{(k)}))]_*^{1/2}} \quad (i \in I_n;\ k = 1,2,3). \quad \text{(SS(k))}$$

Applying the same technique and argumentations presented for the total-step methods, and starting from the initial conditions (42), we can prove that the inequality

$$|\varepsilon_i^{(m+1)}| < \frac{q^{k+2}}{n-1} |\varepsilon_i^{(m)}|^3 \left(\sum_{j=1}^{i-1} |\varepsilon_j^{(m+1)}| + q^{k-1} \sum_{j=i+1}^{n} |\varepsilon_j^{(m)}|^k \right) < \frac{1}{q} \quad (k = 1,2,3) \quad (46)$$

holds for each $m = 0, 1, \dots$ and $i = 1, \dots, n$, assuming that for $i = 1$ the first sum in (46) is neglected. Substituting $|\varepsilon_i^{(m)}| = t_i^{(m)}/q$ in (46), one obtains

$$t_i^{(m+1)} < \frac{(t_i^{(m)})^3}{n-1} \left(\sum_{j=1}^{i-1} t_j^{(m+1)} + \sum_{j=i+1}^{n} (t_j^{(m)})^k \right) \quad (i \in I_n;\ k = 1,2,3). \quad (47)$$

Since the derivation of the inequalities (46) are essentially the same as that given above for the total-step methods, our main attention will be paid to the precise determination of the convergence rate of the single-step methods (SS(k)). This convergence analysis, similar to that presented by Alefeld and Herzberger [1], uses the notion of the R-order of convergence introduced by Ortega and Rheinboldt [7]. The R-order of an iterative process IP with the limit point ζ will be denoted by $O_R((IP), \zeta)$.

Theorem 2. *Assume that the initial conditions* (42) *hold. Then the R-order of convergence of the single-step methods* (SS(k)), *for which the relations* (47) *are valid, is given by*

$$O_R((\text{SS(k)}), \zeta) \geq 3 + \tau_n(k), \quad (48)$$

where $\tau_n(k) > k$ is the unique positive root of the equation

$$\tau^n - k^{n-1}\tau - 3k^{n-1} = 0. \quad (49)$$

Proof: First we note that the condition (42) implies

$$|\varepsilon_i^{(0)}| q = t_i^{(0)} \leq t = \max_{1 \leq i \leq n} t_i^{(0)} < 1. \quad (50)$$

According to this we conclude from (47) that the sequences $\{t_i^{(m)}\}$ $(i \in I_n)$ converge to 0. Hence, the sequences $\{|\varepsilon_i^{(m)}|\}$ are also convergent which means that $z_i^{(m)} \to \zeta_i$ $(i \in I_n)$.

Following Alefeld and Herzberger [1], it can be derived from the relations (47) and (50) that

$$t_i^{(m+1)} \leq t^{r_i^{(m)}} \quad (i = 1, \ldots, n;\ m = 0, 1, \ldots). \tag{51}$$

The column vectors $\mathbf{r}^{(m)} = [r_1^{(m)} \cdots r_n^{(m)}]^T$ are successively computed by

$$\mathbf{r}^{(m+1)} = A_n(k)\mathbf{r}^{(m)} \tag{52}$$

starting with $\mathbf{r}^{(0)} = [1 \cdots 1]^T$. The $n \times n$ matrix $A_n(k)$ in (52) is given by

$$A_n(k) = \begin{bmatrix} 3 & k & & & & & \\ & 3 & k & & \mathbf{O} & & \\ & & 3 & k & & & \\ & & & \ddots & \ddots & & \\ & & \mathbf{O} & & & & \\ & & & & & 3 & k \\ 3 & k & 0 & 0 & \cdots & 0 & 3 \end{bmatrix} \quad (k = 1, 2, 3)$$

(see [10, Sec. 2.3] for more general case). The characteristic polynomial of the matrix $A_n(k)$ is

$$g_n(\lambda; k) = (\lambda - 3)^n - (\lambda - 3)k^{n-1} - 3k^{n-1}.$$

Substituting $\tau = \lambda - 3$, we obtain

$$\phi_n(\tau; k) = g_n(\tau + 3; k) = \tau^n - k^{n-1}\tau - 3k^{n-1}.$$

It is easy to show that the equation

$$\tau^n - k^{n-1}\tau - 3k^{n-1} = 0$$

has the unique positive root $\tau_n(k) > k$. The corresponding (positive) eigenvalue of the matrix $A_n(k)$ is $3 + \tau_n(k)$. The matrix $A_n(k)$ is irreducible and primitive so that it has the unique positive eigenvalue equal to its spectral radius $\rho(A_n(k))$. According to the analysis presented in [1] it can be shown that the lower bound of the R-order of iterative method (SS(k)), for which the inequalities (47) are valid, is given by the spectral radius $\rho(A_n(k))$. Therefore, we have

$$O_R((\mathrm{SS(k)}), \zeta) \geq \rho(A_n(k)) = 3 + \tau_n(k),$$

where $\tau_n(k) > k$ is the unique positive root of Eq. (49). □

The lower bounds of $O_R((\mathrm{SS}), \zeta)$, $O_R((\mathrm{SSN}), \zeta)$ and $O_R((\mathrm{SSH}), \zeta)$ are tabulated for $n = 3(1)10$ in Table 1.

Table 1. The lower bound of the R-order of convergence

Methods\\n	3	4	5	6	7	8	9	10	very large n
(SS):	4.672	4.453	4.341	4.274	4.229	4.196	4.172	4.153	$\to 4$
(SSN):	5.862	5.586	5.443	5.357	5.299	5.257	5.225	5.201	$\to 5$
(SSH):	6.974	6.662	6.503	6.404	6.339	6.291	6.255	6.228	$\to 6$

6. Numerical Results

To demonstrate the convergence rate of the presented iterative methods, we tested a lot of polynomial equations. In order to save all significant digits of the obtained approximations, we implemented the corresponding algorithms on PC PENTIUM III using the programming package *Mathematica* 4.0 with multiple precision arithmetic. Iterative formulas in total step and single step mode have been written in the forms

$$\hat{z}_i = z_i - \frac{nP(z_i)}{P'(z_i) + [(n-1)((n-1)P'(z_i)^2 - nP(z_i)P''(z_i) - P(z_i)^2 f_i(z^{(k)}, z^{(k)}))]_*^{1/2}},$$
$$(i \in I_n;\ k = 1,2,3), \quad \text{(TS(k))}$$

$$\hat{z}_i = z_i - \frac{nP(z_i)}{P'(z_i) + [(n-1)((n-1)P'(z_i)^2 - nP(z_i)P''(z_i) - P(z_i)^2 f_i(\hat{z}, z^{(k)}))]_*^{1/2}},$$
$$(i \in I_n;\ k = 1,2,3), \quad \text{(SS(k))}$$

where $*$ indicates the choice of the proper value of the square root. We have used the following criterion for this selection between the two values of the square root in the last two iterative formulas, proposed by Henrici [5, p. 532]:

The argument of the square root is to be chosen to differ by less than $\pi/2$ from the argument of $P'(z_i)$.

This criterion provides the greater denominator in magnitude (between two values), giving a smaller step $|\hat{z}_i - z_i|$. A great number of numerical examples showed that the above criterion gives quite satisfactory results even for crude initial approximations.

Total step as well as single step methods with the Newton and Halley corrections, considered in this paper, use the already calculated values P, P', P'' at the points $z_1, \ldots, z_n$ so that the convergence speed of the implemented iterative formulas is accelerated with the negligible number of additional operations. Such an approach provided high computational efficiency of the proposed methods.

The performed numerical experiments demonstrated very fast convergence of the modified methods of Laguerre's type. For illustration we present two numerical examples. As a measure of closeness of approximations with regard to the exact zeros, we have calculated Euclid's norm

$$e^{(m)} := \|z^{(m)} - \zeta\|_E = \left(\sum_{i=1}^{n} |z_i^{(m)} - \zeta_i|^2\right)^{1/2}.$$

Example 1. Laguerre-like methods were applied for the simultaneous approximation to the zeros of the polynomial

$$P(z) = z^9 + 3z^8 - 3z^7 - 9z^6 + 3z^5 + 9z^4 + 99z^3 + 297z^2 - 100z - 300.$$

The exact zeros of this polynomial are $-3, \pm1, \pm2i, \pm2i \pm i$. All tested methods started with the following initial approximations:

$$\begin{aligned}
z_1^{(0)} &= -3.2 + 0.2i, & z_2^{(0)} &= -1.2 - 0.2i, & z_3^{(0)} &= 1.2 + 0.2i,\\
z_4^{(0)} &= -0.2 - 2.2i, & z_5^{(0)} &= 0.1 + 1.7i, & z_6^{(0)} &= -1.9 + 1.3i,\\
z_7^{(0)} &= -1.8 - 0.8i, & z_8^{(0)} &= 2.3 + 1.1i, & z_9^{(0)} &= 1.9 - 0.7i.
\end{aligned}$$

The measure of accuracy $e^{(m)}$ $(m = 1, 2, 3)$ is displayed in Table 2 where the denotation $A(-q)$ means $A \times 10^{-q}$.

Example 2. We considered the polynomial

$$P(z) = z^5 - (4 + 5i)z^4 + (6 + 20i)z^3 - (4 + 30i)z^2 + (-15 + 20i)z + 75i,$$

with the zeros $\zeta_{1,2} = 1 \pm 2i$, $\zeta_3 = -1$, $\zeta_4 = 3$ and $\zeta_5 = 5i$. The following complex numbers were chosen as starting approximations to these zeros:

$$\begin{aligned}
z_1^{(0)} &= 1.8 + 1.3i, & z_2^{(0)} &= 1.8 - 1.3i, & z_3^{(0)} &= -1.8 - 0.7i,\\
z_4^{(0)} &= 3.7 + 0.7i, & z_5^{(0)} &= 0.7 + 4.3i.
\end{aligned}$$

In the presented example for the initial approximations we have even $e^{(0)} \approx 2.31$. For this reason, the results of the first iteration are not so good; the values $e^{(1)}$ for the six tested methods belong to the range $[0.037, 0.121]$ (see Table 3). But, in spite of the crude initial approximations, the results of the second and third iteration

Table 2. Euclid's norm of errors; $A(-q)$ means $A \times 10^{-q}$

Methods	(TS)	(TSN)	(TSH)	(SS)	(SSN)	(SSH)
$e^{(1)}$	1.55(−2)	6.16(−3)	2.87(−3)	1.35(−2)	5.15(−3)	2.84(−3)
$e^{(2)}$	1.99(−9)	5.06(−13)	6.07(−18)	1.26(−9)	7.52(−14)	2.98(−18)
$e^{(3)}$	1.94(−36)	1.57(−64)	5.18(−107)	4.95(−38)	7.61(−70)	3.68(−108)

Table 3. Euclid's norm of errors; $A(-q)$ means $A \times 10^{-q}$

Methods	(TS)	(TSN)	(TSH)	(SS)	(SSN)	(SSH)
$e^{(1)}$	1.06(−1)	1.21(−1)	5.74(−2)	9.36(−2)	1.11(−1)	3.70(−2)
$e^{(2)}$	7.69(−7)	3.25(−8)	2.56(−11)	1.80(−7)	9.24(−9)	8.42(−13)
$e^{(3)}$	1.71(−27)	4.27(−41)	3.75(−67)	2.28(−30)	7.48(−47)	5.93(−79)

Table 4. Approximations obtained in the third iteration

	i	Re $(z_i^{(2)})$	Im $(z_i^{(2)})$
	1	0.**9**(26)84845	1.**9**(27)29125
	2	0.**9**(27)84084	−1.**9**(27)70377
(TS)	3	−0.**9**(29)53103	1.99×10^{-29}
	4	2.**9**(32)86230	8.87×10^{-34}
	5	1.15×10^{-33}	4.**9**(33)15682
	1	0.**9**(40)60986	1.**9**(40)83124
	2	0.**9**(44)27715	−2.**0**(44)72575
(TSN)	3	−0.**9**(41)85732	4.22×10^{-42}
	4	3.**0**(42)14543	-2.30×10^{-44}
	5	-4.06×10^{-47}	5.**0**(46)36924
	1	1.**0**(67)55519	1.**9**(67)53245
	2	0.**9**(68)06762	−2.**0**(66)19232
(TSH)	3	−0.**9**(66)68593	3.31×10^{-69}
	4	2.**9**(72)52960	1.97×10^{-44}
	5	-7.22×10^{-74}	4.**9**(73)26273
	1	0.**9**(29)80806	2.**0**(29)12251
	2	0.**9**(34)84904	−1.**9**(34)61506
(SS)	3	−1.**0**(43)43834	2.73×10^{-44}
	4	2.**9**(54)33467	-2.37×10^{-44}
	5	-1.29×10^{-44}	5.**0**(81)11730
	1	0.**0**(46)25791	2.**0**(47)93914
	2	1.**0**(56)62790	−1.**9**(57)74275
(SSN)	3	−1.**0**(67)39568	-5.77×10^{-68}
	4	2.**9**(79)54372	-1.32×10^{-79}
	5	-1.96×10^{-101}	4.**9**(100)76104
	1	1.**0**(78)13737	1.**9**(78)42319
	2	1.**0**(88)98675	−1.**9**(87)14523
(SSH)	3	−0.**9**(103)76017	1.17×10^{-104}
	4	2.**9**(128)80206	-5.65×10^{-131}
	5	1.11×10^{-151}	4.**9**(151)04961

show excellent behaviour of the Laguerre-like simultaneous methods. The measure of accuracy $e^{(m)}$ $(m = 1, 2, 3)$ is displayed in Table 3.

To illustrate very fast convergence of the presented Laguerre-like methods, we give in Table 4 the results of the third iteration. To save a space, the number of repetitions of the digits 9 and 0 (printed in bold) is indicated by the number in the parentheses after the repeating digit.

From Tables 2 and 3 and a number of tested polynomial equations we can conclude that two iterative steps of the presented Laguerre-like methods are usually sufficient in solving most of practical problems when initial approximations are reasonably good and polynomials are well-conditioned. The third iteration demonstrates spectacularly fast convergence producing approximations of very high accuracy, rarely required in practice at present.

Appendix: The Choice of the Sign

As mentioned in Section 2, we have to choose the sign in front of the square root in the iterative formulas (TS(k)) and (SS(k)) so that the correction step $|\hat{z}_i - z_i|$ is smaller in magnitude. A practical criterion for the selection of the proper value of the square root is given in Section 6. Here we discuss this question from a theoretical point of view related to the convergence analysis presented in Sections 4 and 5, assuming that reasonably close initial approximations to the zeros are chosen so that the conditions (17) hold.

Let $\varepsilon = \max_i |\varepsilon_i|$. From (33) we conclude that $v_i^{(k)} = O_M(\varepsilon^{k+2})$, where the symbol O_M denotes the quantity of the same order in magnitude. Then, taking the principal value of the complex square root,

$$[1 + v_i^{(k)}]_*^{1/2} \approx 1 + v_i^{(k)}/2 = 1 + O_M(\varepsilon^{k+2}),$$

so that we have

$$\begin{aligned}[(n-1)w_i^{(k)}]_*^{1/2} &= (n/\varepsilon_i - \delta_{1,i})[1 + v_i^{(k)}]_*^{1/2} \approx (n/\varepsilon_i - \delta_{1,i})(1 + v_i^{(k)}/2) \\ &= n/\varepsilon_i - \delta_{1,i} + O_M(\varepsilon^{k+1}).\end{aligned}$$

Now the iterative formulas (TS(k)) and (SS(k)) can be rewritten in the form

$$\hat{z}_i = z_i - \frac{n\varepsilon_i}{\delta_{1,i}\varepsilon_i \pm (n - \delta_{1,i}\varepsilon_i) + O_M(\varepsilon^{k+2})}. \tag{$*$}$$

Depending on the choice of the sign in $(*)$, let us introduce the notation

$$\begin{aligned} x_i^+ &= \delta_{1,i}\varepsilon_i + (n - \delta_{1,i}\varepsilon_i) + O_M(\varepsilon^{k+2}) = n + O_M(\varepsilon^{k+2}), \\ x_i^- &= \delta_{1,i}\varepsilon_i - (n - \delta_{1,i}\varepsilon_i) + O_M(\varepsilon^{k+2}) = 2\delta_{1,i}\varepsilon_i - n + O_M(\varepsilon^{k+2})\end{aligned}$$

for the denominator in $(*)$. We see that for sufficiently small ε the additional term $O_M(\varepsilon^{k+2})$ has negligible influence to the value of the denominator in (TS(k)) and (SS(k)). For this reason, it is sufficient to compare the absolute value $|2\delta_{1,i}\varepsilon_i - n|$ (the main part of x_i^-) with n (the main part of x_i^+).

Using the bounds (17) and 20) we find

$$\begin{aligned}|2\delta_{1,i}\varepsilon_i - n| &= |2 + 2\varepsilon_i \sum_{j\neq i} \frac{1}{z_i - \zeta_j} - n| \leq n - 2 + 2|\varepsilon_i| \sum_{j\neq i} \frac{1}{|z_i - \zeta_j|} \\ &\leq n - 2 + 2 \cdot \frac{1}{q} \cdot \frac{(n-1)q}{2n} = n - 2 + \frac{n-1}{n} < n - 1 < n.\end{aligned}$$

Therefore, $|x_i^-| < |x_i^+|$, which means that we should choose the sign "+" before the square root in (37).

The choice of the sign $+$ in $(*)$ becomes more obvious if we find the errors $\hat{\varepsilon}_i$ from $(*)$. We obtain

$$\hat{\varepsilon}_i = \frac{\varepsilon_i(2\delta_{1,i}\varepsilon_i - n + O_M(\varepsilon^{k+2}))}{2\delta_{1,i}\varepsilon_i - n + O_M(\varepsilon_i^{k+2})} \quad \text{for } x_i^-, \quad \hat{\varepsilon}_i = \frac{O_M(\varepsilon^{k+3})}{n + O_M(\varepsilon^{k+2})} \quad \text{for } x_i^+.$$

Since $x_i^- \to -n$ and $x_i^+ \to n$ when $\varepsilon \to 0$, from the above expressions we conclude that the choice of the sign " $-$" gives only a linear convergence, while the use of "$+$" provides the very fast convergence (at least $k+3 \geq 4$).

References

[1] Alefeld, G., Herzberger, J.: On the convergence speed of some algorithms for the simultaneous approximation of polynomial zeros. SIAM J. Numer. Anal. *11*, 237–243 (1974).
[2] Alefeld, G., Herzberger, J.: Introduction to interval computations. New York: Academic Press, 1983.
[3] Hansen, E., Patrick, M.: A family of root finding methods. Numer. Math. *27*, 257–269 (1977).
[4] Hansen, E., Patrick, M., Rusnak, J.: Some modifications of Laguerre's method. BIT *17*, 409–417 (1977).
[5] Henrici, P.: Applied and computational complex analysis, vol I. New York: Wiley, 1974.
[6] Milovanović, G. V., Petković, M. S.: On the convergence order of a modified method for simultaneous finding polynomial zeros. Computing *30*, 171–178 (1983).
[7] Ortega, J. M., Rheinboldt, W. C.: Iterative solution of nonlinear equations in several variables. New York: Academic Press, 1970.
[8] Ostrowski, A. M.: Solution of equations in Euclidean and Banach space. New York: Academic Press, 1973.
[9] Parlett, B.: Laguerre's method applied to the matrix eigenvalue problem. Math. Comput. *18*, 464–485 (1964).
[10] Petković, M. S.: Iterative methods for simultaneous inclusion of polynomial zeros. Berlin Heidelberg New York: Springer, 1989.
[11] Petković, M. S., Milovanović, G. V.: A note on some improvements of the simultaneous methods for the determination of polynomial zeros. J. Comput. Appl. Math. *9*, 65–69 (1983).
[12] Petković, M. S., Petković, Lj. D.: Complex interval arithmetic and its applications. Berlin: Wiley-VCH, 1998.
[13] Petković, M. S., Stefanović, L. V.: On some improvements of square root iteration for polynomial complex zeros. J. Comput. Appl. Math. *15*, 13–25 (1986).

M. S. Petković
Faculty of Electronic Engineering
University of Niš
18 000 Niš
Yugoslavia
e-mail: msp@junis.ni.ac.yu

L. Petković
D. Živković
Faculty of Mechanical Engineering
University of Niš
18 000 Niš
Yugoslavia

Computing [Suppl] 15, 211–225 (2001)

Computing

A Smoothing Newton Method for Ball Constrained Variational Inequalities with Applications

L. Qi, Hong Kong, and **G. Zhou,** Sydney

Dedicated to Professor Tetsuro Yamamoto on the occasion of his 65th birthday

Abstract

In this paper we present a smoothing Newton method for solving ball constrained variational inequalities where we only require the function value of the underlying function and its derivative value in the feasible region. Global and superlinear convergence of this smoothing Newton method are established. Preliminary numerical results are reported.

AMS Subject Classifications: 90C33, 90C30, 65H10.

Key Words: Ball constrained variational inequalities, nonsmooth equations, smooth approximation, smoothing Newton method, convergence.

1. Introduction

Consider the variational inequality problem: Find $y^* \in X$ such that

$$(y - y^*)^T F(y^*) \geq 0 \quad \text{for all } y \in X, \tag{1.1}$$

where X is a nonempty closed subset of $\Re^n$ and $F : D \to \Re^n$ is continuously differentiable on some open set D, which contains X. In this paper, unless otherwise stated, we assume that

$$X := \{y \in \Re^n : \|y\| \leq r\}, \tag{1.2}$$

where $r > 0$ and $\|\cdot\|$ is the Euclidean norm. Then (1.1) becomes the ball constrained variational inequality problem. It is well known that solving (1.1) is equivalent to solving the following Robinson's normal equation

$$E(x) := F(\Pi_X(x)) + x - \Pi_X(x) = 0 \tag{1.3}$$

in the sense that if $x^* \in \Re^n$ is a solution of (1.3) then $y^* := \Pi_X(x^*)$ is a solution of (1.1), and conversely if y^* is a solution of (1.1) then $x^* := y^* - F(y^*)$ is a solution of (1.3) [18]. Many algorithms have been proposed to solve the box constrained variational inequality problem, see [1]–[6], [8]–[13], [15], [19], [22], [23] and references therein. To the best of our knowledge, these methods cannot handle the ball constrained variational inequality problem directly.

In this paper we present a smoothing Newton method for solving the ball constrained variational inequality problem. The smoothing parameter t is treated as a variable in the semismooth equation H defined in (3.3), which has the property that if (x^*, t^*) is a zero solution of H then $t^* = 0$ and $\Pi_X(x^*)$ is a solution of (1.1). Moreover, the smoothing parameter t can be controlled in such a way that it converges to zero neither too fast nor too slow by using a particularly designed Newton equation and a line search model.

This paper is organized as follows. In Section 2 we give some smooth approximations to the projection operator Π_X. We present a smoothing method for solving (1.1) in Section 3 and establish its global and superlinear convergence in Section 4. In Section 5, we discuss its applications to the ball constrained convex programming problem and the problem of minimizing a sum of Euclidean norms. Our preliminary numerical results are reported in Section 6.

Some words about our notation. For a continuously differentiable function $F : \Re^n \to \Re^m$, we denote the Jacobian of F at $x \in \Re^n$ by $F'(x)$, whereas the transposed Jacobian as $\nabla F(x)$. In particular, if $m = 1$, the gradient $\nabla F(x)$ is viewed as a column vector.

Let $F : \Re^n \to \Re^m$ be a locally Lipschitzian vector function. By Rademacher's theorem, F is differentiable almost everywhere. Let Ω_F denote the set of points where F is differentiable. Then the B-subdifferential of F at $x \in \Re^n$ is defined to be

$$\partial_B F(x) = \left\{ \lim_{\substack{x^k \to x \\ x^k \in \Omega_F}} \nabla F(x^k)^T \right\}, \tag{1.4}$$

while Clarke's generalized Jacobian of F at x is defined to be

$$\partial F(x) = \operatorname{conv} \partial_B F(x), \tag{1.5}$$

(see [7, 14, 16]). F is called *semismooth* at x if F is directionally differentiable at x and for all $V \in \partial F(x+h)$ and $h \to 0$,

$$F(x+h) - F(x) = Vh + o(\|h\|); \tag{1.6}$$

F is called *p-order semismooth*, $p \in (0, 1]$, at x if F is semismooth at x and for all $V \in \partial F(x+h)$ and $h \to 0$,

$$F(x+h) - F(x) = Vh + O(\|h\|^{1+p}); \tag{1.7}$$

F is called *strongly semismooth* at x if F is 1-order semismooth at x. F is called a (strongly) semismooth function if it is (strongly) semismooth everywhere (see [14, 16]). Here, $o(\|h\|)$ stands for a vector function $e : \Re^n \to \Re^m$, satisfying

$$\lim_{h\to 0}\frac{e(h)}{\|h\|}=0,$$

while $O(\|h\|^2)$ stands for a vector function $e:\Re^n\to\Re^m$, satisfying

$$\|e(h)\|\le M\|h\|^2$$

for all h satisfying $\|h\|\le\delta$, and some $M>0$ and $\delta>0$.

Theorem 1.1. (*see* [10, *Theorem* 19]) *Suppose that the function* $\mathcal{F}:\Re^n\to\Re^m$ *is p-order semismooth at x and the function* $\mathcal{G}:\Re^m\to\Re^l$ *is p-order semismooth at* $\mathcal{F}(x)$. *Then the composite function* $\mathcal{H}=\mathcal{G}\circ\mathcal{F}$ *is p-order semismooth at x.*

Clearly, the function E defined in (1.3) is not continuously differentiable but we can prove that it is semismooth.

Theorem 1.2. *The function E defined in* (1.3) *is semismooth on* $\Re^n$. *Further, if* F' *is Lipschitz continuous then E is strongly semismooth.*

Proof:

$$\Pi_X(x)=\begin{cases}\dfrac{rx}{\|x\|} & \text{if } \|x\|>r,\\ x & \text{if } \|x\|\le r.\end{cases}$$

Then

$$\Pi_X(x)=\frac{rx}{\max\{r,\|x\|\}}=\frac{rx}{r+\max\{0,(\|x\|-r)\}}. \tag{1.8}$$

Since the function $\|\cdot\|$, max functions and linear functions are all strongly semismooth, from Theorem 1.1 E is semismooth on $\Re^n$. Furthermore, if F' is Lipschitz continuous then E is strongly semismooth on $\Re^n$. $\square$

For a convex set $X\subset\Re^n$, $\Pi_X(\cdot)$ is the projection operator onto X. For a vector $x\in\Re^n$, $\|x\|$ represents the Euclidean norm $(\sum_{i=1}^n x_i^2)^{1/2}$. Let I_n denote the $n\times n$ identity matrix. Let $\Re_+=\{\varepsilon\in\Re:\varepsilon\ge 0\}$ and $\Re_{++}=\{\varepsilon\in\Re:\varepsilon>0\}$. Finally, we use $\varepsilon\downarrow 0^+$ to denote the case that a positive scalar ε tends to 0.

2. Smooth Approximations to Π_X

In this section we will present some smooth approximations to the projection operator Π_X and study properties of these smooth approximations.

In [4], Chen and Mangasarian presented a class of smooth approximations to the function $(s)_+=\max\{0,s\}$. Among these smooth approximations, the Chen-Harker-Kanzow-Smale smooth function is most often used. It is defined by

$$\psi(t,s) = \frac{s + \sqrt{s^2 + 4t^2}}{2}, \quad (t,s) \in \Re_{++} \times \Re. \tag{2.1}$$

Proposition 2.1. *(see [4]) $\psi(t,s)$ has the following properties:*

(i) For any given $t > 0$, $\psi(t,s)$ is continuously differentiable;

(ii) $0 \leq \psi(t,s) - (s)_+ \leq t$;

(iii) $\psi(t,s)$ is strictly increasing and strictly convex in s for a fixed $t > 0$;

(iv) $0 < \psi'_s(t,s) < 1$.

Similarly, we can give a class of smooth approximations to the projection operator Π_X defined in (1.8). For simplicity, throughout this paper we only use the following smooth function to approximate Π_X, which is based on the Chen-Harker-Kanzow-Smale smooth function.

$$\phi(t,x) = \frac{rx}{q(t,x)}, \quad (t,x) \in \Re_{++} \times \Re^n, \tag{2.2}$$

where

$$\begin{aligned} q(t,x) &= r + \psi\left(t, \sqrt{\|x\|^2 + t^2} - r\right) \\ &= \frac{1}{2}\left(r + \sqrt{\|x\|^2 + t^2} + \sqrt{\left(\sqrt{\|x\|^2 + t^2} - r\right)^2 + 4t^2}\right). \end{aligned} \tag{2.3}$$

Proposition 2.2. *ϕ has the following properties:*

(i) For any given $t > 0$, $\phi(t,\cdot)$ is continuously differentiable;

(ii) $\phi(t,x) \in \operatorname{int} X$, for any given $t > 0$;

(iii) $|\phi(t,x) - \Pi_X(x)| \leq 2t$;

(iv) For any given $t > 0$,

$$\nabla_x \phi(t,x) = \frac{r}{q(t,x)} I_n - \frac{re(t,x)xx^T}{2q(t,x)^2} \tag{2.4}$$

is symmetric and positive definite, and satisfies $\|\nabla \phi_x(t,x)\| < 1$, where

$$e(t,x) = \frac{1}{\sqrt{\|x\|^2 + t^2}} + \frac{\sqrt{\|x\|^2 + t^2} - r}{\sqrt{\left(\sqrt{\|x\|^2 + t^2} - r\right)^2 + 4t^2}\sqrt{\|x\|^2 + t^2}}; \tag{2.5}$$

(v) *For any given* $x \in \Re^n$ *and* $t > 0$,

$$\nabla_t \phi(t,x) = -\frac{rt}{2q^2(t,x)} \left(e(t,x) + \frac{4}{\sqrt{\left(\sqrt{\|x\|^2 + t^2} - r\right)^2 + 4t^2}} \right) x. \tag{2.6}$$

Proof: It is clear that (i) holds. For any $t > 0$, $q(t,x) > \max\{r, \|x\|\}$, So (ii) holds. By Proposition 2.1,

$$|q(t,x) - \max\{r, \|x\|\}| \leq 2t.$$

Hence,

$$\begin{aligned} \|\phi(t,x) - \Pi_X(x)\| &= \frac{r\|x\| |q(t,x) - \max\{r, \|x\|\}|}{q(t,x) \max\{r, \|x\|\}} \\ &\leq |q(t,x) - \max\{r, \|x\|\}| \\ &\leq 2t. \end{aligned}$$

By simple computation, (iv) and (v) hold. □

3. Algorithm

Define $p : \Re^{n+1} \to \Re^n$ by

$$p(t,x) = \begin{cases} \phi(|t|, x) & \text{if } t \neq 0, \\ \Pi_X(x) & \text{if } t = 0. \end{cases} \tag{3.1}$$

From Proposition 3.1 of [15] and Theorem 1.1 we have

Proposition 3.1. *p is a strongly semismooth function on* $\Re \times \Re^n$.

It follows from Proposition 2.2 that the following proposition holds.

Proposition 3.2. *p has the following properties*:

(i) *For any given* $t \neq 0$, $p(t, \cdot)$ *is continuously differentiable*;

(ii) $p(t,x) \in \operatorname{int} X$, *for any given* $t \neq 0$;

(iii) $|p(t,x) - \Pi_X(x)| \leq 2t$;

(iv) *For any* $x \in \Re^n$ *and* $t > 0$,

$$\nabla_x p(t,x) = \nabla_x \phi(t,x), \quad \nabla_t p(t,x) = \nabla_t \phi(t,x). \tag{3.2}$$

Let $z := (t, x) \in \Re \times \Re^n$ and define $H : \Re \times \Re^n \to \Re^{n+1}$ by

$$H(z) := \begin{pmatrix} t \\ F(p) + x - p \end{pmatrix}, \tag{3.3}$$

where $p = p(t, x)$. From Propositions 3.1 and 3.2, H is continuously differentiable on $(\Re \setminus \{0\}) \times \Re^n$ and strongly semismooth on $\Re \times \Re^n$. Clearly, if $z^* = (t^*, x^*)$ is a solution of $H(z) = 0$ then $\Pi_X(x^*)$ is a solution of (1.1), and conversely if y^* is a solution of (1.1) then $z^* = (0, y^* - F(y^*))$ is a solution of $H(z) = 0$. So solving (1.1) is equivalent to finding the zeros of H.

Choose $\bar{t} \in \Re_{++}$ and $\gamma \in (0, 1)$ such that $\gamma\bar{t} < 1$. Let $\bar{z} := (\bar{t}, 0) \in \Re \times \Re^n$. Define the merit function $\psi : \Re^{n+1} \to \Re_+$ by

$$\psi(z) := \|H(z)\|^2$$

and define $\beta : \Re^{n+1} \to \Re_+$ by

$$\beta(z) := \gamma \min\{1, \psi(z)\}.$$

Algorithm 3.1.

Step 0. Choose constants $\delta \in (0, 1)$, *and* $\sigma \in (0, \frac{1}{2})$. *Let* $t^0 := \bar{t}$, $x^0 \in \Re^n$ *be an arbitrary point and* $k := 0$.

Step 1. If $H(z^k) = 0$ *then stop. Otherwise, let* $\beta_k := \beta(z^k)$.

Step 2. Compute $\Delta z^k := (\Delta t^k, \Delta x^k) \in \Re^n \times \Re^n$ *by*

$$H(z^k) + H'(z^k)\Delta z^k = \beta_k \bar{z}. \tag{3.4}$$

Step 3. Let l_k *be the smallest nonnegative integer* l *satisfying*

$$\psi(z^k + \delta^l \Delta z^k) \leq [1 - 2\sigma(1 - \gamma\bar{t})\delta^l]\psi(z^k). \tag{3.5}$$

Define $z^{k+1} := z^k + \delta^{l_k} \Delta z^k$.

Step 4. Replace k *by* $k + 1$ *and go to Step* 1.

This algorithm was first given in [15] for solving box constrained variational inequality problems. The following results were proved in [15].

Lemma 3.1. *If for some* $\tilde{z} := (\tilde{t}, \tilde{x}) \in \Re_{++} \times \Re^n$, $H'(\tilde{z})$ *is nonsingular, then there exist a closed neighborhood* $\mathcal{N}(\tilde{z})$ *of* $\tilde{z}$ *and a positive number* $\bar{\alpha} \in (0, 1]$ *such that for any* $z = (t, x) \in \mathcal{N}(\tilde{z})$ *and all* $\alpha \in [0, \bar{\alpha}]$, $t \in \Re_{++}$, $H'(z)$ *is invertible and*

$$\psi(z + \alpha\Delta z) \leq [1 - 2\sigma(1 - \gamma\bar{t})\alpha]\psi(z), \tag{3.6}$$

where Δz *is the solution of the following linear equation*:

$$H(z) + H'(z)\Delta z = \beta(z)\bar{z}.$$

Assumption 3.1. *For any $z = (t, x) \in \Re^{n+1}$ with $t > 0$, $H'(z)$ is nonsingular.*

Lemma 3.2. *Suppose that Assumption 3.1 holds. Then Algorithm 3.1 is well defined at the kth iteration and for any $k \geq 0$,*

$$0 < t^{k+1} \leq t^k \leq \bar{t}, \tag{3.7}$$

and

$$t^k \geq \beta(z^k)\bar{t}. \tag{3.8}$$

4. Convergence

Theorem 4.1. *Suppose that Assumption 3.1 holds. Then an infinite sequence $\{z^k = (t^k, x^k)\}$ is generated by Algorithm 3.1, and there exists at least one accumulation point, say $z^* = (0, x^*)$, of $\{z^k\}$ such that z^* is a solution of $H(z) = 0$.*

Proof: It follows from Lemma 3.2 and Assumption 3.1 that an infinite sequence $\{z^k\}$ is generated such that $t^k \geq \beta_k \bar{t}$ for all $k \geq 0$. Let $L(z^0) = \{z \in \Re^{n+1} | \ \psi(z) \leq \psi(z^0)\}$. Clearly, $L(z^0)$ is bounded. So the sequence $\{z^k\}$ is also bounded. Without loss of generality, assume that $\{z^k\}$ converges to $\tilde{z}$. From the design of Algorithm 3.1, $\psi(z^{k+1}) < \psi(z^k)$ for all $k \geq 0$. Hence the two sequences $\{\psi(z^k)\}$ and $\{\beta(z^k)\}$ are monotonically decreasing. Since $\psi(z^k), \beta(z^k) > 0$ $(k \geq 0)$, there exist $\tilde{\psi}, \tilde{\beta} \geq 0$ such that $\psi(z^k) \to \tilde{\psi}$ and $\beta(z^k) \to \tilde{\beta}$ as $k \to \infty$. If $\tilde{\psi} = 0$ then from the continuity of $\psi(\cdot)$ and $\beta(\cdot)$ we obtain $\psi(\tilde{z}) = 0$ and $\beta(\tilde{z}) = 0$. Then we obtain the desired result. Suppose that $\tilde{\psi} > 0$. It is easy to see that $\tilde{\psi} = \psi(\tilde{z})$, $\beta(\tilde{z}) = \tilde{\beta}$ and $\tilde{t} \geq \beta(\tilde{z})\bar{t} > 0$. Then, from Assumption 3.1, $H'(\tilde{z})$ exists and is invertible. Hence, by Lemma 3.1, there exist a closed neighborhood $\mathcal{N}(\tilde{z})$ of $\tilde{z}$ and a positive number $\bar{\alpha} \in (0, 1]$ such that for any $z = (t, x) \in \mathcal{N}(\tilde{z})$ and all $\alpha \in [0, \bar{\alpha}]$ we have $t \in \Re_{++}$, $H'(z)$ is invertible and

$$\psi(z + \alpha \Delta z) \leq [1 - 2\sigma(1 - \gamma \bar{t})\alpha]\psi(z),$$

where Δz is the solution of the following linear equation:

$$H(z) + H'(z)\Delta z = \beta(z)\bar{z}.$$

Therefore, for a nonnegative integer l such that $\delta^l \in (0, \bar{\alpha}]$, we have

$$\psi(z^k + \delta^l \Delta z^k) \leq [1 - 2\sigma(1 - \gamma \bar{t}\ )\delta^l]\psi(z^k)$$

for all sufficiently large k. Then, for every sufficiently large k, $l_k \leq l$ and hence $\delta^{l_k} \geq \delta^l$. Then

$$\psi(z^{k+1}) \leq [1 - 2\sigma(1 - \gamma\bar{t})\delta^{l_k}]\psi(z^k) \leq [1 - 2\sigma(1 - \gamma \bar{t}\ )\delta^l]\psi(z^k)$$

for all sufficiently large k. This contradicts the fact that the sequence $\{\psi(z^k)\}$ converges to $\tilde{\psi} > 0$. So, we complete our proof. □

Lemma 4.1. *For any* $z = (t, x) \in \Re_{++} \times \Re^n$,

$$H'(z) := \begin{pmatrix} 1 & 0 \\ F'(p)\nabla p_t(t,x) & F'(p)\nabla p_x(t,x) + I_n - \nabla p_x(t,x) \end{pmatrix}, \tag{4.1}$$

where $p = p(t,x)$. *If* $F'(p)$ *is positive semidefinite then* $H'(z)$ *is nonsingular.*

Proof: From Proposition 2.2, there exists an $n \times n$ matrix Q such that

$$\nabla p_x(t,x) = Q \operatorname{diag}(\lambda_i) Q^T,$$

where $0 < \lambda_i < 1$, and $QQ^T = I_n$. Then,

$$F'(p)\nabla p_x(t,x) + I_n - \nabla p_x(t,x) = Q[Q^T F'(p) Q \operatorname{diag}(\lambda_i) + \operatorname{diag}(1 - \lambda_i)]Q^T.$$

Since $F'(p)$ is positive semidefinite, $Q^T F'(p) Q \operatorname{diag}(\lambda_i) + \operatorname{diag}(1 - \lambda_i)$ is nonsingular. Then $F'(p)\nabla p_x(t,x) + I_n - \nabla p_x(t,x)$ is nonsingular. So $H'(z)$ is nonsingular. This completes the proof. □

The theorem below follows from Lemma 4.1 and Theorem 4.1.

Theorem 4.2. *Suppose that* $F'(y)$ *is positive semidefinite for any* $y \in X$. *Then an infinite sequence* $\{z^k\}$ *is generated by Algorithm* 3.1, *and there exists at least one accumulation point, say* z^*, *of* $\{z^k\}$ *such that* z^* *is a solution of* $H(z) = 0$.

Let $z^* = (0, x^*)$ and define

$$A(z^*) = \{\lim H'(t^k, x^k) : t^k \downarrow 0^+ \text{ and } x^k \to x^*\}. \tag{4.2}$$

Clearly, $A(z^*) \subseteq \partial_B H(z^*)$.

Lemma 4.2. *If all* $V \in A(z^*)$ *are nonsingular, then there is a neighborhood* $N(z^*)$ *of* z^* *and a constant* C *such that for any* $z = (t,x) \in N(z^*)$ *with* $t \neq 0$, $H'(z)$ *is nonsingular and*

$$\|(H'(z))^{-1}\| \leq C.$$

Proof: If the conclusion is not true, then there is a sequence $\{z^k = (t^k, x^k)\}$ with all $t^k \neq 0$ such that $z^k \to z^*$, and either all $H'(z^k)$ are singular or $\|(H'(z^k))^{-1}\| \to +\infty$. Since H is locally Lipschitzian, ∂H is bounded in a neighborhood of z^*. By passing to a subsequence, we may assume that $H'(z^k) \to V$. Then V must be singular, a contradiction to the assumption of this lemma. □

Theorem 4.3. *Suppose that Assumption* 3.1 *is satisfied and* z^* *is an accumulation point of the infinite sequence* $\{z^k\}$ *generated by Algorithm* 3.1. *Assume that all*

$V \in A(z^*)$ are nonsingular. Then the whole sequence $\{z^k\}$ converges to z^* superlinearly, i.e.,

$$\|z^{k+1} - z^*\| = o(\|z^k - z^*\|). \tag{4.3}$$

Furthermore, if F' is Lipschitz continuous, then the whole sequence $\{z^k\}$ converges to z^* quadratically, i.e.,

$$\|z^{k+1} - z^*\| = O(\|z^k - z^*\|^2). \tag{4.4}$$

Proof: First, from Theorem 4.1, z^* is a solution of $H(z) = 0$. Then, from Lemma 4.2, there is a scalar C such that for all z^k sufficiently close to z^*,

$$\|H'(z^k)^{-1}\| \leq C. \tag{4.5}$$

Because H is semismooth, for z^k sufficiently close to z^*, we have

$$\begin{aligned}\|z^k + \Delta z^k - z^*\| &= \|z^k + H'(z^k)^{-1}[-H(z^k) + \beta_k \bar{z}] - z^*\| \\ &= O(\|H(z^k) - H(z^*) - H'(z^k)(z^k - z^*)\| + \beta_k \bar{t}\ ) \\ &= O(\|H(z^k) - H(z^*) - H'(z^k)(z^k - z^*)\| + \beta_k \bar{t}\ ) \\ &= o(\|z^k - z^*\|) + O(\psi(z^k)),\end{aligned} \tag{4.6}$$

and

$$\psi(z^k) = \|H(z^k)\|^2 = O(\|z^k - z^*\|^2). \tag{4.7}$$

Therefore, from (4.6) and (4.7), for all z^k sufficiently close to z^*,

$$\|z^k + \Delta z^k - z^*\| = o(\|z^k - z^*\|). \tag{4.8}$$

By (4.8), for any $\epsilon \in (0, \frac{1}{2})$, there is a $k(\epsilon)$ such that for all $k \geq k(\epsilon)$,

$$\|z^k + \Delta z^k - z^*\| \leq \epsilon \|z^k - z^*\|. \tag{4.9}$$

By (3.4) and (4.5),

$$\|\Delta z^k\| = \|H'(z^k)^{-1}[-H(z^k) + \beta_k \bar{z}]\| \leq C\|H(z^k)\| + \bar{t}C[\psi(z^k)]^{\frac{1}{2}} = (1 + \bar{t}\ )C\|H(z^k)\|.$$

Then

$$\|z^k - z^*\| = \|\Delta z^k\| + \|z^k + \Delta z^k - z^*\| \leq (1 + \bar{t}\ )C\|H(z^k)\| + \epsilon\|z^k - z^*\|. \tag{4.10}$$

Thus,

$$\|z^k - z^*\| \leq 2(1 + \bar{t}\ )C\|H(z^k)\|. \tag{4.11}$$

Hence, for all z^k sufficiently close to z^*,

$$\psi(z^k + \Delta z^k) = \|H(z^k + \Delta z^k)\|^2 = O(\|z^k + \Delta z^k - z^*\|^2) = o(\|z^k - z^*\|^2)$$
$$= o(\|H(z^k) - H(z^*)\|^2) = o(\psi(z^k)). \tag{4.12}$$

Therefore, for all z^k sufficiently close to z^* we have

$$z^{k+1} = z^k + \Delta z^k,$$

which, together with (4.8) and (4.12), proves (4.3). If F' is Lipschitz continuous then H is strongly semismooth. Similarly, we can prove (4.4) holds. □

Lemma 4.3. *If $F'(\Pi_X(x^*))$ is positive definite then all $V \in A(z^*)$ are nonsingular.*

Proof: For any $V \in A(z^*)$, there exists a sequence $\{z^k = (t^k, x^k)\}$ such that

$$V = \begin{pmatrix} 1 & 0 \\ F'(\Pi_X(x^*))P_t^* & F'(\Pi_X(x^*))P_x^* + I_n - P_x^* \end{pmatrix},$$

where

$$P_t^* = \lim_{\substack{t^k \downarrow 0+ \\ x^k \to x^*}} \nabla p_t(t^k, x^k),$$

and

$$P_x^* = \lim_{\substack{t^k \downarrow 0+ \\ x^k \to x^*}} \nabla p_x(t^k, x^k).$$

From Proposition 2.2, P_x^* is symmetric, positive semidefinite and $\|P_x^*\| \le 1$. Then there exists an $n \times n$ matrix Q such that

$$P_x^* = Q\,\mathrm{diag}(\lambda_i)Q^T,$$

where $0 \le \lambda_i \le 1$, and $QQ^T = I_n$. Hence,

$$F'(\Pi_X(x^*))P_x^* + I_n - P_x^* = Q[Q^T F'(\Pi_X(x^*))Q\,\mathrm{diag}(\lambda_i) + \mathrm{diag}(1 - \lambda_i)]Q^T.$$

Since $F'(\Pi_X(x^*))$ is positive definite, $Q^T F'(\Pi_X(x^*))Q\,\mathrm{diag}(\lambda_i) + \mathrm{diag}(1 - \lambda_i)$ is nonsingular. So $F'(\Pi_X(x^*))P_x^* + I_n - P_x^*$ and V are nonsingular. □

From Lemma 4.3 and Theorem 4.3, we have

Theorem 4.4. *Suppose that $z^* = (0, x^*)$ is an accumulation point of the infinite sequence $\{z^k\}$ generated by Algorithm 3.1. Assume that $F'(\Pi_X(x^*))$ is positive* *definite. Then the whole sequence* $\{z^k\}$ *converges to* z^* *superlinearly. Furthermore, if* F' *is Lipschitz continuous, then* $\{z^k\}$ *converges to* z^* *quadratically.*

5. Applications

In this section, we will apply the algorithm proposed in Section 3 to solve the ball constrained convex programming problem and the problem of minimizing a sum of Euclidean norms.

The ball constrained convex programming problem. Consider the ball constrained convex programming problem

$$\min_{x \in X} f(x), \tag{5.1}$$

here X is a ball and $f : X \to \Re$ is a twice differentiable convex function. It is well known that solving (5.1) is equivalent to finding $x^* \in X$ such that

$$(x - x^*)^T \nabla f(x^*) \geq 0 \quad \text{for all } x \in X. \tag{5.2}$$

So we can use Algorithm 3.1 to solve (5.1).

The MSN problem. Consider the problem of minimizing a sum of Euclidean norms:

$$\min_{x \in \Re^n} \sum_{i=1}^{m} \|b_i - A_i^T x\|, \tag{5.3}$$

where $b_1, b_2, \ldots, b_m \in \Re^d$ are column vectors in the Euclidean d-space, $A_1, A_2, \ldots, A_m \in \Re^{n \times d}$ are n-by-d matrices. Let $A = [A_1, A_2, \ldots, A_m]$, $b^T = [b_1^T, \ldots, b_m^T]$ and

$$f(x) = \sum_{i=1}^{m} f_i(x), \tag{5.4}$$

where

$$f_i(x) = \|b_i - A_i^T x\|, \quad \text{for } i = 1, \ldots, m. \tag{5.5}$$

Problem (5.3) is a convex programming problem, but its objective function f is not differentiable at any point x where $b_i - A_i^T x = 0$ for some i. Problem (5.3) arises in many applications, such as the VLSL design, the Euclidean facilities location problem and the Steiner minimal tree problem under a given topology, see [17], [20] and [21].

Lemma 5.1. (*see* [17]) *The vector* x^* *is a solution to* (5.3) *if and only if* (x^*, y^*), $y^* = (y_1^*, \ldots, y_m^*) \in \Re^{md}$, *is a solution to the following system:*

$$\begin{cases} Ay = 0, & \\ \|y_i\| \le 1, & i = 1, \dots, m, \\ (b_i - A_i^T x) - \|b_i - A_i^T x\| y_i = 0, & i = 1, \dots, m. \end{cases} \tag{5.6}$$

Let $B = \{s \in \Re^d : \|s\| \le 1\}$, $u = (x, y) \in \Re^n \times \Re^{md}$, and

$$X = \Re^n \times B \times \cdots \times B \subset \Re^{n+md}. \tag{5.7}$$

Define $F : \Re^{n+md} \to \Re^{n+md}$ by

$$F(u) = \begin{pmatrix} Ay \\ A^T x - b \end{pmatrix}. \tag{5.8}$$

Lemma 5.2. *(x^*, y^*) is a solution to (5.6) if and only if $u^* = (x^*, y^*)$ satisfies that*

$$(u - u^*)^T F(u^*) \ge 0 \quad \text{for all } u \in X, \tag{5.9}$$

where X is defined in (5.7).

Proof: From Lemma 2.3 [17], (5.6) is equivalent to

$$\begin{cases} Ay = 0, & \\ y_i - \Pi_B(y_i + (b_i - A_i^T x)) = 0, & i = 1, \dots, m, \end{cases} \tag{5.10}$$

i.e.,

$$u - \Pi_X(u - F(u)) = 0,$$

which is clearly equivalent to (5.9). □

Define $P(t, u) : \Re^{n+md+1} \to \Re^{n+md}$ by

$$P(t, u) = \begin{pmatrix} x \\ p(t, y_1) \\ \vdots \\ p(t, y_m) \end{pmatrix},$$

where for $i = 1, \dots, m$, $p(t, y_i)$ is defined in (3.1) with $r = 1$. Then we can use Algorithm 3.1 to solve (5.3).

6. Numerical Experiments

In this section we present some preliminary computational results. Extensive computational study of the algorithm will be given in a separate paper.

Algorithm 3.1 was implemented in MATLAB. All experiments were carried out on a DEC Alpha Server 8200. Throughout the computational experiments, we used the following parameters:

$$\delta = 0.5, \quad \sigma = 0.0005, \quad \bar{t} = 0.002, \quad \text{and} \quad \gamma = 0.5.$$

We terminated our iteration when one of the following conditions was satisfied

$$k > 50; \quad \psi(z^k) \leq 1e-8; \quad \text{or} \quad ls > 20.$$

Here ls was the number of line search iterations at each step.

The first experiment presents the performance on the problem (5.1) with the function $f(x) = \frac{1}{2}x^T Ax + g^T x$ and $X = \{x \in \Re^n : \|x\| \leq r\}$, where $A = UDU^T$ is an $n \times n$ matrix, with D being a diagonal matrix with diagonal elements selected randomly from a uniform distribution on $(0, 1)$ and $U = I_n - \frac{2}{\|u\|^2} uu^T$. Each element of $u \in \Re^n$ and $g \in \Re^n$ is randomly generated. The starting point is $x^0 = 0$. The numerical results are presented in Table 1, where r is the radius of the ball X, **Ite** denotes the number of iterations, **LS** denotes the total number of line search and **CPU** denotes the CPU time in second for solving the problem.

The second experiment presents the performance on the problem (5.3) with $n = 2$, $d = 2$, $A_i = I_2$ for $i = 1, 2, \ldots, m$ and each element of b being selected randomly from a uniform distribution on $(-10, 10)$. The starting point is $x^0 = 0$. The numerical results are presented in Table 2. Table 3 gives detailed results for the above problem with $m = 500$, which show the superlinear rate of convergence.

Table 1. Numerical results for problem (5.1)

n	r	Ite	LS	CPU
50	0.1	4	6	0.07
50	1	4	6	0.08
50	10	10	12	0.15
100	0.1	4	6	0.97
100	1	4	7	1.07
100	10	8	10	1.67
150	0.1	4	6	5.10
150	1	4	6	4.58
150	10	7	9	7.03
200	0.1	4	6	14.35
200	1	4	6	14.75
200	10	7	10	24.07

Table 2. Numerical results for problem (5.3)

m	Ite	LS	CPU
100	4	16	1.25
200	5	18	3.82
300	6	25	10.37
500	5	17	16.53
800	6	25	59.12
1000	7	27	93.57
1500	7	25	191.88
2000	7	26	357.47

Table 3. Verification of superlinear convergence

k	$\psi(z^k)$	t^k	ls
1	4.62e + 02	2.00e − 02	12
2	3.42e + 02	1.50e − 02	2
3	2.63e − 01	1.00e − 02	1
4	7.03e − 04	2.63e − 03	1
5	1.13e − 07	7.03e − 06	1
6	2.81e − 15	1.13e − 09	1

Acknowledgements

This paper is dedicated to Professor Tetsuro Yamamoto on the occasion of his 65th birthday. Professor Yamamoto is a devoted mathematician in computational mathematics. The first author would like to thank Professor Tetsuro Yamamoto for his help and friendship of many years. L. Qi's work is supported by the Australian Research Council and the Research Grant Council of Hong Kong. G. Zhou's work is supported by the Australian Research Council.

References

[1] Billups, S. C., Dirkse, S. P., Ferris, M. C.: A comparison of algorithms for large-scale mixed complementarity problems. Comput. Optim. Appl. *7*, 3–25 (1997).

[2] Burke, J. V., Xu, S.: A non-interior predictor-corrector path following algorithm for the monotone linear complementarity problem. Math. Programm. *87*, 113–130 (2000).

[3] Chen, B., Chen, X.: A global and local superlinear continuation-smoothing method for $P_0 + R_0$ and monotone NCP. SIAM J. Optim. *9*, 624–645 (1999).

[4] Chen, C., Mangasarian, O. L.: A class of smoothing functions for nonlinear and mixed complementarity problems. Comput. Optim. Appl. *5*, 97–138 (1996).

[5] Chen, X., Qi, L., Sun, D.: Global and superlinear convergence of the smoothing Newton method and its application to general box constrained variational inequalities. Math. Comput. *67*, 519–540 (1998).

[6] Chen, X., Ye, Y.: On homotopy-smoothing methods for variational inequalities. SIAM J. Control Optim. *37*, 589–616 (1999).

[7] Clarke F. H.: Optimization and nonsmooth analysis. New York: Wiley, 1983.

[8] De Luca, T., Facchinei, F., Kanzow, C.: A semismooth equation approach to the solution of nonlinear complementarity problems. Math. Programm. *75*, 407–439 (1996).

[9] Facchinei, F., Kanzow, C.: A nonsmooth inexact Newton method for the solution of large-scale nonlinear complementarity problems. Math. Programm. *76*, 493–512 (1997).

[10] Fischer, A.: Solution of monotone complementarity problems with locally Lipschitzian functions. Math. Programm. *76*, 513–532 (1997).

[11] Hotta, K., Yoshise, A.: Global convergence of a class of non-interior-point algorithms using Chen-Harker-Kanzow functions for nonlinear complementarity problems. Math. Programm. *86*, 105–133 (1999).

[12] Jiang, H., Qi, L.: A new nonsmooth equations approach to nonlinear complementarity problems. SIAM J. Control Optim. *35*, 178–193 (1997).

[13] Qi, H.-D.: A regularized smoothing Newton method for box constrained variational inequality problems with P_0 functions. SIAM J. Optim. *10*, 315–330 (2000).

[14] Qi, L.: Convergence analysis of some algorithms for solving nonsmooth equations. Math. Oper. Res. *18*, 227–244 (1993).

[15] Qi, L., Sun, D., Zhou, G.: A new look at smoothing Newton methods for nonlinear complementarity problems and box constrained variational inequalities. Math. Programm. *87*, 1–35 (2000).

[16] Qi, L., Sun, J.: A nonsmooth version of Newton's method. Math. Programm. *58*, 353–367 (1993).

[17] Qi, L., Zhou, G.: A smoothing Newton method for minimizing a sum of Euclidean norms. SIAM J. Optim. *11*, 389–410 (2000).

[18] Robinson, S. M.: Normal maps induced by linear transformation. Math. Oper. Res. *17*, 691–714 (1992).

[19] Sun, D.: A regularization Newton method for solving nonlinear complementarity problems. Appl. Math. Optim. *40*, 315–339 (1999).

[20] Xue, G., Ye, Y.: An efficient algorithm for minimizing a sum of Euclidean norms with applications. SIAM J. Optim. *7*, 1017–1036 (1997).

[21] Xue, G., Ye, Y.: An efficient algorithm for minimizing a sum of P-norms. SIAM J. Optim. *10*, 551–579 (2000).

[22] Yamashita, N., Fukushima, M.: Modified Newton methods for solving semismooth reformulations of monotone complementarity problems. Math. Programm. *76*, 273–284 (1997).

[23] Zhou, G., Sun, D., Qi, L.: Numerical experiments for a class of squared smoothing Newton methods for box constrained variational inequality problems. In: Reformulation: nonsmooth, piecewise smooth, semismooth and smoothing methods (Fukushima, M., Qi, L., eds.), pp. 421–441. Boston, Kluwer, 1999.

L. Qi
Department of Applied Mathematics
The Hong Kong Polytechnic University
Hung Hom, Kowloon
Hong Kong
e-mail: maqilq@polyu.edu.hk

G. Zhou
School of Mathematics
The University of New South Wales
Sydney 2052
Australia
e-mail: zhou@maths.unsw.edu.au

Computing [Suppl] 15, 227–238 (2001)

Computing

An Explicit Inversion Formula for Tridiagonal Matrices

T. Tsuchiya and **Q. Fang**, Matsuyama

Dedicated to Professor Tetsuro Yamamoto on the occasion of his 65th birthday

Abstract

Discretizing two-point boundary value problems on an interval by finite difference method, we obtain a certain type of tridiagonal coefficient matrices. In this paper we give an explicit inversion formula for such tridiagonal matrices using Yamamoto-Ikebe's inversion formula.

AMS Subject Classifications: 65F05.

Key Words: Tridiagonal matrices, explicit inversion formula.

1. Introduction

Let $(a,b) \subset \mathbb{R}$ be a one-dimensional bounded interval. Let $p \in C^2[a,b]$ be a function such that $p(x) \geq \delta > 0$ with some positive constant δ. Also, let $q, r \in L^\infty(a,b)$ be functions. Suppose that we are discretizing the following two-point boundary value problem by finite difference method: for given $f(x)$ find $u(x)$ such that

$$-(p(x)u'(x))' + q(x)u'(x) + r(x)u(x) = f(x) \text{ in } (a,b), \quad u(a) = u(b) = 0.$$

Then, we obtain a finite dimensional linear equation $P\vec{x} = \vec{f}$, where $P = HA$, $\vec{f} := (f(x_1), \ldots, f(x_n))^t$, $H := \mathrm{diag}(2(h_1+h_2)^{-1}, \ldots, 2(h_n+h_{n+1})^{-1})$, and

$$A := \begin{pmatrix} \tilde{b}_1 & \tilde{c}_1 & & & & \\ \tilde{a}_2 & \tilde{b}_2 & \tilde{c}_2 & & & \\ & \tilde{a}_3 & \ddots & \ddots & & \\ & & \ddots & \ddots & \ddots & \\ & & & \tilde{a}_{n-1} & \tilde{b}_{n-1} & \tilde{c}_{n-1} \\ & & & & \tilde{a}_n & \tilde{b}_n \end{pmatrix}, \quad \tilde{a}_i \neq 0, \ \tilde{c}_i \neq 0 \quad \text{for } \forall\, i. \tag{1.1}$$

Here,

$$a = x_0 < x_1 < \cdots < x_i < \cdots < x_{n+1} = b, \quad h_i := x_i - x_{i-1}, \quad h := \max h_i \tag{1.2}$$

is triangulation (or partition) of the interval $[a, b]$. For the tridiagonal matrix (1.1), Yamamoto-Ikebe [6] (also see [2, 4, 5]) give the following explicit inversion formula.

Theorem 1.1 (Yamamoto-Ikebe [6]). *Let A be a nonsingular tridiagonal matrix defined in* (1.1). *Define the two sequences $\{u_i\}$ and $\{v_i\}$ as follows:*

$$u_0 := 0, \quad u_1 := h_1, \quad u_i := -\frac{1}{\tilde{c}_{i-1}}(\tilde{a}_{i-1}u_{i-2} + \tilde{b}_{i-1}u_{i-1}), \quad i = 2, 3, \ldots, n+1,$$

$$v_{n+1} := 0, \quad v_n := h_2, \quad v_i := -\frac{1}{\tilde{a}_{i+1}}(\tilde{b}_{i+1}v_{i+1} + \tilde{c}_{i+1}v_{i+2}), \quad i = n-1, n-2, \ldots, 0,$$

where the constants h_1, h_2, $\tilde{a}_1$, and $\tilde{c}_n$ may be chosen arbitrarily, but may not be zero. Then, $A^{-1} = (\alpha_{ij})$ is given by

$$\alpha_{ij} := \begin{cases} -\frac{u_i v_j}{\tilde{a}_1 h_1 v_0} \prod_{k=2}^{j} \frac{\tilde{c}_{k-1}}{\tilde{a}_k} & (i \leq j) \\ -\frac{u_j v_i}{\tilde{a}_1 h_1 v_0} \prod_{k=2}^{j} \frac{\tilde{c}_{k-1}}{\tilde{a}_k} & (i \geq j), \end{cases}$$

where empty products are understood to be 1. *We remark here that $v_0 \neq 0$ since A is nonsingular.*

Using the Yamamoto-Ikebe formula, we can show, for example, the following. Let A be the matrix defined by

$$A := \begin{pmatrix} a_1 + a_2 & -a_2 & & & \\ -a_2 & a_2 + a_3 & -a_3 & & \\ & -a_3 & \ddots & \ddots & \\ & & \ddots & \ddots & -a_n \\ & & & -a_n & a_n + a_{n+1} \end{pmatrix}, \tag{1.3}$$

where $a_i > 0$ for $i = 1, 2, \ldots, n+1$. Then we have

Theorem 1.2 (Yamamoto [5]). *The matrix A defined by* (1.3) *is nonsingular and $A^{-1} = (\alpha_{ij})$ is written by*

$$\alpha_{ij} := \begin{cases} \left(\sum_{k=1}^{n+1} \frac{1}{a_k}\right)^{-1} \left(\sum_{k=1}^{i} \frac{1}{a_k}\right) \left(\sum_{k=j+1}^{n+1} \frac{1}{a_k}\right) & (i \leq j), \\ \left(\sum_{k=1}^{n+1} \frac{1}{a_k}\right)^{-1} \left(\sum_{k=1}^{j} \frac{1}{a_k}\right) \left(\sum_{k=i+1}^{n+1} \frac{1}{a_k}\right) & (i \geq j). \end{cases} \tag{1.4}$$

Note that the matrix A defined by (1.3) is associated with approximation of the two-point boundary value problem

$$-(p(x)u'(x))' = f(x) \text{ in } (a,b), \quad u(a) = u(b) = 0 \tag{1.5}$$

in the following manner. With the triangulation (1.2) of $[a,b]$ we discretize Eq. (1.5) by Shortley-Weller finite difference method

$$-\frac{p_{i+1/2}\frac{u_{i+1}-u_i}{h_{i+1}} - p_{i-1/2}\frac{u_i-u_{i-1}}{h_i}}{\frac{h_{i+1}+h_i}{2}} = f_i, \quad i = 1,2,\dots,n,$$

where $p_{i+1/2} := p(x_i + h_{i+1}/2)$ and $p_{i-1/2} := p(x_i - h_i/2)$. Then we obtain the coefficient matrix of the form HA, where

$$H := \operatorname{diag}(2(h_1+h_2)^{-1},\dots,2(h_n+h_{n+1})^{-1})$$

and A defined by (1.3) with $a_i := p_{i-1/2}/h_i$. We immediately notice that Yamamoto's explicit inversion formula (1.4) is associated with an approximation of the Green function $G(x,y)$ of (1.5) defined by

$$G(x,y) := \begin{cases} \left(\int_a^b \frac{ds}{p(s)}\right)^{-1} \int_a^x \frac{ds}{p(s)} \int_y^b \frac{ds}{p(s)} & (x \le y) \\ \left(\int_a^b \frac{ds}{p(s)}\right)^{-1} \int_a^y \frac{ds}{p(s)} \int_x^b \frac{ds}{p(s)} & (x \ge y). \end{cases}$$

For example,

$$\sum_{k=1}^{n+1} \frac{1}{a_k} = \sum_{k=1}^{n+1} \frac{h_k}{p_{k-1/2}}$$

is an approximation of $\int_a^b (ds/p(s))$ by the mid-point rule.

As well as finite difference approximation, finite element and finite volume approximations of (1.6) may also be understood through Yamamoto's inversion formula (see Fang-Tsuchiya-Yamamoto [1]). By this observation the following results have been shown: Let U, $\hat{U}$, and $\bar{U}$ be finite difference, piecewise linear finite element, and finite volume solutions on the partition (1.2), respectively. Let also u be the exact solution of (1.6). Then, with certain regularity assumptions of p and f in (1.6), the following hold:

(1) Even if the partition (1.2) is not uniform, we have $u(x_i) - U(x_i) = \mathcal{O}(h^2)$ at each nodal point x_i ([1]).

(2) Let K be a positive constant. If a nodal point x_i is close enough to one of the end points, that is, $|x_i - y| \le Kh$, $(y = a$ or $b)$ for sufficiently small h, we have $u(x_i) - U(x_i) = \mathcal{O}(h^3)$ ([7]).

(3) Suppose that we have $\|u - U\| = \mathcal{O}(h^2)$ with a norm $\|\cdot\|$. Then we would have $\|u' - U'\| = \mathcal{O}(h)$ for the derivatives u' and U'. However, if we define the "recovered derivative" or "averaged derivative" $D_hU(x_i)$ by

$$D_h U(x_i) := \frac{h_i}{h_i + h_{i+1}} \left(\frac{U(x_{i+1}) - U(x_i)}{h_{i+1}} \right) + \frac{h_{i+1}}{h_i + h_{i+1}} \left(\frac{U(x_i) - U(x_{i-1})}{h_i} \right)$$

at each nodal point x_i, then we have $u'(x_i) - D_h U(x_i) = \mathcal{O}(h^2)$ with the non-uniform partition (1.2) ([8]).

(4) For the piecewise linear finite element solution $\hat{U}$, the well-known estimate $u(x_i) - \hat{U}(x_i) = \mathcal{O}(h^2)$ is obtained easily ([1]). Also all other known error bounds such as $\|u - \hat{U}\|_{W^{1,p}} = \mathcal{O}(h)$, $1 \le p \le \infty$ are obtained systematically ([3]).

(5) For $\hat{U}$, the "recovered derivative" $D_h \hat{U}(x_i)$ is defined at each nodal point as well, and $u'(x_i) - D_h \hat{U}(x_i) = \mathcal{O}(h^2)$ with the non-uniform partition (1.2) ([3]).

(6) For the finite volume solution $\bar{U}$, we have $u(x_i) - \bar{U}(x_i) = \mathcal{O}(h^2)$ with the non-uniform partition (1.2) ([1]).

Gathering all, we may say that, using Yamamoto's explicit inversion formula (1.4), we have been obtaining new insights to a universal understanding of finite difference, finite element, and finite volume methods applied to the two-point boundary value problem (1.5).

Naturally, we would like to extend our analysis to more general cases. In this paper, as a first step to the goal, we try to find an explicit inversion formula of matrices defined by

$$A := \begin{pmatrix} a_1 + b_1 + a_2 & -a_2 & & & \\ -a_2 & a_2 + b_2 + a_3 & -a_3 & & \\ & -a_3 & \ddots & \ddots & \\ & & \ddots & \ddots & -a_n \\ & & & -a_n & a_n + b_n + a_{n+1}, \end{pmatrix} \tag{1.6}$$

which is associated with discretization of two-point boundary value problem of the following equation:

$$-(p(x)u'(x))' + r(x)u(x) = f(x) \text{ in } (a, b), \quad u(a) = u(b) = 0. \tag{1.7}$$

Numerical experiments given in Section 4 strongly suggest that, by using the newly obtained explicit inversion formula, a similar error analysis of finite difference method applied to (1.7) may be developed.

2. Lemmas for the Sequences u_i and v_i

We now apply Theorem 1.1 to the above matrix A defined by (1.6). By a straightforward calculation we obtain several u_i such as

$$u_0 = 0, \quad u_1 = a_1 h_1 \frac{1}{a_1}, \quad u_2 = a_1 h_1 \left[\frac{1}{a_1} + \frac{1}{a_2} + \frac{1}{a_1} b_1 \frac{1}{a_2}\right],$$

$$u_3 = a_1 h_1 \left[\frac{1}{a_1} + \frac{1}{a_2} + \frac{1}{a_3} + \frac{1}{a_1} b_1 \left(\frac{1}{a_2} + \frac{1}{a_3}\right) + \left(\frac{1}{a_1} + \frac{1}{a_2}\right) b_2 \frac{1}{a_3} + \frac{1}{a_1} b_1 \frac{1}{a_2} b_2 \frac{1}{a_3}\right],$$

$$\begin{aligned} u_4 = a_1 h_1 \Bigg[& \frac{1}{a_1} + \frac{1}{a_2} + \frac{1}{a_3} + \frac{1}{a_4} + \frac{1}{a_1} b_1 \left(\frac{1}{a_2} + \frac{1}{a_3} + \frac{1}{a_4}\right) \\ & + \left(\frac{1}{a_1} + \frac{1}{a_2}\right) b_2 \left(\frac{1}{a_3} + \frac{1}{a_4}\right) + \left(\frac{1}{a_1} + \frac{1}{a_2} + \frac{1}{a_3}\right) b_3 \frac{1}{a_4} \\ & + \frac{1}{a_1} b_1 \frac{1}{a_2} b_2 \left(\frac{1}{a_3} + \frac{1}{a_4}\right) + \frac{1}{a_1} b_1 \left(\frac{1}{a_2} + \frac{1}{a_3}\right) b_3 \frac{1}{a_4} \\ & + \left(\frac{1}{a_1} + \frac{1}{a_2}\right) b_2 \frac{1}{a_3} b_3 \frac{1}{a_4} + \frac{1}{a_1} b_1 \frac{1}{a_2} b_2 \frac{1}{a_3} b_3 \frac{1}{a_4}\Bigg]. \end{aligned}$$

Therefore, we infer a formula of general u_i in the following manner. First, we define the order $\preceq$ among symbols $S := \{a_1, \ldots, a_{n+1}, b_1, \ldots, b_n\}$ by

$$a_1 \preceq b_1 \preceq a_2 \preceq b_2 \preceq \cdots \preceq b_n \preceq a_{n+1}.$$

As usual, if $\alpha \preceq \beta$ and $\alpha \neq \beta$ for $\alpha, \beta \in S$, we write $\alpha \prec \beta$. Note that the order $\preceq$ is for the *symbols* a_i and b_i, *not* for the *values* of a_i and b_i. For $1 \leq k \leq n+1$, define B_0^k by

$$B_0^k := \frac{1}{a_1} + \cdots + \frac{1}{a_k},$$

and B_j^k for $1 \leq j \leq k-1$ by

$$B_j^k := \sum \left(\sum_{a_1 \preceq a_l \prec b_{n_1}} \frac{1}{a_l}\right) b_{n_1} \left(\sum_{b_{n_1} \prec a_l \prec b_{n_2}} \frac{1}{a_l}\right) b_{n_2} \cdots b_{n_j} \left(\sum_{b_{n_j} \prec a_l \preceq a_k} \frac{1}{a_l}\right),$$

where the left-most summation is taken for all combinations of $a_1 \prec b_{n_1} \prec b_{n_2} \prec \cdots \prec b_{n_j} \prec a_k$ (that is, $1 \leq n_1 < n_2 < \cdots < n_j \leq k-1$). Then, the following lemma is valid.

Lemma 2.1. *Let the matrix A be defined by* (1.6). *Let the sequence* $\{u_i\}$ *be defined as in Theorem* 1.1 *for A with* $\tilde{a}_1 := -a_1$ *and* $\tilde{c}_n := -a_{n+1}$. *Then, we have*

$$u_k = a_1 h_1 \left(B_0^k + B_1^k + \cdots + B_{k-1}^k\right) \tag{2.1}$$

for $k = 1, 2, \ldots, n+1$

Proof: The proof is by induction. By the above calculation, (2.1) holds for $k = 1, 2$. Suppose that (2.1) holds for $1, \ldots, k-1$.

The formula for u_i in Theorem 1.1 for the matrix A defined by (1.6) is written as

$$\begin{aligned} u_k &= \frac{1}{a_k}\left((-a_{k-1})u_{k-2} + (a_{k-1} + b_{k-1} + a_k)u_{k-1}\right) \\ &= u_{k-1} + \frac{a_{k-1}}{a_k}(u_{k-1} - u_{k-2}) + \frac{b_{k-1}}{a_k}u_{k-1}, \quad 2 \le k \le n+1. \end{aligned}$$

From the induction hypothesis, the above equation is written using B_j^n by

$$\begin{aligned} u_k = a_1 h_1 \Big[& B_0^{k-1} + B_1^{k-1} + \cdots + B_{k-3}^{k-1} + B_{k-2}^{k-1} \\ & + \frac{a_{k-1}}{a_k}\left(B_0^{k-1} - B_0^{k-2} + B_1^{k-1} - B_1^{k-2} + \cdots + B_{k-3}^{k-1} - B_{k-3}^{k-2} + B_{k-2}^{k-1}\right) \\ & + \frac{b_{k-1}}{a_k}\left(B_0^{k-1} + B_1^{k-1} + \cdots + B_{k-3}^{k-1} + B_{k-2}^{k-1}\right)\Big]. \end{aligned} \tag{2.2}$$

From the definition of B_j^n, $n = k-1, k-2$, we notice in (2.2) that

$$B_0^{k-1} + \frac{a_{k-1}}{a_k}\left(B_0^{k-1} - B_0^{k-2}\right) = B_0^k, \quad \frac{b_{k-1}}{a_k}B_{k-2}^{k-1} = B_{k-1}^k. \tag{2.3}$$

We divide B_j^{k-1} into two sums. Define

$$C_j^{k-1} := \sum \left(\sum_{a_1 \preceq a_l \prec b_{n_1}} \frac{1}{a_l} \right) b_{n_1} \cdots b_{n_{j-1}} \left(\sum_{b_{n_{j-1}} \prec a_l \prec b_{k-2}} \frac{1}{a_l} \right) b_{k-2} \frac{1}{a_{k-1}},$$

where the left-most summation is taken for all combinations of $a_1 \prec b_{n_1} \prec \cdots \prec b_{n_{j-1}} \prec b_{k-2}$ (that is, $1 \le n_1 < \cdots < n_{j-1} < k-2$). Also, define $D_j^{k-1} := B_j^{k-1} - C_j^{k-1}$.

We now consider the term

$$B_j^{k-1} + \frac{a_{k-1}}{a_k}(B_j^{k-1} - B_j^{k-2}) = D_j^{k-1} + \frac{a_{k-1}}{a_k}(D_j^{k-1} - B_j^{k-2}) + C_j^{k-1} + \frac{a_{k-1}}{a_k}C_j^{k-1}$$

for $1 \le j \le k-3$. Writing D_j^{k-1} and B_j^{k-2} as

$$D_j^{k-1} = \sum \left(\sum_{a_1 \preceq a_l \prec b_{n_1}} \frac{1}{a_l} \right) b_{n_1} \left(\sum_{b_{n_1} \prec a_l \prec b_{n_2}} \frac{1}{a_l} \right) b_{n_2} \cdots b_{n_j} \left(\frac{1}{a_{n_j+1}} + \cdots + \frac{1}{a_{k-2}} + \frac{1}{a_{k-1}} \right),$$

$$B_j^{k-2} = \sum \left(\sum_{a_1 \preceq a_l \prec b_{n_1}} \frac{1}{a_l} \right) b_{n_1} \left(\sum_{b_{n_1} \prec a_l \prec b_{n_2}} \frac{1}{a_l} \right) b_{n_2} \cdots b_{n_j} \left(\frac{1}{a_{n_j+1}} + \cdots + \frac{1}{a_{k-2}} \right),$$

(where the summations of the above equations are taken all combinations of $a_1 \prec b_{n_1} \prec \cdots \prec b_{n_j} \prec b_{k-2}$, (that is, $1 \le n_1 < \cdots < n_j \le k-3$)), we notice that the number of terms in D_j^{k-1} and B_j^{k-2} are equal. Moreover, we obtain

$$\frac{a_{k-1}}{a_k}(D_j^{k-1} - B_j^{k-2}) = \sum \left(\sum_{a_1 \preceq a_l \prec b_{n_1}} \frac{1}{a_l} \right) b_{n_1} \left(\sum_{b_{n_1} \prec a_l \prec b_{n_2}} \frac{1}{a_l} \right) b_{n_2} \cdots b_{n_j} \frac{1}{a_k}$$

and

$$D_j^{k-1} + \frac{a_{k-1}}{a_k}(D_j^{k-1} - B_j^{k-2}) = \sum \left(\sum_{a_1 \preceq a_l \prec b_{n_1}} \frac{1}{a_l} \right) b_{n_1} \cdots b_{n_j} \left(\frac{1}{a_{n_j+1}} + \cdots + \frac{1}{a_{k-1}} + \frac{1}{a_k} \right), \tag{2.4}$$

where $n_j + 1 \le k - 2$, and the summations of the above equations are taken all combinations of $a_1 \prec b_{n_1} \prec \cdots \prec b_{n_j} \prec b_{k-2}$, (that is, $1 \le n_1 < \cdots < n_j \le k-3$). Since

$$\frac{a_{k-1}}{a_k} C_j^{k-1} = \sum \left(\sum_{a_1 \preceq a_l \prec b_{n_1}} \frac{1}{a_l} \right) b_{n_1} \cdots b_{n_{j-1}} \left(\sum_{b_{n_{j-1}} \prec a_l \prec b_{k-2}} \frac{1}{a_l} \right) b_{k-2} \frac{1}{a_k},$$

we have

$$C_j^{k-1} + \frac{a_{k-1}}{a_k} C_j^{k-1} = \sum \left(\sum_{a_1 \preceq a_l \prec b_{n_1}} \frac{1}{a_l} \right) b_{n_1} \cdots b_{n_{j-1}} \left(\sum_{b_{n_{j-1}} \prec a_l \prec b_{k-2}} \frac{1}{a_l} \right) b_{k-2} \left(\frac{1}{a_{k-1}} + \frac{1}{a_k} \right), \tag{2.5}$$

where the summations of the above equations are taken all combinations of $a_1 \prec b_{n_1} \prec \cdots \prec b_{n_j} \prec b_{k-2}$, (that is, $1 \le n_1 < \cdots < n_j \le k-3$). Combining (2.4) and (2.5) we obtain

$$B_j^{k-1} + \frac{a_{k-1}}{a_k}(B_j^{k-1} - B_j^{k-2}) = D_j^k, \quad 1 \le j \le k-3. \tag{2.6}$$

From the definition we clearly have

$$\frac{b_{k-1}}{a_k} B_j^{k-1} = C_{j+1}^k, \quad 0 \le j \le k-3. \tag{2.7}$$

Combining (2.6) and (2.7) we get

$$B_j^{k-1} + \frac{a_{k-1}}{a_k}(B_j^{k-1} - B_j^{k-2}) + \frac{b_{k-1}}{a_k} B_{j-1}^{k-1} = B_j^k, \quad 1 \le j \le k-3. \tag{2.8}$$

At last, noticing $(b_{k-1}/a_k)B_{k-3}^{k-1} = C_{k-2}^k$ and $B_{k-2}^{k-1} + (a_{k-1}/a_k)B_{k-2}^{k-1} = D_{k-2}^k$, we obtain

$$B^{k-1}_{k-2} + \frac{a_{k-1}}{a_k} B^{k-1}_{k-2} + \frac{b_{k-1}}{a_k} B^{k-1}_{k-3} = B^k_{k-2}. \tag{2.9}$$

Inserting (2.3), (2.8), and (2.9) into (2.2) we finally obtain

$$u_k = a_1 h_1 (B^k_0 + B^k_1 + \cdots + B^k_{k-1}),$$

which completes the proof. □

Next, we consider the sequence v_i in Theorem 1.1. By a straightforward calculation we obtain

$$v_{n+1} = 0, \quad v_n = a_{n+1} h_2 \frac{1}{a_{n+1}}, \quad v_{n-1} = a_{n+1} h_2 \left[\frac{1}{a_n} + \frac{1}{a_{n+1}} + \frac{1}{a_n} b_n \frac{1}{a_{n+1}} \right],$$

$$\begin{aligned} v_{n-2} = a_{n+1} h_2 \Bigg[&\frac{1}{a_{n-1}} + \frac{1}{a_n} + \frac{1}{a_{n+1}} + \frac{1}{a_{n-1}} b_{n-1} \left(\frac{1}{a_n} + \frac{1}{a_{n+1}} \right) \\ &+ \left(\frac{1}{a_{n-1}} + \frac{1}{a_n} \right) b_2 \frac{1}{a_{n+1}} + \frac{1}{a_{n-1}} b_{n-1} \frac{1}{a_n} b_n \frac{1}{a_{n+1}} \Bigg], \end{aligned}$$

$$\begin{aligned} v_{n-3} = a_{n+1} h_2 \Bigg[&\frac{1}{a_{n-2}} + \frac{1}{a_{n-1}} + \frac{1}{a_n} + \frac{1}{a_{n+1}} + \frac{1}{a_{n-2}} b_{n-2} \left(\frac{1}{a_{n-1}} + \frac{1}{a_n} + \frac{1}{a_{n+1}} \right) \\ &+ \left(\frac{1}{a_{n-2}} + \frac{1}{a_{n-1}} \right) b_{n-1} \left(\frac{1}{a_n} + \frac{1}{a_{n+1}} \right) + \left(\frac{1}{a_{n-2}} + \frac{1}{a_{n-1}} + \frac{1}{a_n} \right) b_n \frac{1}{a_{n+1}} \\ &+ \frac{1}{a_{n-2}} b_{n-2} \frac{1}{a_{n-1}} b_{n-1} \left(\frac{1}{a_n} + \frac{1}{a_{n+1}} \right) + \frac{1}{a_{n-2}} b_{n-2} \left(\frac{1}{a_{n-1}} + \frac{1}{a_n} \right) b_n \frac{1}{a_{n+1}} \\ &+ \left(\frac{1}{a_{n-2}} + \frac{1}{a_{n-1}} \right) b_{n-1} \frac{1}{a_n} b_n \frac{1}{a_{n+1}} + \frac{1}{a_{n-2}} b_{n-2} \frac{1}{a_{n-1}} b_{n-1} \frac{1}{a_n} b_n \frac{1}{a_{n+1}} \Bigg]. \end{aligned}$$

Indeed, we have a formula for v_i which is very similar to (2.1). For $1 \le k \le n+1$, define

$$B^{k,n+1}_0 := \frac{1}{a_k} + \cdots + \frac{1}{a_{n+1}},$$

and, for $1 \le j \le n+1-k$, by

$$B^{k,n+1}_j := \sum \left(\sum_{a_k \preceq a_l \prec b_{n_1}} \frac{1}{a_l} \right) b_{n_1} \left(\sum_{b_{n_1} \prec a_l \prec b_{n_2}} \frac{1}{a_l} \right) b_{n_2} \cdots b_{n_j} \left(\sum_{b_{n_j} \prec a_l \preceq a_{n+1}} \frac{1}{a_l} \right),$$

where the left-most summation is taken for all combinations of $a_k \prec b_{n_1} \prec b_{n_2} \prec \cdots \prec b_{n_j} \prec a_{n+1}$ (that is, $k \le n_1 < n_2 < \cdots < n_j \le n$). Then, we can prove the following lemma in exactly the same manner as before.

Lemma 2.2. *Let the matrix A be defined by* (1.6). *Let the sequence* $\{v_i\}$ *be defined as in Theorem* 1.1 *for A with* $\tilde{a}_1 := -a_1$ *and* $\tilde{c}_n := -a_{n+1}$. *Then, we have*

$$v_k = a_{n+1}h_2(B_0^{k+1,n+1} + B_1^{k+1,n+1} + \cdots + B_{n-k}^{k+1,n+1})$$

for $k = 0, 1, 2, \ldots, n$.

Note that, from the definition of B_j^k and $B_j^{k,n+1}$, we have $B_j^{n+1} = B_j^{1,n+1}$ for $0 \le j \le n$. In particular, we have

$$B_0^{n+1} + B_1^{n+1} + \cdots + B_n^{n+1} = B_0^{1,n+1} + B_1^{1,n+1} + \cdots + B_n^{1,n+1}.$$

3. An Explicit Inversion Formula

Using the lemmas proved in the previous section, we now apply Theorem 1.1 to the matrix (1.6) and have the following explicit inversion formula.

Theorem 3.1. *Let the matrix A be of the form*

$$A := \begin{pmatrix} a_1+b_1+a_2 & -a_2 & & & \\ -a_2 & a_2+b_2+a_3 & -a_3 & & \\ & -a_3 & \ddots & \ddots & \\ & & \ddots & \ddots & -a_n \\ & & & -a_n & a_n+b_n+a_{n+1} \end{pmatrix}$$

with $a_i > 0$, $1 \le i \le n+1$. *Define* $B_0^{1,k}$, $B_0^{k+1,n+1}$ *by*

$$B_0^{1,k} := \frac{1}{a_1} + \cdots + \frac{1}{a_k}, \quad B_0^{k+1,n+1} := \frac{1}{a_{k+1}} + \cdots + \frac{1}{a_{n+1}},$$

and $B_i^{1,k}$, $B_j^{k+1,n+1}$ *for* $1 \le i \le k-1$ *and* $1 \le j \le n-k$, *respectively, by*

$$B_j^{1,k} := \sum \left(\sum_{a_1 \preceq a_l \prec b_{n_1}} \frac{1}{a_l} \right) b_{n_1} \left(\sum_{b_{n_1} \prec a_l \prec b_{n_2}} \frac{1}{a_l} \right) b_{n_2} \cdots b_{n_j} \left(\sum_{b_{n_j} \prec a_l \preceq a_k} \frac{1}{a_l} \right),$$

$$B_j^{k+1,n+1} := \sum \left(\sum_{a_{k+1} \preceq a_l \prec b_{n_1}} \frac{1}{a_l} \right) b_{n_1} \left(\sum_{b_{n_1} \prec a_l \prec b_{n_2}} \frac{1}{a_l} \right) b_{n_2} \cdots b_{n_j} \left(\sum_{b_{n_j} \prec a_l \preceq a_{n+1}} \frac{1}{a_l} \right),$$

where the left-most summations are taken for all combinations of $a_1 \prec b_{n_1} \prec b_{n_2} \prec \cdots \prec b_{n_j} \prec a_k$ (*that is,* $1 \le n_1 < n_2 < \cdots < n_j \le k-1$) *and* $a_{k+1} \prec b_{n_1} \prec b_{n_2} \prec \cdots \prec b_{n_j} \prec a_{n+1}$ (*that is,* $k+1 \le n_1 < n_2 < \cdots < n_j \le n$), *respectively. Also, define*

$$\beta := B_0^{1,n+1} + B_1^{1,n+1} + \cdots + B_n^{1,n+1}.$$

Then, if $\beta \neq 0$, A is nonsingular, and $A^{-1} = (g_{ij})$ is written by

$$g_{ij} := \begin{cases} \beta^{-1}\left(\sum_{k=0}^{i-1} B_k^{1,i}\right)\left(\sum_{k=0}^{n-j} B_k^{j+1,n+1}\right) & (i \leq j), \\ \beta^{-1}\left(\sum_{k=0}^{j-1} B_k^{1,j}\right)\left(\sum_{k=0}^{n-i} B_k^{i+1,n+1}\right) & (i \geq j). \end{cases}$$

4. Numerical Experiments

In this section we give numerical examples which suggest strongly that, by using the explicit inversion formula given in Theorem 3.1, a systematic error analysis of the Shortley-Weller finite difference method applied to (1.7) may be possible.

Now, we consider the following two-point boundary value problem:

$$-(p(x)u')' + r(x)u = f(x) \quad \text{in } I := (0,1), \qquad u(0) = u(1) = 0, \tag{4.1}$$

where $p(x) := e^{1-x}$, $r(x) =: e^{1-x}$, and $f(x) := (1+x)e^{1-x} + 1 - x$ so that the exact solution is $u(x) = x(1 - e^{x-1})$.

Here, we use the following partition of I. Let h, $0 < h < 1$ be given. Then, we divide

$(0, 1/8)$ with the uniform mesh size $h/2$,
$(1/8, 1/4)$ with the uniform mesh size $h/8$,
$(1/4, 1/2)$ with the uniform mesh size h,
$(1/2, 1)$ with the uniform mesh size $h/2$,

respectively. We discretize Eq. (4.1) by the Shortley-Weller finite difference method. Then, the coefficient matrix is of form HA, where $H := \mathrm{diag}(2(h_1 + h_2)^{-1}, \ldots, 2(h_n + h_{n+1})^{-1})$ and A is defined by (1.6) with $a_i := p(x_{i-1/2})/h_i$ and $b_i := r(x_i)$.

Table 1 shows the numerical result. In the table "max-norm" stands for $\max\{|u(x_i) - U(x_i)| : x_i \text{ are nodal points}\}$, where u is the exact solution and U is the finite difference solution of (4.1), respectively. Also, "n" stands for the number

Table 1. The numerical result for (4.1)

h	n	max-norm	$\frac{\text{max-norm}}{h^2}$
$1/2^3$	23	7.09e − 5	4.54e − 3
$1/2^4$	47	1.78e − 5	4.55e − 3
$1/2^5$	95	4.44e − 6	4.55e − 3
$1/2^6$	191	1.11e − 6	4.55e − 3
$1/2^7$	383	2.78e − 7	4.55e − 3
$1/2^8$	767	6.94e − 8	4.55e − 3
$1/2^9$	1535	1.74e − 8	4.55e − 3

Table 2. The numerical result for (4.2)

h	n	max-norm	$\frac{\text{max-norm}}{h^2}$
$1/2^3$	23	7.42e − 5	4.75e − 3
$1/2^4$	47	1.86e − 5	4.76e − 3
$1/2^5$	95	4.65e − 6	4.77e − 3
$1/2^6$	191	1.16e − 6	4.77e − 3
$1/2^7$	383	2.91e − 7	4.77e − 3
$1/2^8$	767	7.27e − 8	4.77e − 3
$1/2^9$	1535	1.82e − 8	4.77e − 3

Table 3. The numerical result for (4.3)

h	n	max-norm	$\frac{\text{max-norm}}{h^2}$
$1/2^3$	23	1.61e − 2	1.03e + 0
$1/2^4$	47	3.89e − 3	9.96e − 1
$1/2^5$	95	9.65e − 4	9.89e − 1
$1/2^6$	191	2.41e − 4	9.87e − 1
$1/2^7$	383	6.02e − 5	9.86e − 1
$1/2^8$	767	1.50e − 5	9.86e − 1
$1/2^9$	1535	3.76e − 6	9.86e − 1

of nodes. We see clearly that $u(x_i) - U(x_i) = \mathcal{O}(h^2)$ with the non-uniform partition.

Then second example is the equation

$$-(p(x)u')' + r(x)u = f(x) \quad \text{in } I, \qquad u(0) = u(1) = 0, \tag{4.2}$$

where $p(x) := e^{1-x}$, $r(x) := (x-0.5)e^{1-x}$, and $f(x) = (1 + x(x-0.5))e^{1-x} + 1 - x(x-0.5)$ so that the exact solution is $u(x) = x(1 - e^{x-1})$. Note that $r(x)$ changes its sign in I in this example. We use the same partition of I as above. Table 2 shows the numerical result. Clearly, the error bound $u(x_i) - U(x_i) = \mathcal{O}(h^2)$ holds even in the case that "$r(x) \geq 0, \forall x \in I$" does not hold.

The third example is the equation

$$-u'' - 2\pi^2 u = f(x) \quad \text{in } I, \quad u(0) = u(1) = 0, \tag{4.3}$$

where $f(x) := -\pi^2 \sin(\pi x)$ so that the exact solution is $u(x) = \sin(\pi x)$. Hence, $r(x)$ is negative in I in this example. We use the same partition of I as above. Table 3 shows the numerical result. We see that $u(x_i) - U(x_i) = \mathcal{O}(h^2)$ even in this example.

References

[1] Fang, Q., Tsuchiya, T., Yamamoto, T.: Finite difference, finite element and finite volume methods applied to two-point boundary value problems. J. Comp. Appl. Math. (to appear).

[2] Meurant, G.: A review on the inverse of symmetric tridiagonal and block tridiagonal matrices. SIAM J. Matrix Anal. Appl. *13*, 707–728 (1992).

[3] Tsuchiya, T., Yoshida, K.: An application of Yamamoto's explicit inversion formula for tridiagonal matrices to finite element error analysis (in preparation).
[4] Yamamoto, T.: A new insight of the Shortley-Weller approximation for Dirichlet problems. In: Dagstuhl Proceedings of the Seminar 'Symbolic Algebraic Methods and Verification Methods' (Alefeld, G., et al., eds.), pp. 245–253. Berlin Heidelberg New York Tokyo: Springer, 2001.
[5] Yamamoto, T.: Inversion formulas for tridiagonal matrices with applications to boundary value problems. In: Numer. Funct. Anal. Optim. *22* (in press).
[6] Yamamoto, T., Ikebe, Y.: Inversion of band matrices. Linear Algebra Appl. *24*, 105–111 (1979).
[7] Yoshida, K.: Error analysis of the Shortley-Weller finite difference method applied to two-point boundary value problems (in preparation).
[8] Yoshida, K., Tsuchiya, T.: Recovered derivatives for the Shortley-Weller finite difference approximation (submitted).

T. Tsuchiya
Q. Fang
Department of Mathematical Sciences
Faculty of Science
Ehime University
Matsuyama 790-8577
Japan
e-mails: tsuchiya@math.sci.ehime-u.ac.jp
fang@math.sci.ehime-u.ac.jp

Computing [Suppl] 15, 239–249 (2001)

Computing

On the Rate of Convergence of the Levenberg-Marquardt Method*

N. Yamashita and **M. Fukushima,** Kyoto

Dedicated to Professor Tetsuro Yamamoto on the occasion of his 65th birthday

Abstract

We consider a rate of convergence of the Levenberg-Marquardt method (LMM) for solving a system of nonlinear equations $F(x) = 0$, where F is a mapping from R^n into R^m. It is well-known that LMM has a quadratic rate of convergence when $m = n$, the Jacobian matrix of F is nonsingular at a solution x and an initial point is chosen sufficiently close to x. In this paper, we show that if $\|F(x)\|$ provides a local error bound for the system of nonlinear equations, then a sequence generated by LMM converges to the solution set quadratically.

AMS Subject Classifications: 65H10, 49M15, 90C33.

Key Words: Levenberg-Marquardt method, error bound, quadratic convergence, linear complementarity problem.

1. Introduction

We consider the system of nonlinear equations

$$F(x) = 0, \tag{1}$$

where F is a mapping from R^n to R^m. Throughout the paper, we assume that F is continuously differentiable and the system (1) has a nonempty solution set, which we denote by X^*.

The Levenberg-Marquardt method (LMM) is classical but still one of the most popular solution methods for (1) [1, 3, 6]. LMM is a kind of Newton-type method and it adopts the solution d^k of the following system of linear equations as a search direction:

$$(F'(x^k)^T F'(x^k) + \mu_k I)d = -F'(x^k)^T F(x^k), \tag{2}$$

* This work was supported in part by a Grant-in-Aid for Scientific Research from the Ministry of Education, Science, Sports and Culture of Japan.

where μ_k is a positive parameter. Since $(F'(x^k)^T F'(x^k) + \mu_k I)$ is a positive definite matrix, the system (2) has the unique solution d^k, which is a nice property of LMM compared with the pure Newton's method or the Gauss-Newton method. Moreover, the search direction d^k is a descent direction of the merit function $\phi : R^n \to R$ defined by

$$\phi(x) := \frac{1}{2}\|F(x)\|^2.$$

Therefore, combining LMM with Armijo step size based on the merit function ϕ, we may expect that such LMM converges globally. In particular we can show that any accumulation point x^* of a sequence generated by such LMM is a stationary point of ϕ, i.e.,

$$0 = \nabla\phi(x^*) = F'(x^*)^T F(x^*),$$

and hence x^* is a solution of (1) if $F'(x^*)$ is nonsingular. As to a rate of convergence, it is well-known that LMM has a superlinear convergence property if $F'(x^*)$ is nonsingular and μ_k is updated by an appropriate rule.

However, the nonsingularity of $F'(x^*)$ seems too stringent a requirement for the purpose of ensuring superlinear convergence of LMM. Recently, algorithms that enjoy superlinear convergence without the nonsingularity condition on $F'(x^*)$ have been proposed [9, 10] for the nonlinear complementarity problem, which is in some sense more general than the system of equations. To guarantee the superlinear convergence, those algorithms require a local error bound assumption instead of the nonsingularity condition. A local error bound for the system (1) is defined as follows.

Definition 1.1. *Let N be a subset of R^n such that $X^* \cap N \neq \emptyset$. We say that $\|F(x)\|$ provides a local error bound on N for the system* (1) *if there exists a positive constant c such that*

$$c \operatorname{dist}(x, X^*) \leq \|F(x)\| \quad \forall x \in N. \tag{3}$$

We note that, if $F'(x^*)$ is nonsingular at a solution x^* of (1), then x^* is an isolated solution and $\|F(x)\|$ provides a local error bound on some neighborhood of x^*. Note that the converse is not necessarily true. For example, let us consider the mapping $F : R^2 \to R^2$ defined by $F(x_1, x_2) = (e^{x_1} - 1, 0)^T$. The solution set of (1) is given by $X^* = \{x \in R^2 | x_1 = 0\}$, and hence $\operatorname{dist}(x, X^*) = |x_1|$. Then it is not difficult to see that (3) holds for any $c \in (0, 1)$ provided N is chosen as $N = \{x \in R^2 | -a < x_1 < a\}$ with sufficiently small $a > 0$. However $F'(x)$ is clearly singular for any $x \in X^*$. Hence we see that a local error bound condition is weaker than the nonsingularity of $F'(x^*)$. The purpose of this paper is to reveal that the key condition for superlinear convergence of a Newton-type method is the local error bound rather than the nonsingularity. More specifically, we show that LMM has a quadratic rate of convergence under a local error bound condition. Furthermore, applying this result to the system of nonlinear equations equivalent to the linear complementarity problem, we show that the LMM for the linear complementarity problem enjoys more favorable convergence property compared with other existing algorithms.

The paper is organized as follows. In Section 2, we show that LMM with unit step size has the quadratic rate of convergence under a local error bound condition. In Section 3, we consider the LMM with Armijo step size rule based on the merit function ϕ and show its global convergence. In Section 4, we apply the results obtained in Sections 2 and 3 to the linear complementarity problem.

2. The Quadratic Rate of Convergence of LMM

In this section, we discuss local convergence properties of LMM without line search, i.e., the next iterate x^{k+1} is given by

$$x^{k+1} := x^k + d^k,$$

where d^k is the solution of the system (2) of linear equations.

Most of the previous local convergence analysis of LMM was based on the fact that (2) is almost equivalent to the Newton equation near a solution where the Jacobian of F is nonsingular. Here we will take another viewpoint, i.e., we consider an unconstrained minimization problem equivalent to (2). By using a property of the minimization problem, we will investigate a rate of convergence of LMM. To this end, define the quadratic function $\theta^k : R^n \to R$ by

$$\theta^k(d) = \|F'(x^k)d + F(x^k)\|^2 + \mu_k\|d\|^2.$$

The function θ^k is strictly convex. Therefore, the unconstrained minimization problem

$$\min \theta^k(d) \tag{4}$$

is equivalent to (2), which can be easily verified from the first-order optimality condition for (4).

Next, we make some assumptions that will guarantee the quadratic rate of convergence of LMM.

Assumption 2.1. *For some $x^* \in X^*$, the following conditions (a) and (b) hold:*

(a) There exist constants $b \in (0,1)$ and $c_1 \in (0,\infty)$ such that

$$\|F'(y)(x-y) - (F(x) - F(y))\| \le c_1\|x-y\|^2$$
$$\forall x, y \in N(x^*, b) := \{x \mid \|x - x^*\| \le b\}.$$

(b) $\|F(x)\|$ provides a local error bound on $N(x^*, b)$ for the system (1), i.e., there exists a constant $c_2 \in (0,\infty)$ such that

$$c_2 \operatorname{dist}(x, X^*) \leq \|F(x)\| \quad \forall x \in N(x^*, b).$$

Assumption 2.1(a) holds when F is continuously differentiable and F' is Lipschitz continuous. Note that, by Assumption 2.1(a), there exists a positive constant L such that

$$\|F(x) - F(y)\| \leq L\|x - y\| \quad \forall x, y \in N(x^*, b). \tag{5}$$

Assumption 2.1(b) plays an important role in the paper. It is well-known [7] that the local error bound condition holds when F is a piecewise linear function.

We need a further assumption concerning the choice of parameters $\{\mu_k\}$.

Assumption 2.2. *$\mu_k := \|F(x^k)\|^2$ for all k.*

Under these assumptions, we show the following key lemmas. For all k, $\bar{x}^k$ denotes a vector in X^* such that

$$\|x^k - \bar{x}^k\| = \operatorname{dist}(x^k, X^*).$$

Below, x^* refers to the solution of (1) as specified in Assumption 2.1.

Lemma 2.1. *Suppose that Assumptions* 2.1 *and* 2.2 *hold. If $x^k \in N(x^*, b/2)$, then the solution d^k of the linear system* (2) *satisfies*

$$\|d^k\| \leq c_3 \operatorname{dist}(x^k, X^*),$$

$$\|F'(x^k)d^k + F(x^k)\| \leq c_4 \operatorname{dist}(x^k, X^*)^2,$$

where $c_3 = \sqrt{c_1^2 + c_2^2}/c_2$ and $c_4 = \sqrt{c_1^2 + L^2}$.

Proof: Since d^k solves the minimization problem (4), we get

$$\theta^k(d^k) \leq \theta^k(\bar{x}^k - x^k). \tag{6}$$

Furthermore, since $x^k \in N(x^*, b/2)$, we get

$$\|\bar{x}^k - x^*\| \leq \|\bar{x}^k - x^k\| + \|x^* - x^k\| \leq \|x^* - x^k\| + \|x^* - x^k\| \leq b,$$

and hence $\bar{x}^k \in N(x^*, b)$. It then follows from the definition of θ^k, the inequality (6), and Assumption 2.1(a) that

$$\|d^k\|^2 \leq \frac{1}{\mu_k}\theta^k(d^k) \leq \frac{1}{\mu_k}\theta^k(\bar{x}^k - x^k) = \frac{1}{\mu_k}\left(\|F'(x^k)(\bar{x}^k - x^k) + F(x^k)\|^2 + \mu_k\|\bar{x}^k - x^k\|^2\right)$$
$$\leq \frac{1}{\mu_k}\left(c_1^2\|x^k - \bar{x}^k\|^4 + \mu_k\|x^k - \bar{x}^k\|^2\right) \leq \left(\frac{c_1^2\|x^k - \bar{x}^k\|^2}{\mu_k} + 1\right)\|x^k - \bar{x}^k\|^2. \quad (7)$$

Since

$$\mu_k = \|F(x^k)\|^2 \geq c_2^2\|x^k - \bar{x}^k\|^2$$

by Assumption 2.1(b) and Assumption 2.2, the inequality (7) yields

$$\|d^k\| \leq \sqrt{\frac{c_1^2 + c_2^2}{c_2^2}}\|x^k - \bar{x}^k\|.$$

Next we show the second inequality. In a way similar to proving the first inequality, we can show

$$\|F'(x^k)d^k + F(x^k)\|^2 \leq \theta^k(d^k) \leq \theta^k(\bar{x}^k - x^k) \leq c_1^2\|x^k - \bar{x}^k\|^4 + \mu_k\|x^k - \bar{x}^k\|^2.$$

Since

$$\mu_k = \|F(x^k)\|^2 = \|F(x^k) - F(\bar{x}^k)\|^2 \leq L^2\|x^k - \bar{x}^k\|^2$$

by the Lipschitz property (5) and Assumption 2.2, we then have

$$\|F'(x^k)d^k + F(x^k)\| \leq \sqrt{c_1^2 + L^2}\|x^k - \bar{x}^k\|^2. \quad \square$$

The next lemma shows that $\mathrm{dist}(x^k, X^*)$ converges to 0 quadratically, as long as the iterates $\{x^k\}$ lie sufficiently near x^*.

Lemma 2.2. *If $x^k, x^{k-1} \in N(x^*, b/2)$, then*

$$\mathrm{dist}(x^k, X^*) \leq c_5 \,\mathrm{dist}(x^{k-1}, X^*)^2 \quad (8)$$

holds, where $c_5 = (c_1c_3^2 + c_4)/c_2$.

Proof: Since $x^k, x^{k-1} \in N(x^*, b/2)$ and $x^k = x^{k-1} + d^{k-1}$, it follows from Assumption 2.1 (a), (b) and Lemma 2.1 that

$$\begin{aligned}
c_2 \,\mathrm{dist}(x^k, X^*) &= c_2\mathrm{dist}(x^{k-1} + d^{k-1}, X^*) \leq \|F(x^{k-1} + d^{k-1})\| \\
&\leq \|F'(x^{k-1})d^{k-1} + F(x^{k-1})\| + c_1\|d^{k-1}\|^2 \leq c_4 \,\mathrm{dist}(x^{k-1}, X^*)^2 \\
&\quad + c_1c_3^2 \,\mathrm{dist}(x^{k-1}, X^*)^2 \\
&= (c_1c_3^2 + c_4)\mathrm{dist}(x^{k-1}, X^*)^2.
\end{aligned}$$

Therefore, setting $c_5 = (c_1c_3^2 + c_4)/c_2$ gives (8). □

From this lemma, we see that $\{\text{dist}(x^k, X^*)\}$ converges to 0 quadratically if $x^k \in N(x^*, b/2)$ for all k. Next, we show that the latter condition holds, provided the initial point x^0 is chosen sufficiently close to x^*. Let

$$r := \min\left\{\frac{b}{2+4c_3}, \frac{1}{2c_5}\right\}. \tag{9}$$

Lemma 2.3. *If $x^0 \in N(x^*, r)$, then we have $x^k \in N(x^*, b/2)$ for all k.*

Proof: We show that, for each k, if $x^l \in N(x^*, b/2), l = 1, \ldots, k$, then $x^{k+1} \in N(x^*, b/2)$. Then, since $x^0 \in N(x^*, r) \subseteq N(x^*, b/2)$, it shows the lemma.

We consider two cases $k = 0$ and $k \geq 1$: When $k = 0$, it follows from Lemma 2.1 that

$$\begin{aligned}\|x^1 - x^*\| &\leq \|x^0 + d^0 - x^*\| \leq \|x^0 - x^*\| + \|d^0\| \leq r + c_3 \,\text{dist}(x^0, X^*)\\ &\leq r + c_3\|x^0 - x^*\| \leq (1 + c_3)r.\end{aligned} \tag{10}$$

Since $(1 + c_3)r \leq b/2$, we have $x^1 \in N(x^*, b/2)$. Next consider the case $k \geq 1$. Then, since $x^l \in N(x^*, b/2)$ for $l = 0, 1, \ldots, k$, we obtain from Lemma 2.2 that, for $l = 1, \ldots, k$,

$$\text{dist}(x^l, X^*) \leq c_5 \,\text{dist}(x^{l-1}, X^*)^2 \leq \cdots \leq c_5^{2^l - 1}\|x^0 - x^*\|^{2^l} \leq r\left(\frac{1}{2}\right)^{2^l - 1},$$

where the last inequality follows from $\|x^0 - x^*\| \leq r$ and $r \leq \frac{1}{2c_5}$. Therefore, from Lemma 2.1, we have for $l = 1, \ldots, k$

$$\|d^l\| \leq c_3 \,\text{dist}(x^l, X^*) \leq c_2 r\left(\frac{1}{2}\right)^{2^l - 1} \leq c_3 r\left(\frac{1}{2}\right)^{2l-1}. \tag{11}$$

It then follows that

$$\begin{aligned}\|x^{k+1} - x^*\| &\leq \|x^1 - x^*\| + \sum_{l=1}^{k}\|d^l\| \leq (1 + c_3)r + c_3 r\sum_{l=1}^{k}\left(\frac{1}{2}\right)^{2l-1}\\ &\leq (1 + c_3)r + c_3 r\sum_{l=1}^{\infty}\left(\frac{1}{2}\right)^{l} \leq (1 + 2c_3)r \leq \frac{b}{2},\end{aligned}$$

where the second inequality follows from (10) and the last inequality follows from $r \leq \frac{b}{2+6c_4}$. Consequently we have $x^{k+1} \in N(x^*, b/2)$. □

By using Lemmas 2.2 and 2.3, we show our main theorem.

Theorem 2.1. *Suppose that Assumptions* 2.1 *and* 2.2 *hold. Suppose also that* $\{x^k\}$ *is generated by LMM with* $x^0 \in N(x^*, r)$. *Then* $\{\mathrm{dist}(x^k, X^*)\}$ *converges to* 0 *quadratically. Moreover the sequence* $\{x^k\}$ *converges to a solution* $\hat{x} \in X^* \cap N(x^*, b/2)$.

Proof: The first part of the theorem directly follows from Lemmas 2.2 and 2.3. So, we only show that the sequence $\{x^k\}$ converges to a solution $\hat{x} \in X^* \cap N(x^*, b/2)$. Since $\{\mathrm{dist}(x^k, X)\}$ converges to 0 and $x^k \in N(x^*, b/2)$ for all k, it suffices to show that $\{x^k\}$ converges. Since (11) implies

$$\|d^k\| \le c_3 r \left(\frac{1}{2}\right)^{2k-1}$$

for all $k \ge 1$, we have for any positive integers p, q such that $p \ge q$

$$\|x^p - x^q\| \le \sum_{i=q}^{p-1} \|d^i\| \le \sum_{i=q}^{\infty} \|d^i\| \le c_3 r \sum_{i=q}^{\infty} \left(\frac{1}{2}\right)^{2i-1} = \frac{1}{3} c_3 r \left(\frac{1}{2}\right)^{2q-3}.$$

This means that $\{x^k\}$ is a Cauchy sequence, and hence it converges. □

3. Global Convergence

In the previous section, we have investigated the rate of convergence of LMM with unit step size. In this section, we consider LMM with Armijo step size rule, and discuss its global convergence. To determine a step size, we use the merit function

$$\phi(x) = \frac{1}{2}\|F(x)\|^2.$$

We consider the following algorithm.

LMM with line search

Step 0: Choose parameters $\alpha, \beta, \gamma \in (0,1)$ and an initial point $x^0 \in R^n$. Set $\mu^0 := \|F(x^0)\|^2$ and $k := 0$.

Step 1: If x^k satisfies a termination criterion, then stop.

Step 2: Find the solution d^k of the system of linear equations

$$(F'(x^k)^T F'(x^k) + \mu_k I)d = -F'(x^k)^T F(x^k).$$

If d^k satisfies

$$\|F(x^k + d^k)\| \le \gamma \|F(x^k)\|, \tag{12}$$

then $x^{k+1} := x^k + d^k$ and go to Step 4. Otherwise go to Step 3.

Step 3: Let m_k be the smallest nonnegative integer m such that

$$\phi(x^k + \beta^m d^k) - \phi(x^k) \leq \alpha\beta^m \nabla\phi(x^k)^T d^k. \tag{13}$$

Set $x^{k+1} := x^k + \beta^{m_k} d^k$.

Step 4: Set $\mu^{k+1} := \|F(x^{k+1})\|^2$ and $k := k+1$. Go to Step 1.

For the above algorithm, we establish the following global convergence theorem. To this end, we assume that $\nabla\phi(x^k) \neq 0$ for all k, so the algorithm generates an infinite sequence $\{x^k\}$.

Theorem 3.1. *Let $\{x^k\}$ be generated by LMM with line search. Then any accumulation point of the sequence $\{x^k\}$ is a stationary point of ϕ. Moreover, if an accumulation point x^* of the sequence $\{x^k\}$ is a solution of the system* (1), *then the sequence* $\{\text{dist}(x^k, X^*)\}$ *converges to* 0 *quadratically, provided Assumption* 2.1 *with this particular x^* and Assumption* 2.2 *hold.*

Proof: Since $\nabla\phi(x^k) \neq 0$ implies $d^k \neq 0$, we have

$$\nabla\phi(x^k)^T d^k = (F'(x^k)^T F(x^k))^T d^k = -((F'(x^k)^T F'(x^k) + \mu_k I)d^k)^T d^k < 0.$$

It then follows from (12) and (13) that $\{\phi(x^k)\}$ is monotonically decreasing, and hence μ_k is also monotonically decreasing and has a limit. If $\mu_k \to 0$, then $F(x^k) \to 0$, which implies that any accumulation point of $\{x^k\}$ is a solution of (1), and hence it is a stationary point of ϕ. If $\lim_{k\to\infty} \mu_k = \bar{\mu} > 0$, then we have

$$\nabla\phi(x^k)^T d^k = -((F'(x^k)^T F'(x^k) + \mu_k I)d^k)^T d^k < -\bar{\mu}\|d^k\|^2.$$

Hence, we can show that any accumulation point is a stationary point of ϕ by using the standard arguments. This completes the proof of the first part of the theorem.

Next we show the last part. Since the accumulation point x^* is a solution of (1), there exists a $\bar{k}$ such that

$$\|F(x^{\bar{k}})\| \leq \frac{c_2^2 \gamma}{c_5 L} \tag{14}$$

and

$$\|x^{\bar{k}} - x^*\| \leq r,$$

where r is the positive constant specified by (9). Now let $\{y^k\}$ be a sequence generated by LMM with unit step size and $y^0 := x^{\bar{k}}$. Then, by Theorem 2.1, $\text{dist}(y^l, X^*)$ converges to 0 quadratically. Therefore, it suffices to show that $x^{\bar{k}+l} = y^l$ for all l, i.e., $\{y^l\}$ satisfies

$$\|F(y^{l+1})\| \le \gamma \|F(y^l)\|$$

for all l. Note that, from Lemmas 2.2 and 2.3, we have

$$\operatorname{dist}(y^{l+1}, X^*) \le c_5 \operatorname{dist}(y^l, X^*)^2$$

for all l. Let $\bar{y}^{l+1}$ be a point in X^* such that $\|y^{l+1} - \bar{y}^{l+1}\| = \operatorname{dist}(y^{l+1}, X^*)$. It then follows from the Lipschitz property (5) and Assumption 2.1(b) that

$$\begin{aligned} \|F(y^{l+1})\| &= \|F(y^{l+1}) - F(\bar{y}^{l+1})\| \le L \operatorname{dist}(y^{l+1}, X^*) \\ &\le c_5 L \operatorname{dist}(y^l, X^*)^2 \le \frac{c_5 L \|F(y^l)\|}{c_2^2} \|F(y^l)\|. \end{aligned} \tag{15}$$

Since (14) implies

$$\frac{c_5 L \|F(y^0)\|}{c_2^2} \le \gamma,$$

we have from (15) and $\gamma < 1$ that

$$\frac{c_5 L \|F(y^l)\|}{c_2^2} \le \gamma,$$

and hence

$$\|F(y^{l+1})\| \le \gamma \|F(y^l)\|$$

for all l. This completes the proof. □

4. Application to the Linear Complementarity Problem

In this section, we apply the results obtained in the previous sections to the linear complementarity problem LCP(M, q), which is to find a vector $x \in R^n$ such that

$$x \ge 0, \quad Mx + q \ge 0, \quad x^T(Mx + q) = 0,$$

where M is an $n \times n$ matrix and q is an n-dimensional vector.

Until now, various reformulations of LCP(M, q) to an equivalent system of equations have been proposed. In this section, we consider the following reformulation based on the Fischer-Burmeister function [4]:

$$H(x) := \begin{pmatrix} \psi(x_1, (Mx+q)_1) \\ \vdots \\ \psi(x_n, (Mx+q)_n) \end{pmatrix} = 0, \tag{16}$$

where $\psi : R^2 \to R$ is the Fischer-Burmeister function defined by

$$\psi(a,b) := \sqrt{a^2+b^2} - a - b.$$

Since this function is not differentiable, the previous results cannot be applied to the system (16) directly. However, using the generalized Jacobian $\partial H(x)$ in the sense of [2], we can modify LMM so that it can deal with the nonsmooth system (16). More specifically, we adopt, as a search direction, the solution d^k of the linear system

$$((J^k)^T J^k + \mu_k I)d = -\nabla\phi(x^k),$$

where $J^k \in \partial H(x^k)$. Note that $\phi(x) := \frac{1}{2}\|H(x)\|^2$ is differentiable. Though the function H is not differentiable everywhere, H enjoys some nice properties related with Assumption 2.1. In particular, H provides a local error bound on $\{x \in R^n \mid \|H(x)\| \le \varepsilon\}$ for LCP(M,q), where ε is a sufficiently small positive constant [5], and if $(x_i, F_i(x)) \ne (0,0)$ for all i, then H is twice continuously differentiable at x. Moreover, any stationary point of ϕ is a solution of LCP(M,q) when M is a P_0 matrix.

Using these properties and the results obtained in the previous sections, we can establish the following theorem. We omit the proof, since it is a rather straightforward modification of that of Theorem 3.1.

Theorem 4.1. *Suppose that M is a P_0 matrix. Then any accumulation point of the sequence generated by LMM with line search is a solution of LCP(M,q). Moreover, if an accumulation point x^* of the generated sequence is a nondegenerate solution, i.e., $x_i^* + F_i(x^*) > 0$ for all i, then the sequence converges to the solution set of LCP (M,q) quadratically.* □

For LCP(M,q), some existing algorithms, such as interior point methods and proximal point algorithms, are known to have a quadratic rate of convergence when one of the following conditions holds [5, 8–11]:

(a) M is positive semidefinite and one of the accumulation points is a nondegenerate solution. (Note that the nondegenerate condition is not needed for superlinear convergence of the algorithm proposed in [8])

(b) M is a P_0 matrix and a kind of nonsingularity holds at one of the accumulation points.

In this section, we have shown that LMM has a quadratic rate of convergence when M is a P_0 matrix and one of the accumulation points is a nondegenerate solution, which is a new result for LCP(M,q).

References

[1] Bertsekas, D. P.: Nonlinear programming. Belmont: Athena Scientific 1995.

[2] Clarke, F. H.: Optimization and nonsmooth analysis. New York: Wiley 1983.

[3] Facchinei, F., Kanzow, C.: A nonsmooth inexact Newton method for the solution of large-scale nonlinear complementarity problems. Math. Programm. *76*, 493–512 (1997).

[4] Fischer, A.: An NCP-function and its use for the solution of complementarity problems. In: Recent advances in nonsmooth optimization (Du, D. Z., Qi, L., Womersley, R. S., eds.), pp. 88–105. Singapore: World Scientific 1995.

[5] Fischer, A.: On the local superlinear convergence of a Newton-type method for LCP under weak conditions. Optim. Methods Software *6*, 83–107 (1995).

[6] Fletcher, R.: Practical methods of optimization. New York: Wiley 1987.

[7] Robinson, S. M.: Some continuity properties of polyhedral multifunctions. Math. Programm. Study *14*, 206–214 (1981).

[8] Sturm, J. F.: Superlinear convergence of an algorithm for monotone linear complementarity problems, when no strictly complementary solution exists. Math. Oper. Res. *24*, 72–94 (1999).

[9] Tseng, P.: Error bounds and superlinear convergence analysis of some Newton-type methods in optimization. In: Nonlinear optimization and related topics (Di Pillo, G., Giannessi, F., eds.), pp. 445–462. Boston: Kluwer, 2000.

[10] Yamashita, N., Fukushima, M.: The proximal point algorithm with genuine superlinear convergence for the monotone complementarity problem. SIAM J. Optim. *11*, 364–379 (2001).

[11] Yamashita, N., Imai, J., Fukushima, M.: The proximal point algorithm for the P_0 complementarity problem. In: Applications and algorithms of complementarity (Ferris, M. C., Mangasarian, O. L., Pang, J. S., eds.), pp. 361–379. Boston: Kluwer, 2000.

N. Yamashita
M. Fukushima
Department of Applied Mathematics and Physics
Graduate School of Informatics
Kyoto University
Kyoto 606-8501, Japan
e-mail: nobuo@amp.i.ac.jp

GPSR Compliance

The European Union's (EU) General Product Safety Regulation (GPSR) is a set of rules that requires consumer products to be safe and our obligations to ensure this.

If you have any concerns about our products, you can contact us on ProductSafety@springernature.com

In case Publisher is established outside the EU, the EU authorized representative is:

Springer Nature Customer Service Center GmbH
Europaplatz 3
69115 Heidelberg, Germany

Batch number: 10683525

Printed by Printforce, the Netherlands